CCNA® Cisco® Certified Network Associate Wireless Study Guide

(Exam 640-721)

CCNA® Cisco® Certified Network Associate Wireless Study Guide

(Exam 640-721)

Henry Chou
Michael Kang

New York Chicago San Francisco Lisbon London Madrid
Mexico City Milan New Delhi San Juan Seoul Singapore Sydney Toronto

The McGraw-Hill Companies

Cataloging-in-Publication Data is on file with the Library of Congress

McGraw-Hill books are available at special quantity discounts to use as premiums and sales promotions, or for use in corporate training programs. To contact a representative, please e-mail us at bulksales@mcgraw-hill.com.

CCNA® Cisco® Certified Network Associate Wireless Study Guide (Exam 640-721)

1234567890 WFR WFR 109876543210

ISBN: Book p/n 978-0-07-170149-5 and CD p/n 978-0-07-170150-1
of set 978-0-07-170152-5
MHID: Book p/n 0-07-170149-4 and CD p/n 0-07-170150-8
of set 0-07-170152-4

Sponsoring Editor
Tim Green

Editorial Supervisor
Jody McKenzie

Project Editor
Rachel Gunn

Acquisitions Coordinator
Meghan Riley

Technical Editor
Tom Carpenter

Copy Editor
Sally Engelfried

Proofreader
Nancy Bell

Indexer
Jack Lewis

Production Supervisor
Jim Kussow

Composition
Apollo Publishing Service

Illustration
Apollo Publishing Service

Art Director, Cover
Jeff Weeks

Cover Designer
Peter Grame

This book is dedicated to my wife, Amy Chou, for her unconditional support and enduring love, and to my parents, Dr. Chen-Kung Chou and Dr. Sheau-Farn Yeh, for their inspiration and constant guidance.

—Henry

This book is dedicated to my parents, Alan and Eugenia Kang, for their continuing love and support, as well as to my brother and best friend, David Kang. Without you, I would not be who I am today.

—Michael

ABOUT THE AUTHORS

H. Henry Chou is currently a Solutions Architect with Cisco Advanced Services. A network professional for over 10 years, Henry has served as a Systems Engineer and Network Consulting Engineer and has worked closely on the design, configuration, deployment, and support of large LAN, WAN, wireless, voice, and security infrastructure for service providers, enterprise, commercial, and federal, state, and local government customers, including many Fortune 100 companies.

Henry has Masters and Bachelors of Science degrees in Industrial and Enterprise Systems Engineering from the University of Illinois at Urbana-Champaign. Henry's professional certifications include the CCNA, CCNP, CCDP, CCIP, CCSP, and CCIE (Routing and Switching) #10315.

Michael N. Kang is a Senior Network Consulting Engineer with Cisco Advanced Services. An Information Technology professional for over 12 years, Michael has focused on Cisco Networking since 2001 and has served as a Systems Engineer and Network Consulting Engineer specializing in enterprise-class wireless networking. Michael's work has included full-scale wireless LAN design and deployment for a variety of notable clients in health care, sports and entertainment, hospitality, small government, retail, technology, and finance. In his time in the industry, Michael has successfully led numerous wireless engagements encompassing secure corporate and public wireless LAN access, voice over wireless LAN, point-of-sales, and outdoor wireless mesh and bridging applications.

Michael holds a Bachelors of Business Administration in Management Information Systems from the University of Texas at Austin, where he graduated with highest honors. Michael's professional certifications include the CCNA, CCDA, CCNP, and CCDP as well as the CWNA and CWSP with the CWNP program.

About the Technical Editor

Tom Carpenter is a technical experts' expert. He teaches in-depth courses on Microsoft technologies, wireless networking and security, and professional development skills such as project management, team leadership, and communication skills for technology professionals. Tom holds CWNA, CWSP and CWTS certifications with the CWNP program and is also a Microsoft Certified Partner. He is married to his lovely wife, Tracy, and lives with her and their four children, Faith, Rachel, Thomas, and Sarah in Ohio.

About LearnKey

LearnKey provides self-paced learning content and multimedia delivery solutions to enhance personal skills and business productivity. LearnKey claims the largest library of rich streaming-media training content that engages learners in dynamic media-rich instruction complete with video clips, audio, full motion graphics, and animated illustrations. LearnKey can be found on the Web at www.LearnKey.com.

CONTENTS AT A GLANCE

CONTENTS

Part V
Maintaining and Troubleshooting a Cisco Wireless Network

Part VI
Appendixes

ACKNOWLEDGMENTS

I would like to thank Tim Green for giving me the opportunity to work on this book and work with an exceptional support team from McGraw-Hill to make this book possible.

I would also like to thank my co-author for his continued perseverance during this entire process.

I'd also like to thank our technical editor, Tom Carpenter, for his honest and constructive feedback. Thanks for your thorough review of the technical details.

—Henry

I would like to thank Timothy Green, Meghan Riley, and Tom Carpenter for their professional guidance and support during this challenging yet rewarding experience.

I would also like to extend my sincere thanks to my co-author for enabling me to participate in this process and for his incredible patience and effort in making this book a reality.

—Michael

PREFACE

IEEE 802.11-based wireless LAN technology has quickly become one of the hottest technologies of the past decade and will continue to be viable for years to come. Whether you are studying to become CCNA Wireless certified or are simply trying to gain a better understanding of Cisco Unified Wireless Network components, operations, configuration, and integration with an existing Enterprise network, you will find a wealth of information consisting of theory and practical real-world knowledge presented in this book.

The objective of this study guide is to prepare you for the Cisco IUWNE (640-721) exam by familiarizing you with the technology or body of knowledge tested on the exam. Because the primary focus of the book is to help you pass the test, we don't always cover every aspect of the related technology. Some aspects of the technology are covered only to the extent necessary to help you understand what you need to know to pass the exam, but we hope this book will serve you as a valuable professional resource after your exam.

In This Book

This book is organized to serve as an in-depth review for the Cisco IUWNE (640-721) exam for both experienced Cisco professionals and newcomers to Cisco wireless LAN networking technologies. Each chapter covers a major aspect of the exam, with an emphasis on the "why" as well as the "how to" of working with and supporting Cisco Unified Wireless LAN solutions as a network administrator or engineer.

On the CD

For more information on the CD-ROM, please see the appendix about the CD-ROM at the back of the book.

Exam Readiness Checklist

At the end of the introduction you will find an Exam Readiness Checklist. This table has been constructed to allow you to cross-reference the official exam objectives with the objectives as they are presented and covered in this book.

The checklist also allows you to gauge your level of expertise on each objective at the outset of your studies. This should allow you to check your progress and make sure you spend the time you need on more difficult or unfamiliar sections. References have been provided for the objective exactly as the vendor presents it, the section of the study guide that covers that objective, and a chapter reference.

In Every Chapter

We've created a set of chapter components that call your attention to important items, reinforce important points, and provide helpful exam-taking hints. Take a look at what you'll find in every chapter:

- Every chapter begins with **Certification Objectives**—what you need to know in order to pass the section on the exam dealing with the chapter topic. The objective headings identify the objectives within the chapter, so you'll always know an objective when you see it!
- **Exam Watch** notes call attention to information about, and potential pitfalls in, the exam. These helpful hints are written by authors who have taken the exams and received their certification—who better to tell you what to worry about? They know what you're about to go through!

exam watch

On the exam, pay close attention to questions on the client traffic flow from different types of roaming. The client traffic flow will help you understand the difference in different types of roaming. The wired network design (VLAN and subnet) usually holds the key to questions about types of roaming in a wireless LAN.

- **On the Job** notes describe the issues that come up most often in real-world settings. They provide a valuable perspective on certification- and product-related topics. They point out common mistakes and address questions that have arisen from on the job discussions and experience.
- **Inside the Exam** sidebars highlight some of the most common and confusing problems that students encounter when taking a live exam. Designed to anticipate what the exam will emphasize, getting inside the exam will help

ensure you know what you need to know to pass the exam. You can get a leg up on how to respond to those difficult to understand questions by focusing extra attention on these sidebars.

- The **Certification Summary** is a succinct review of the chapter and a restatement of salient points regarding the exam.

- The **Two-Minute Drill** at the end of every chapter is a checklist of the main points of the chapter. It can be used for last-minute review.
- The **Self-Test** offers questions similar to those found on the certification exams. The answers to these questions, as well as explanations of the answers, can be found at the end of each chapter. By taking the Self-Test after completing each chapter, you'll reinforce what you've learned from that chapter while becoming familiar with the structure of the exam questions.

Some Pointers

Once you've finished reading this book, set aside some time to do a thorough review. You might want to return to the book several times and make use of all the methods it offers for reviewing the material:

1. *Reread all the Two-Minute Drills, or have someone quiz you.* You can also use the drills as a way to do a quick cram before the exam. You might want to make some flash cards out of 3 × 5 index cards that have the Two-Minute Drill material on them.
2. *Reread all the Exam Watch notes and Inside the Exam elements.* Remember that these notes are written by authors who have taken the exam and passed. They know what you should expect—and what you should be on the lookout for.
3. *Retake the Self-Tests.* Taking the tests right after you've read the chapter is a good idea because the questions help reinforce what you've just learned. However, it's an even better idea to go back later and do all the questions in the book in one sitting. Pretend that you're taking the live exam. When you go through the questions the first time, you should mark your answers on a separate piece of paper. That way, you can run through the questions as many times as you need to until you feel comfortable with the material.

INTRODUCTION

Over the past several years, wireless LAN technology has evolved from basic Layer 2 bridging to a complex architecture supporting Layer 2 through 7 applications with high user and infrastructure densities in the enterprise, commercial, and public access spaces. As a result of this evolution, a rapidly increasing number of companies and service providers are deploying large wireless LANs in order to extend the reach of the access layer network beyond that of traditional access networks, composed of routers and switches. A wireless LAN is no longer a "nice to have" component of the network; it is now a crucial part of business operations and a mission-critical element of the network infrastructure in many organizations.

As enterprises and service providers build out larger and more complex wireless LANs, the need for experienced and knowledgeable professionals to design, configure, and support the wireless LAN also grows significantly. The Cisco Certified Network Associate (CCNA) Wireless program is designed for network professionals to demonstrate their competency in wireless LAN and the Cisco Unified Wireless Network solution. A CCNA Wireless certificate demonstrates one's knowledge and ability to install, configure, and administer a Cisco Unified Wireless LAN for midsized-to-large corporate networks.

The Goal of This Book

The goal of this book is to assist you in preparing for the Cisco 640-721 IUWNE Exam. This book includes detailed theoretical explanations of IUWNE topics and provides figures to illustrate the concepts and configuration examples. The theoretical explanations include background information, technical standards, and relevant technologies. Each chapter includes a self-test section to assess your knowledge of the subject presented.

Furthermore, this book includes information that goes beyond the scope of the certification exam and the course. The book shares the authors' combined over 20 years of real-world consulting experience from working with clients ranging from commercial to service providers to Fortune 50 customers and provides tips, information, and practical best practices that are not covered on the 640-721 IUWNE Exam or most similar CCNA-Wireless study guides.

Who Should Read This Book?

In addition to network professionals, system administrators, and RF engineers preparing for the CCNA Wireless exam, this book is an excellent reference book for engineers and IT professionals who desire to learn about the Cisco Unified Wireless Network solution and IEEE 802.11-based wireless LAN networks. The book caters to individuals at various knowledge levels, and the discussions of practical and technical concepts are designed to be easy to follow. The book begins with RF fundamentals in Chapter 1 and increases in technical depth in subsequent chapters while building on information presented in previous chapter(s). That said, this book is also intended to be flexible, so you can easily move between chapters to concentrate on specific subject(s) of your choice.

At a minimum, the reader should have a basic understanding of modern computer networking, TCP/IP, and the OSI Reference Model. No previous certification is required for reading this book, but CCNA, CCENT or an equivalent level of knowledge is strongly recommended for individuals to take full advantage of this book.

The knowledge presented in this book can also be used for CCNP Wireless and CCIE Wireless certification exams. The examples in this book are demonstrated on the Cisco 4400 and WiSM family of wireless LAN controllers. As a result, candidates who are preparing for the CCNP Wireless and CCIE Wireless exams may find this book useful for preparing for these more advanced Cisco wireless certifications.

Knowing the Test

Cisco certification tests are computer-based testing. The candidate will take the test in front of computer terminals operated by Pearson-Vue, which provides a secure testing environment in a number of facilities around the world. During the test, Cisco does not allow the candidates to go back and review a test question after an answer is entered and the candidate has moved on to the next question. The candidate can change the answer to a question as many times as desired as long as the Next button is not clicked. Once the candidate moves to the next question, the answer is locked in and the question cannot be reviewed again.

exam watch

It is critical to remember that you are not allowed to go back to review a question once you move on to the next question. Cisco used to allow candidates to go back and review answered questions and change answers, but that option is no longer available. Once you skip a question, you miss it. Unlike many other tests, previewing all questions before answering is not an option for Cisco computer-based certification tests.

Strategies for Exam Preparation

There are many different strategies for preparing for this test. One is to take the official Cisco 640-721 IUWNE training course from a Cisco Certified Training Partner. If you have taken the training course already, use this book to reinforce concepts you have learned from the classroom and use the self-test questions in the end of each chapter to assess what you need to study. You may not need to read the entire chapter, but rather to focus on appropriate sections. Use the practice exam included in this book to prepare you for the final studying before the exam. You will fill the gaps in knowledge and be ready for the exam.

You can take and pass the Cisco 640-721 IUWNE test without taking any courses. However, Cisco does recommend that you take the recommended courses for all exams, as Cisco believes that the training courses are the best way to learn about Cisco technologies and products. For those who have not taken the Cisco 640-721 IUWNE training course, this book may provide the necessary information for passing the exam. Read and understand the concepts and information covered by this book and thoroughly understand the practice tests. In addition, we suggest that you try to get some hands-on experience working with Cisco access points and Wireless LAN Controllers, so you feel comfortable with the user interfaces and configuration tasks required to build a wireless LAN based on Cisco Unified Wireless Network solution.

If you are already an experienced RF and wireless LAN engineer and have a fair amount of professional experience, here are some suggestions for your test preparations. Use this book and review each chapter. Assess your knowledge level against test objectives and use this book to assess what you need to study. It may not be necessary for you to read each entire chapter, but rather focus on appropriate sections in the chapters. Also, use the practice tests to familiarize yourself with the test and question format, and use the book for study in the days leading up to your exam for final review of subjects and information covered by the exam.

Strategies for the Exam

There are many different ways to study for an exam. Our suggestion is simple and straightforward: use your time wisely and effectively. This book has provided an approach that allows readers with little to no RF knowledge to understand the basics of 802.11 wireless LAN. Each chapter contains sections that correspond to different exam objectives covered in the CCNA-Wireless exam. Each topic is explained in detail and with informative figures and examples. Within the chapter, we have also included on the job scenarios and tips which will help you understand how information you learn from the chapter is applied in the real world. We have found that this real-world experience is invaluable to exam preparation and can provide an added advantage in understanding the material. As you prepare for the CCNA Wireless exam, and other advanced Cisco Wireless certifications, we recommend that you practice with real Cisco wireless LAN gear.

In addition to focused studying and practice, you should not be burned out mentally. Here are some additional tips for preparing yourself before, during, and after the exam.

One day before the exam

Verify and confirm with Pearson-Vue your exam time, exam center, and, most importantly, the exam itself. Make sure you're registered for the correct exam. Make sure you have the directions to get to the test center. A good night of sleep is a must before your exam. Do not attempt to do heavy studying on the day before the exam.

On the day of the exam

Make sure you have consumed food prior to the exam, but do not take the exam right after eating. It's a scientific fact that our brains function better when fueled. But eating right before the exam may cause indigestion and could be distracting for test taking. The test center will give you a pen and a marker and scratch paper, so there is no need to bring your own. Leave all heavy books at home and wear comfortable clothing. Allow plenty of time for travel and try to arrive at the test center early.

During the exam

Pace yourself. The exam consists of 65 questions, and you have 90 minutes to complete the exam. The computer terminal will show time remaining and questions remaining. There should be plenty of time to complete each question. Ensure you feel comfortable with the answer to the question before you move on to the next question. As mentioned earlier, there is no going back once you click the Next button. If you do not know the answer, use the process of elimination. Try to find an answer that is best for the question. Use the pen and paper provided by the test center to help you with this process. If you cannot answer a question, make a guess and move on to the next question. You should not leave any questions unanswered. Also, once you have moved on to the next question, try to stay calm and put the previous questions in the past.

Retesting

The exam can be taken multiple times. Even if you are struggling, use this exam to your advantage by remembering the topics that are causing you trouble. Cisco makes you wait five business days before rescheduling for the same test. Continue with the study and focus on the areas that need most attention. When you're ready, simply contact Pearson-Vue to schedule your next attempt. You can track your certification progress at www.cisco.com/go/certifications/login.

Exam Readiness Checklist

Official Objective	Study Guide Coverage	Ch #	Pg #	Beginner	Intermediate	Advanced
Describe WLAN fundamentals						
Describe basics of spread spectrum technology (modulation, DSS, OFDM, MIMO, Channels reuse and overlap, Rate-shifting, CSMA/CA)	Spread Spectum Technology	1	4			
Describe the impact of various wireless technologies (Bluetooth, WiMAX, ZigBee, cordless phone)	The Impact of Various Wireless Technologies	3	96			
Describe wireless regulatory bodies, standards and certifications (FCC, ETSI, 802.11a/b/g/n, WiFi Alliance)	Wireless Regulatory Bodies, Wireless LAN Standards and Certifications	2	46, 47			
Describe WLAN RF principles (antenna types, RF gain/loss, EIRP, refraction, reflection, ETC)	Wireless LAN RF Principles	1	19			
Describe networking technologies used in wireless (SSID --> WLAN_ID --> Interface --> VLAN, 802.1q trunking)	Wireless Regulatory Bodies	2	46			
Describe wireless topologies (IBSS, BSS, ESS, Point-to-Point, Point-to-Multipoint, basic Mesh, bridging)	Wireless Topologies and Frame Types	2	61			
Describe 802.11 authentication and encryption methods (Open, Shared, 802.1X, EAP, TKIP, AES)	Wireless Authentication, Wireless Encryption, Authentication Sources	5	139, 151, 158			
Describe frame types (associated/unassociated, management, control, data)	Wireless Topologies and Frame Types	2	61			
Install a basic Cisco wireless LAN						
Describe the basics of the Cisco Unified Wireless Network architecture (Split MAC, LWAPP, stand-alone AP versus controller-based AP, specific hardware examples)	The Basics of the Cisco Unified Wireless Network Architecture	6	176			
Describe the Cisco Mobility Express Wireless architecture (Smart Business Communication System—SBCS, Cisco Config Agent—CCA, 526WLC, 521AP—stand-alone and controller-based)	Cisco Mobility Express Wireless Architecture	7	226			
Describe the modes of controller-based AP deployment (local, monitor, HREAP, sniffer, rogue detector, bridge)	The Modes of Controller-based AP Deployment	6	184			

Exam Readiness Checklist *(continued)*

Official Objective	Study Guide Coverage	Ch	Pg #	Beginner	Intermediate	Advanced
Describe controller-based AP discovery and association (OTAP, DHCP, DNS, Master-Controller, Primary-Secondary-Tertiary, n+1 redundancy)	Controller-based AP Discovery and Association	6	191			
Describe roaming (Layer 2 and Layer 3, intra-controller and inter-controller, mobility groups)	Roaming	6	200			
Configure a WLAN controller and access points WLC: ports, interfaces, WLANs, NTP, CLI and Web UI, CLI wizard, LAG AP: Channel, Power	Configuration of Cisco Wireless LAN Components	8	260			
Configure the basics of a stand-alone access point (no lab) (Express setup, basic security)	Standalone Access Point Configuration	8	250			
Describe RRM	Radio Resource Management (RRM) Concepts	8	287			
Install wireless clients						
Describe client OS WLAN configuration (Windows, Apple, and Linux)	Operating System Client Configuration	10	342			
Install Cisco ADU	Cisco Client Configuration	10	358			
Describe basic CSSC	Cisco Client Configuration	10	358			
Describe CCX versions 1 through 5	Cisco Compatible eXtensions (CCX)	10	374			
Implement basic WLAN security						
Describe the general framework of wireless security and security components (authentication, encryption, MFP, IPS)	The General Framework of Wireless Security and Security Components	4	116			
Describe and configure authentication methods (Guest, PSK, 802.1X, WPA/WPA2 with EAP-TLS, EAP-FAST, PEAP, LEAP)	Wireless Authentication	5	139			
Describe and configure encryption methods (WPA/WPA2 with TKIP, AES)	Wireless Encryption	5	151			
Describe and configure the different sources of authentication (PSK, EAP-local or -external, Radius)	Authentication Sources	5	158			

Exam Readiness Checklist (continued)

Official Objective	Study Guide Coverage	Ch	Pg #	Beginner	Intermediate	Advanced
Operate basic WCS						
Describe key features of WCS and Navigator (versions and licensing)	The Wireless Control System (WCS) and Navigator	9	302			
Install/upgrade WCS and configure basic administration parameters (ports, O/S version, strong passwords, service vs. application)	Installing and Configuring the Wireless Control System (WCS)	9	307			
Configure controllers and APs (using the Configuration tab not templates)	The Wireless Control System (WCS) and Navigator	9	302			
Configure and use maps in the WCS (add campus, building, floor, maps, position AP)	Installing and Configuring the Wireless Control System (WCS)	9	307			
Use the WCS monitor tab and alarm summary to verify the WLAN operations	Monitoring with the Wireless Control System	9	329			
Conduct basic WLAN maintenance and troubleshooting						
Identify basic WLAN troubleshooting methods for controllers, access points, and clients methodologies	Wireless LAN Troubleshooting Methods	12	426			
Describe basic RF deployment considerations related to site survey design of data or VoWLAN applications, Common RF interference sources such as devices, building material, AP location Basic RF site survey design related to channel reuse, signal strength, cell overlap	Cisco Wireless Network Deployment Considerations	11	410			
Describe the use of WLC show, debug and logging	Use of Wireless LAN Controller Show, Debug, and Logging Commands	12	429			
Describe the use of the WCS client troubleshooting tool	Use of the WCS Client Troubleshooting Tool	12	440			
Transfer WLC config and O/S using maintenance tools and commands	Cisco Wireless Network Maintenance Tasks	11	396			
Describe and differentiate WLC WLAN management access methods (console port, CLI, telnet, ssh, http, https, wired versus wireless management)	Cisco Wireless Network Administration and Access Methods	11	388			

Part I

Wireless LAN Fundamentals

CHAPTERS

1

Radio Frequency Basics

CERTIFICATION OBJECTIVES

1.01 Spread Spectrum Technology

1.02 Wireless WLAN RF Principles

✓ Two-Minute Drill

Q&A Self Test

One of the key elements of working with wireless LANs is understanding the radio frequency environment. Wireless LAN technology is distinguished from traditional LAN by one major characteristic: the medium by which data is transmitted. In a traditional Local Area Network, data is transmitted over copper and fiber cabling. Wireless LANs, by contrast, transmit data over the air in the form of radio frequency waves. Many of the design and implementation considerations for deploying and supporting a wireless LAN lie in the radio frequency (RF) environment. As you begin to work more with wireless LAN technology, it will become more apparent that wireless LAN represents a true convergence of expertise. An effective wireless LAN engineer needs to understand the basics of Cisco routing and switching and also have the ability to understand the fundamentals of the radio frequency upon which it runs. The purpose of this chapter is to establish a foundation for the technologies that a typical 802.11 wireless LAN runs on.

CERTIFICATION OBJECTIVE 1.01

Spread Spectrum Technology

When the discussion about transmitting data using radio frequency technology begins, the first question is, how? A part of the answer to this question lies in the purpose of advancing networking technology. As traditional wired networks have evolved, one element has remained constant: the ability to increase the amount and speed of data that is transmitted over a given medium. This medium can be copper cable, fiber optic phone lines, or in this case, radio frequency waves. *Spread spectrum* technology simply refers to one of several methods by which data is converted to electrical signals for transmission over airwaves.

The impetus behind spread spectrum's wide use is twofold. First of all, spread spectrum technology is naturally resistant to interference which, as you will read in later chapters, is the largest obstacle to the proper operation of all varieties of wireless technology. The reason for this is that spread spectrum technology runs with a wide bandwidth (or portion of the radio frequency utilized). A very simplified way of looking at this requires you to understand that, within a radio frequency range, there are a given number of channels on which data can be transmitted. An example of this would be your typical citizen's band (CB) radio, which allows you to switch between certain channels. These channels have a very narrow bandwidth (known as *narrowband transmissions*) and can be very easily distorted by external sources of interference (you may have noticed when using a CB radio that there is a large amount of crosstalk and lost messages when too many people are sharing the same channel). The narrowband transmissions you use in each channel/station on the CB radio utilize only the minimal amount of bandwidth in the radio frequency range to transmit data. The same is not true of spread spectrum transmissions because the channels (or stations) in spread spectrum technology are much wider and often provide substantially more bandwidth than is absolutely necessary to transmit the data. Although this may at first seem like an inefficient use of the bandwidth, it is this very characteristic that gives the data transmission on a spread spectrum network the ability to resist interference sources and even share the channel with other devices. This will be discussed in more detail when unlicensed frequency bands (such as 2.4 GHz and 5 GHz) are covered.

The second advantage of spread spectrum technology lies in its inherent security. This is because spread spectrum, while running on a wide bandwidth, also has the characteristic of utilizing low peak power. In the CB radio example, the narrowband transmissions of a CB need very high peak power to make them distinguishable from the undesirable radio noise in the frequency range. Because the resulting signal is easily distinguished from noise, it can be easily picked up by most radio receivers. Spread spectrum's low peak power makes it characteristically very similar to radio noise with the difference being that it is not an undesirable transmission. This means that receivers have to run on a specific technology that allows them to distinguish spread spectrum signals from normal radio noise. Although this is important to know, it is also important to be aware that this security has been dramatically diminished by the rapidly growing number of receivers that can run a specific technology (IEEE 802.11a/b/g/n being a key example). A comparison of spread spectrum and narrowband transmissions can be seen in Figure 1-1.

Spread spectrum technology is characterized by wide bandwidth and low peak power. This allows it to resist interference and provides it with a certain level of security over traditional narrowband transmissions.

FIGURE 1-1 Comparison of spread spectrum and narrowband transmissions

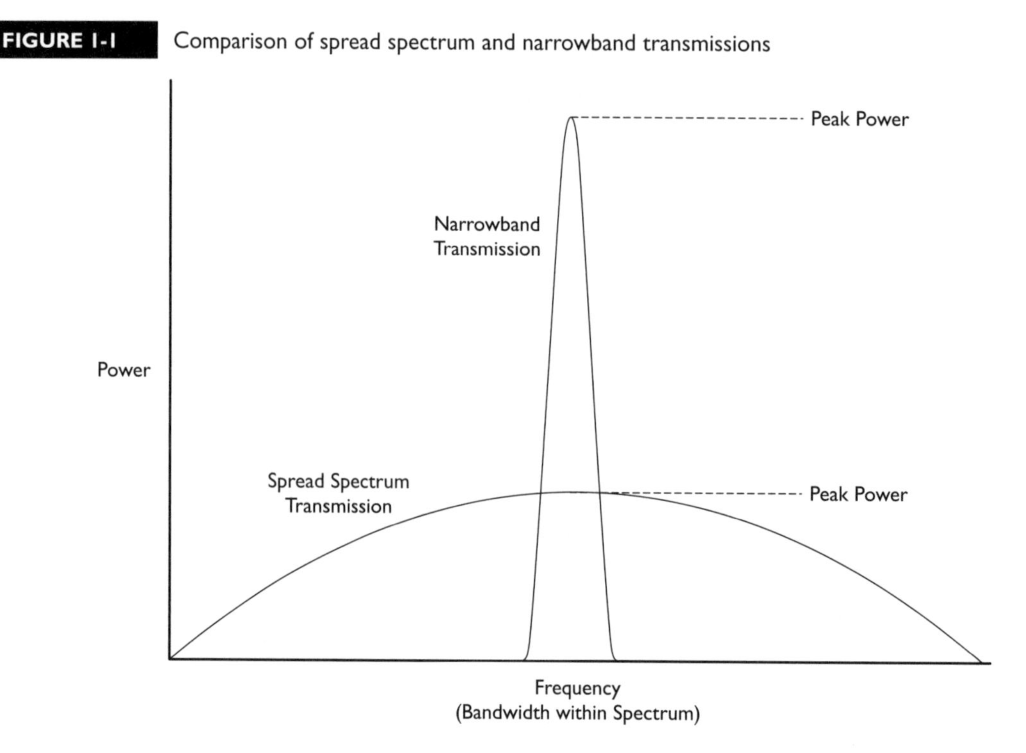

Modulation

The key to understanding the purpose and application of modulation lies in what is desired in radio frequency transmissions.

The goals of radio frequency signal transmissions are to send

- As much data as possible
- As far as possible
- As fast as possible

Modulation is simply the process by which data is placed on a signal medium. In the case of wireless technology, the medium happens to be a radio wave. This is accomplished by altering the waveform (modulation) in order to send an encoded message. This altered waveform is also known as a *carrier signal* and is later decoded (demodulated) by the receiver. An example of basic modulation is music, where sound waves are altered to create melodies using changes in volume, timing, and pitch. Much like in music, an encoded carrier signal has the same alterable properties, which are seen in all waveforms (Figure 1-2):

- **Amplitude** Sometimes recognized known as volume, amplitude represents the size of the wave variation or height of a wave.
- **Phase** Better known as timing, phase represents the speed at which the wave moves along the medium.
- **Frequency** Recognized as pitch, frequency represents the number of wave iterations within a given amount of time. This is directly impacted by phase and the wavelength (or width of a wave). Frequency declines with increasing wavelength and/or decreasing phase, and vice versa.

FIGURE 1-2 Characteristics of waveforms

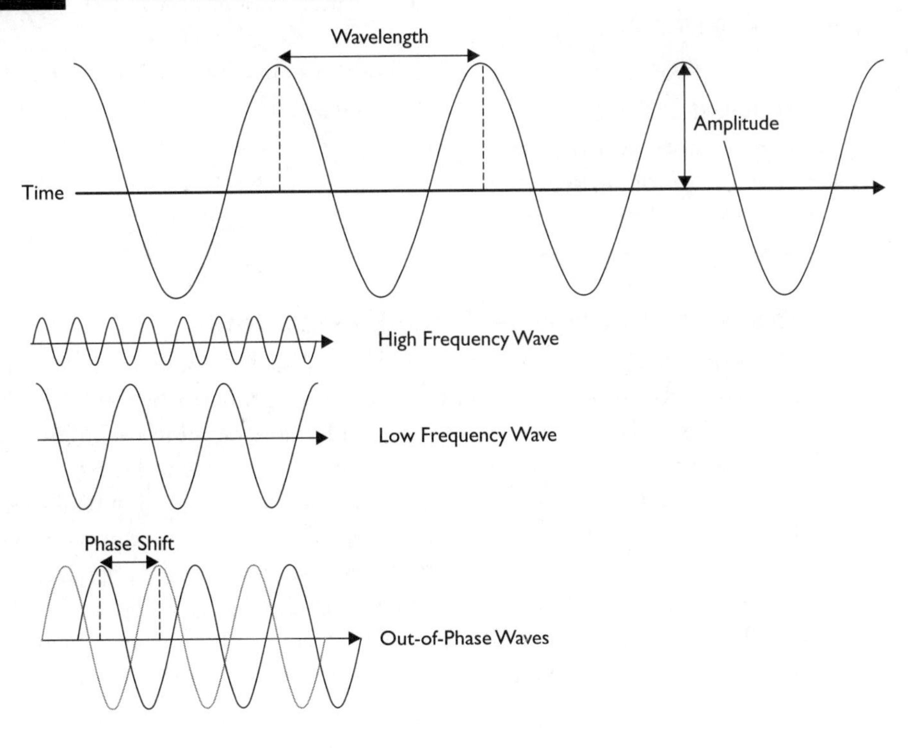

Going back to the goals of radio technology, you can see that, without modulation, the only way to send more data is to increase the frequency spectrum or amount of signal bandwidth that is used. This is because a wider signal bandwidth allows more "space" for data to be transmitted. There are, however, three laws of radio dynamics inherent in all radio frequency transmissions involving wireless devices:

- Higher data rate = shorter transmission range
- Higher power output = increased range, lower battery life on devices
- Higher frequency radios = higher data rates, shorter ranges

Because of these laws, there are inherent limitations in each frequency range, usually related to the inverse relationship between data rate and distance. Modulation plays the role of being the second method upon which more data can be sent over a given frequency. Simply put, more complex modulation allows for a greater amount of data to be sent within a specific frequency range.

Modulation is a key method of sending more data over a given frequency. More complex modulation means higher data rates and throughput.

There are numerous different modulation types, but the ones we will concentrate on are directly related to the operation of 802.11 wireless networks. They are DSSS, OFDM, and MIMO which are discussed in the following sections.

DSSS (Direct Sequence Spread Spectrum)

Direct Sequence Spread Spectrum (DSSS) is a popular modulation type used in the vast majority of wireless networks. It is used primarily by the IEEE 802.11 and 802.11b standard, which will be covered in more detail in the following chapter. DSSS is preferred because it allows high data rates and is relatively easy to implement. Because DSSS works under the principles of spread spectrum technology, there is a surplus of bandwidth for the amount of data that is being transmitted. DSSS takes advantage of this by spreading the data signal over what is a relatively wide 22 MHz frequency which, in the case of wireless, represents a single channel inside the 2.4 GHz band. DSSS accomplishes this spreading of data by breaking the bits of data into pieces called *chips*. This process is also known as *encoding*.

Chips are basically a series of redundant bits used to represent a single data bit. This combined series, known as a *chip stream*, also has an associated *chipping sequence* or *chipping code* for each bit of data. A chipping sequence is simply a unique binary translation of a single bit of data. The data bit is converted into multiple chips and sent in parallel over the widened frequency range. There can be different numbers of chips that are created for each bit, known as the *chipping rate*. The rates mandated by the FCC are a minimum of 10 chips at the 1- and 2-Mbps rates and 8 chips at the 5.5- and 11-Mbps rates. The IEEE 802.11 standard specifies a minimum of 11 chips per bit, which is also known as the *Barker Code*. The best way to understand the concept of DSSS modulation is to use a basic example of a data bit being encoded into a chipping sequence, transmitted over the wireless medium, and then decoded by the receiver:

1. In a basic DSSS modulation scenario, a transmitter—a laptop, wireless phone, or PDA—sends a series of data bits (the string "101" for instance) over the 22 MHz radio frequency channel that it is associated to via an access point.
2. The individual data bits 1 and 0 are each encoded to a chip stream which, depending on the data rate, can be 8 or 11 bits. In this example, we will use 11 bits where a 1 has an associated chipping sequence of 00110011011 and 0 has an associated code of 11001100100. Keep in mind that for each of the 3 bits in "101" there are 11 chips being sent in parallel over the 22 MHz channel.
3. The receiver (the wireless access point in the case of 802.11b) decodes the chip streams (11001100100, 00110011011, and 11001100100) to receive the "101" string.
4. The "101" string is sent to the wired network (or connected device) as a regular series of data bits.

One important thing to note in this example is that each chip stream, because it is spread over a wider frequency range than a single bit, is more resistant to narrowband interference that may occur during transmission. When interference occurs during a transmission, certain chips (bits in the chip stream) may become inverted from 1 to 0, but the integrity of the chip stream is maintained as long as no more than five of these chips are altered. If, for example, the 11001100100 string ("1") was altered to 01001100100, there would only be one chip in the string that is altered. It would take a large number of altered chips—such as the nine

exam watch
DSSS modulation resists narrowband interference by taking individual data bits and spreading them over a 22 MHz frequency band, to ensure data integrity and successful transmission.

comparatively different chips in 00110011011 ("0")—to alter the actual meaning of the overall string. Therefore, the transmitted and decoded value would remain 1 up until five or more of the bits/chips were altered by interference.

Within the realm of DSSS, there are three different modulation types that are used in IEEE 802.11b transmissions at different data rates. It should be noted that BPSK and QPSK are 802.11 standards. These modulation types are as follows:

- **Binary Phase Shift Keying (BPSK)** Uses one phase to represent a binary 1 and another to represent a binary 0 for a total of two bits of binary data. Utilized to transmit data at 1 Mbps.
- **Quadrature Phase Shift Keying (QPSK)** The carrier undergoes four changes in phase and can thus represent four binary bits of data. This is utilized to transmit data at 2 Mbps.
- **Complementary Code Keying (CCK)** Uses a complex set of functions known as complementary codes to send more data. This is utilized to transmit data at 5.5 and 11 Mbps.

OFDM (Orthogonal Frequency Division Multiplexing)

OFDM (Orthogonal Frequency Division Multiplexing) is a modulation method utilized by IEEE 802.11a and 802.11g to achieve higher data rates beyond 11 Mbps all the way up to 54 Mbps. In addition to high data rates, OFDM is also more resistant to interference because it breaks a single data stream into multiple evenly spaced frequency bands known as *subcarriers*. The subcarriers, which each carry a piece of the data stream, transmit independently of one another. To understand how OFDM works, it is helpful to break down the individual definitions.

Frequency Division Multiplexing is, simply put, the process by which a single signal is sent in a stream by a single carrier, similar to water coming out of a faucet. FDM is a very commonly used technology, but all you need to know is that it represents sending data as a single stream over a single channel.

Orthogonal refers to the concept of individual events running independently of each other. Simply put, if something affects or disrupts one event, it will not necessarily disrupt the other because of this independence.

Orthogonal Frequency Division Multiplexing is a variation on Frequency Division Multiplexing, where the single stream is split into independent (orthogonal) subcarriers. This is analogous to placing a showerhead over the flow of water, thereby splitting it into multiple independent streams of water. This yields two major benefits:

1. If there is a blockage of one of the subcarriers (streams of water), the majority of the stream is still sent by the other subcarriers. This is effectively how OFDM is able to resist narrowband interference and still successfully transmit a message.
2. Although each of the subcarriers carries less data (water), there are multiple subcarriers simultaneously delivering data, which yields higher throughput and increased overall data rates.

OFDM runs on 20 MHz wide channels with 52 evenly divided subcarriers that overlap but still run independently of one another. Each of the subcarriers is 312.6 KHz wide.

MIMO (Multiple-Input Multiple-Output)

MIMO (Multiple-Input Multiple-Output) is a modulation type utilized by IEEE 802.11n. MIMO refers to the use of multiple antennas for transmitting and receiving signals. By doing this, MIMO can substantially increase both data throughput and wireless coverage range without needing additional frequency bandwidth or transmit power to the access point radio. In traditional wireless technology, your data rate and signal strength depends heavily on the ratio of signal-to-noise ratio or SNR (which is covered later in this chapter). The greater the SNR, the better the throughput and quality of wireless association. As a wireless device moves farther from the transmitter (the access point), the SNR decreases and the amount of information that can be picked up by the receiver (the associated wireless client device) also decreases.

MIMO provides several advantages over traditional modulation in large part because it utilizes technologies to increase SNR at the receiver end:

Transmit beamforming By utilizing more than one antenna on the transmitter to send data, multiple radio signals containing the same data can be sent simultaneously and synched so that the waves to a single receiving antenna do not arrive out of phase with one another. This combined radio frequency wave, the product of two or more separate transmitting antennas, provides for a higher SNR to the receiver and provides higher data rates at greater distances from the transmitter. This synching process requires an 802.11n client and extensive communication between the receiver and transmitter and involves extensive fine-tuning that is performed by the transmitter. Any physical movement by the receiver requires the synching process to occur again.

Spatial diversity Another concept which will be described later in this chapter is *multipath*. As signals are transmitted, they bounce off various metal objects and can reflect off them several times on their way to a receiver. In a traditional wireless network, a single signal can arrive at a receiver multiple times (as the original signal or various reflections of the signal which show up nanoseconds later). These reflected signals can degrade the original signal. MIMO sends multiple signals or *spatial streams* from different antennas that will naturally travel over different paths due to the minor difference in location of the antennas. Multiple antennas at the receiver can individually receive these disparate spatial streams and combine them with each other after decoding has occurred. This effectively increases SNR to the receiver and takes advantage of the previously troublesome behavior of multipath interference.

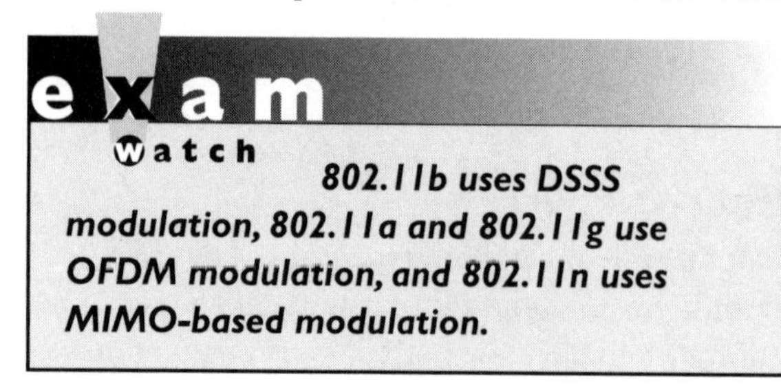

802.11b uses DSSS modulation, 802.11a and 802.11g use OFDM modulation, and 802.11n uses MIMO-based modulation.

Unlicensed Frequency Bands

To review what has been covered thus far, spread spectrum technology is a method by which data is converted into electrical signals. Modulation is the process used to actually convert that data. Before we can cover concepts such as channelization, data rates, collision avoidance, and different 802.11 standards (in the following chapter), it is important to understand the actual radio medium as it pertains to wireless networking.

The radio frequency spectrum has countless ranges, frequencies, and applications, but the three that we will concentrate on are the unlicensed frequency bands directly related to wireless LAN applications, shown in Figure 1-3. These spectrums are 900 MHz, 2.4 GHz, and 5 GHz.

FIGURE 1-3 Unlicensed frequency bands

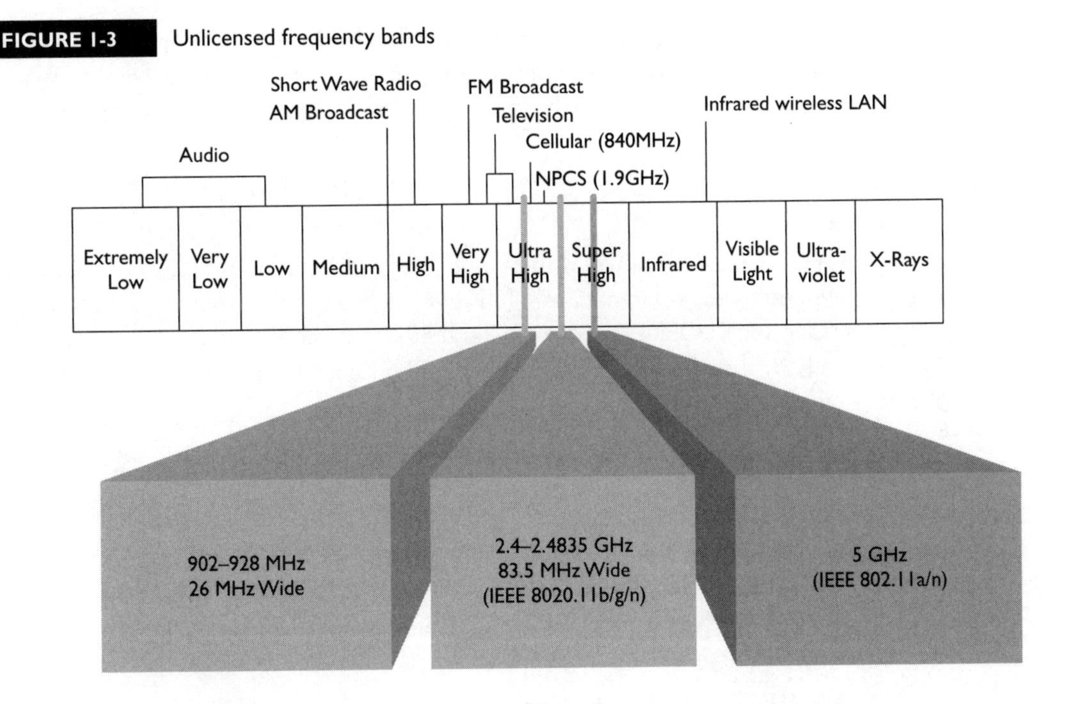

900 MHz

The 900 MHz spectrum is familiar to most of us precisely because it is the range used for a large number of cordless phones in the market. 900 MHz is specifically the spectrum range between 902 and 928 MHz; it is characterized by high range and low bandwidth (limited to 1 Mbps) and is therefore not used in the 802.11 standard. This is due to the relatively narrow bandwidth of the channels and range.

2.4 GHz

The 2.4 GHz range is the most widely used frequency band and specifically covers the range between 2.4000 and 2.4835 GHz. In addition to being utilized by 802.11b/g/n, the 2.4 GHz band is also used by cordless phones, Bluetooth, microwave ovens, and a variety of other devices that will be covered in some detail in Chapter 3. The 2.4 GHz range is characterized by shorter range than 900 MHz, but the capability for substantially higher bandwidth (1, 2, 5.5, or 11 Mbps for 802.11b with DSSS modulation, and up to 54 Mbps for 802.11g with OFDM modulation).

In the United States, the 2.4 GHz range is divided into 11 overlapping 22 MHz-wide channels. These wider channels are the same ones that we referred to in the DSSS discussion and will be further covered in the "Channelization" section of this chapter.

5 GHz

The 5 GHz range, also known as the Unlicensed National Information Infrastructure (U-NII) radio band, is a frequency band that has begun to see increased use since the introduction of OFDM modulation and 802.11a technology. It is also used extensively by the 802.11n standard. The 5 GHz range is actually broken down into 4 ranges, delineated by U-NII:

- **U-NII 1 (Lower U-NII),** 5.15–5.25 GHz This range requires use of an integrated or built-in antenna with maximum radio transmit power limited to 50mW. Indoor use only.
- **U-NII 2 (Middle U-NII),** 5.25–5.35 GHz This range allows an external antenna with maximum radio transmit power limited to 250mW (milliwatts). Indoor and outdoor use.
- **U-NII-2E (U-NII 2 Extended),** 5.470–5.725 GHz This range is subject to the same restrictions and applications as U-NII 2 and was added by the FCC in 2003 to allow for global support. It is also known as *U-NII 2 Worldwide*.
- **U-NII 3 (Upper U-NII),** 5.725–5.825 GHz Sometimes referred to as *U-NII/ISM*. This range allows for an external antenna with maximum radio transmit power limited to 1W (watt). Specifically assigned for outdoor use and used extensively for outdoor applications such as wireless mesh backhaul, wireless ISPs, and outdoor bridging. This frequency range is not available in Europe.

The 5 GHz range is characterized by shorter range than 2.4 GHz and 900 MHz but capability for higher data throughput (up to 54 Mbps with OFDM modulation). In the United States, the 5 GHz range is divided into 23 nonoverlapping 20 MHz wide channels. These channels are the same ones we referred to in the OFDM discussion.

Channelization

The primary benefit of wireless networking is the ability for client devices to move from one location to another while maintaining a connection to the wireless network. In describing spread spectrum transmissions for the purposes of wireless networking, the topic of channels becomes essential to enabling client mobility or roaming. The following sections discuss the applications and limitations of channels in the radio frequency spectrum.

Channel Overlap

In the discussion of the 2.4 GHz band, we can see that, in the United States, there are eleven 22 MHz-wide channels inside a band with only 5 MHz spacing between the center frequencies of each channel. As you move from one channel to another within the 2.4 GHz band, you will see that there are only a few combinations of two channels that do not overlap with each other and only one combination where there can be three nonoverlapping channels. These three channels are 1, 6, and 11. The issue of avoiding channel overlap and its associated interference (known as adjacent channel interference) is a key concern with wireless RF site surveys and wireless deployments. It is also the primary reason why the vast majority of 802.11b/g devices that you see use one of those three nonoverlapping channels. Figure 1-4 shows channel overlap in the 2.4 GHz frequency band.

FIGURE 1-4 Channel overlap in the 2.4 GHz band

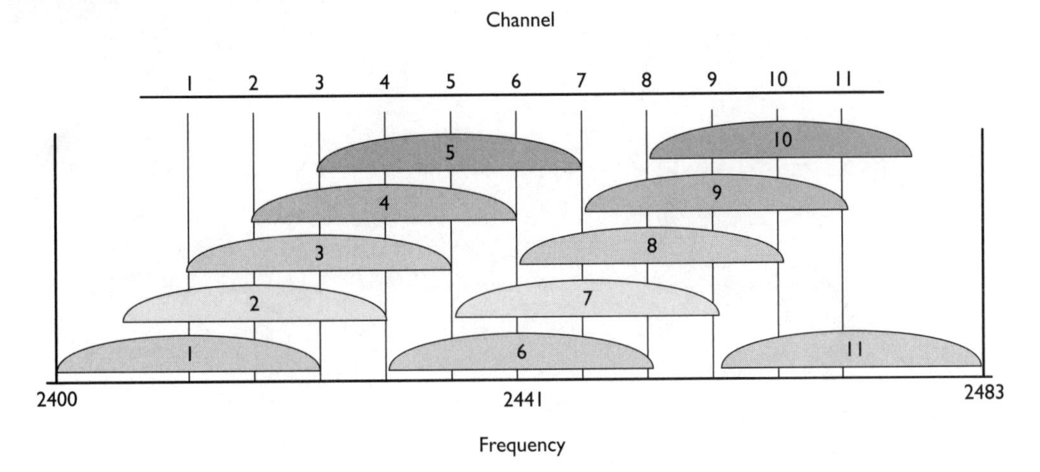

Channel Reuse

Because most enterprise wireless deployments require more than three access points, the question of deploying them involves considering not only outside interference, but also same-channel interference—called *co-channel interference*—from additional APs that are deployed. An example of co-channel interference would be two APs on channel 1 in the same location. Due to the ever-present issue of adjacent channel interference, it is simply not practical to deploy APs using channels other than 1, 6, and 11. The only solution is to perform what is called *channel reuse*, which simply refers to the judicious distancing and placement of access points so that co-channel and adjacent channel interference is minimized. Channel reuse is performed extensively in manual site surveys and is done automatically in the Dynamic Channel Assignment feature of Cisco's Radio Resource Management feature that is covered later in this book. Figure 1-5 gives a quick overview of how channel reuse is applied with APs in the 2.4 GHz (802.11b/g) and 5 GHz (802.11a) bands.

Dynamic Rate Shifting

In the sections covering modulation and unlicensed frequency bands used by wireless technology, there have been multiple mentions of data rates that can be achieved with certain frequencies and modulations. The concept of dynamic rate

FIGURE 1-5 Application of channel reuse

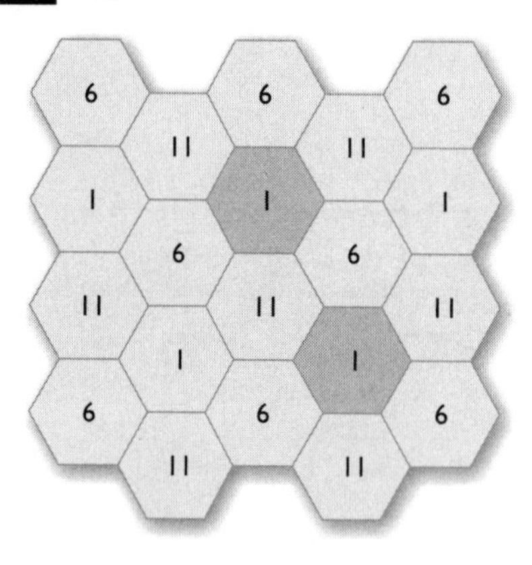

2.4 GHz 802.11b/g

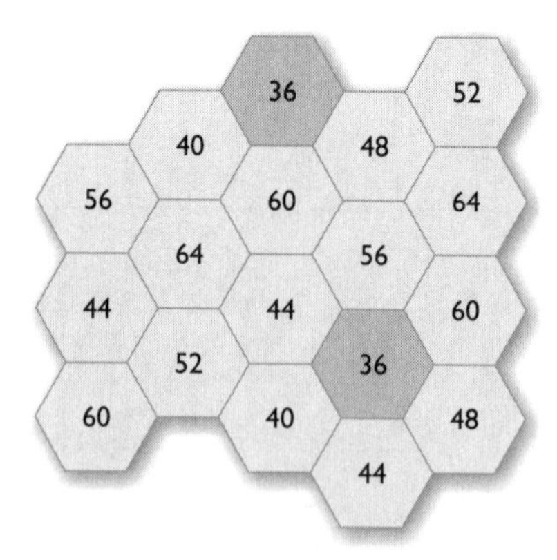

5 GHz 802.11a
(UNII-1 and UNII-2)

shifting is simply this: Provided that the data rates are enabled and available, a wireless client (receiver) device will drop to a lower data rate without dropping that client's connection as it moves farther away from an access point. In the example of 802.11b, the rates will go from 11 to 5.5, then 2, then 1 Mbps as a client moves farther from an access point. The name of the feature that enables this behavior is *dynamic rate shifting*. There is no "in-between" data rate that a client will drop to, and the point at which the client's data rate shifts is determined by the distance the client is from the access point, the RSSI (received signal strength indicator) reading that is picked up by the client, and the SNR (signal-to-noise ratio) that is received by the client. 802.11a, 802.11g, and 802.11n also perform dynamic rate shifting, albeit with a different set of signal strengths and thresholds. Figure 1-6 gives a quick overview of dynamic rate shifting on 802.11b with approximate range comparisons.

FIGURE 1-6

Dynamic rate shifting with 802.11a and 802.11b/g radios

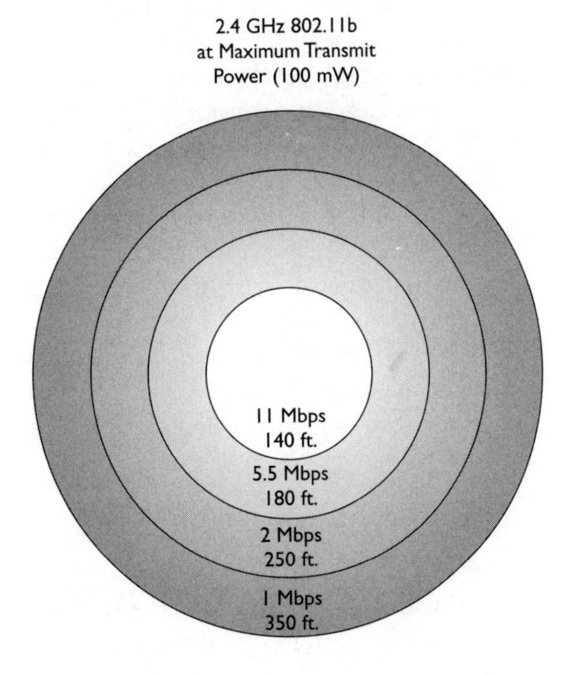

on the Job

One of the best ways to control the effective coverage cell size of a wireless access point is to disable lower data rates. This is often beneficial as it ensures that a specific data rate is maintained throughout your environment. By maintaining a higher data rate and using more access points with smaller cell sizes, a given environment will also be able to support a higher density of wireless LAN users. In wireless implementations where voice over wireless LAN is used, the disabling of lower data rates is often considered a requirement.

Collision Avoidance with CSMA/CA

With all of the recent advances in wireless LAN technology over the years, one thing has remained the same: the role of wireless in a traditional Cisco network infrastructure. Because radio frequency is and always has been a shared medium, a wireless access point is functionally a bridge. Collisions in a wireless LAN are a frequent occurrence due to multiple transmitting and receiving sources and no method of broadcast suppression. Collision avoidance measures are a must on all wireless networks and this is accomplished by CSMA/CA *(Carrier Sense Multiple Access with Collision Avoidance)*. This concept is not to be confused with the similar CSMA/CD (Carrier Sense Multiple Access with Collision Detect) that is seen on traditional wired data networks.

In CSMA/CA, a device that needs to transmit has to first listen to the radio channel for a set interval of time to monitor for activity on the channel. If the channel appears to have no activity (idle state), the device is then able to transmit. If there is activity sensed on the channel (busy state) the device has to hold back its transmission until the channel is detected as clear. Along with the data that is transmitted, a signal is sent to the other devices on the same channel to hold back their transmissions. To accomplish this, there are two different packets that are used to facilitate CSMA/CA:

- **Request-to-Send (RTS)** The RTS frame is sent by the device that needs to transmit.
- **Clear-to-Send (CTS)** The CTS frame is sent by the receiving device when it is ready to receive the transmission.

When either an RTS or CTS is sent, nodes that are not intended recipients are instructed to stop sending data for a given period of time in order to prevent collisions. The concepts of RTS and CTS are also covered in Chapter 2.

CERTIFICATION OBJECTIVE 1.02

Wireless LAN RF Principles

In the implementation of a wireless network, there is a great deal of planning involved to ensure a successful and robust deployment. Without an understanding of the following basic RF principles, performing essential tasks such as site surveys, RF analysis, and wireless site audits becomes a very daunting task. Understanding the following principles in addition to the spread spectrum technology concepts covered in this chapter will enable you to make the best use of Cisco wireless features and properly design and deploy a wireless network from the beginning.

Distortion of an RF Signal

For traditional wired network connectivity, the obstacles that you have to deal with on the physical layer are miscabled ports, faulty hardware, and poor quality cabling and physical connection cuts. For successful wireless connectivity, there are a whole host of additional challenges to face because of the unpredictability of the radio frequency medium and the imperfect nature of the indoor and outdoor environments where wireless is deployed. These factors are discussed in this section.

Reflection

Reflection (Figure 1-7) occurs when an RF signal "bounces" off an object and continues to travel in a different direction. Reflection is also characterized by the fact that the disruptive object has a surface that is larger than the wave and the signal itself largely stays intact. The best way to visualize this is to imagine a laser pointer or light source and a mirror. When you direct the pointer at the mirror, it

FIGURE 1-7

Reflection

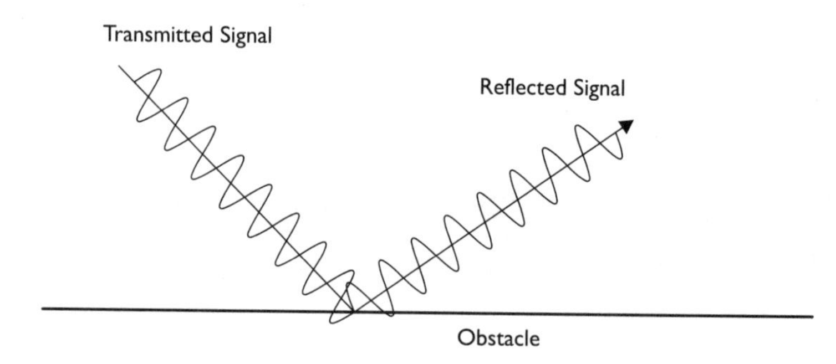

reflects the beam and, depending on the angle, will reflect it into another direction. Reflection can occur on a large variety of objects such as walls, furniture, doors, and even water, but it is most often seen with metal objects such as machinery, filing cabinets, and building materials.

Multipath

Multipath (Figure 1-8) is a variation on reflection where a signal bounces off several objects. Multipath creates a very substantial RF problem in that it can alter or degrade the wave and cause it to go out of phase, disrupting or even cancelling out the original signal. This cancelling out of the original signal can result in coverage "dead spots." In addition to this, the delays associated with reflected signals can cause the same signal to arrive at a receiver several times. In the MIMO discussion earlier in this chapter, we touched on how multiple antenna technology could actually take advantage of this phenomenon. The best method for mitigating the negative effects of multipath is to utilize *antenna diversity* on the access point. Antenna diversity is covered in the next section.

FIGURE 1-8

Multipath

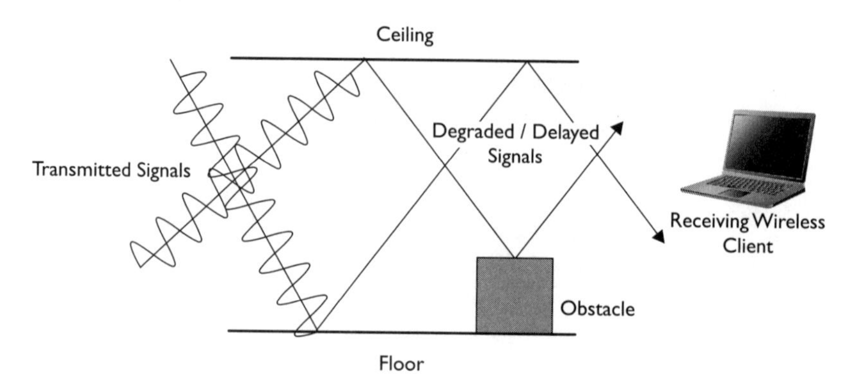

FIGURE 1-9

Scattering

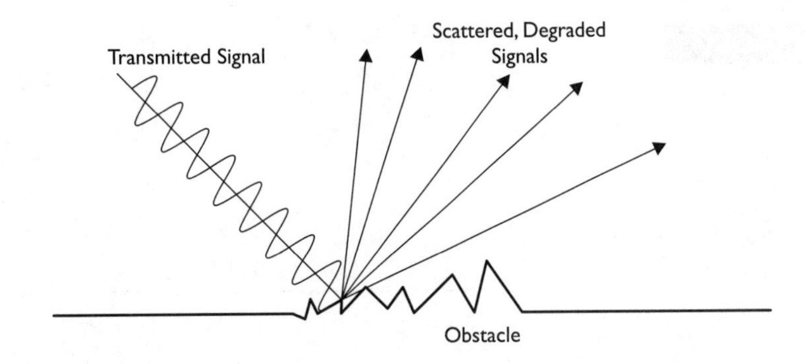

Scattering

Scattering (Figure 1-9) is another phenomenon similar to reflection except for the fact that it involves the bouncing of an original signal off an object that has surfaces that are smaller than the wave. When scattering occurs, a single original signal bounces off the object but reflects off in many different directions. Going back to the laser pointer analogy, imagine if the beam bounced off a disco ball or an uneven glass surface of some kind. Scattering can occur on uneven surfaces, trees, dust particles in the air, and even certain forms of precipitation. Because each of the bounced signals are portions of the original signal and could even be disrupted, scattering can lead to the weakening and even destruction of an RF transmission.

Refraction

Refraction (Figure 1-10) occurs when a signal passes through an object or area of varying density, resulting in a change of direction. In the laser pointer example, refraction would occur if you directed the beam at a glass of water or a prism. Although the object will let part of the signal pass through, it disrupts the direction of the signal and, depending on the material, will reflect some percentage of the signal as well. Refraction is usually prevalent on outdoor links where the RF signal has to pass through areas of changing temperature, humidity, and air density. The net damage of refraction is that it diverts a signal from its intended destination and possibly weakens it as well.

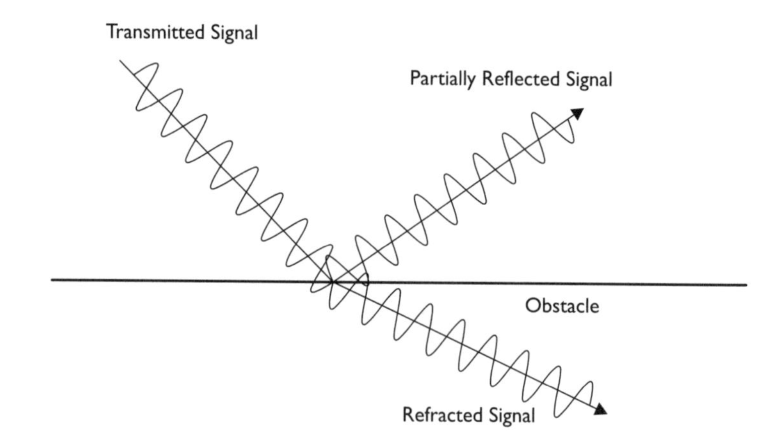

FIGURE 1-10

Refraction

Diffraction

Diffraction, which is often confused with refraction, is the bending of an RF signal around an obstacle. A good way to think of diffraction is to imagine what happens when you put your finger into a pool after dropping a pebble into it. The waves of the water from the pebble bend around your finger and flow in a different direction. Diffraction causes the same problems that refraction does in terms of weakening the signal and altering its direction.

Absorption

Absorption (Figure 1-11) is the most basic of RF distortions. In short, absorption is where an RF signal is absorbed by the object or medium that it hits, effectively weakening and sometimes destroying the signal. The best example of absorption is the muffling or elimination of audible sounds by a wall, object, or other medium. Absorption is one of the major factors that one must consider when placing access points in a particular building or area as it can substantially decrease the effective coverage range of an AP. Walls, human bodies, plants, certain types of glass and wood, and even water can absorb RF signals.

FIGURE 1-11

Absorption

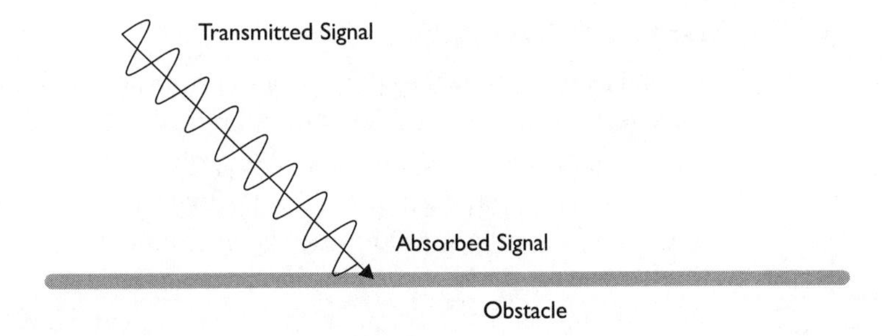

Antenna-Related RF Concepts

The key element of a wireless access point or any radio transmitter is the antenna that it uses. An antenna is functionally a device that allows you to transmit a signal and also shape the RF energy (transmit power) to suit the environment that you are providing wireless coverage for. When planning and troubleshooting a wireless network, having knowledge of each different antenna's capabilities and purpose is very important. Before we discuss antennas, it is important to talk about characteristics of antennas.

Polarization

Polarization refers to the direction in which an electromagnetic wave is emitted by an antenna. In simpler terms, polarization defines how an antenna is oriented in relationship to the ground. There are two major kinds of polarization when dealing with wireless antennas:

- **Vertical** The antenna is oriented perpendicular to the ground with the metal part radiating the signal to the sides. Vertical orientation is typically used for omnidirectional antennas.
- **Horizontal** The antenna is oriented parallel to the ground with the metal part radiating the signal up and down. Horizontal orientation is sometimes seen in client adapters but is generally less favorable.

Antenna Diversity

Antenna diversity refers to the use of multiple antennas on an access point to mitigate the effects of multipath-related interference. When an access point has diversity antennas, only one antenna at a given point in time is transmitting and receiving a signal. In a properly installed AP with diversity, the antennas are polarized parallel to each other and a short distance apart from one another. The access point has intelligence to decide which of the antennas to use at a given time based on signal strength indicators. A common misconception about access points with multiple antennas on the same radio is that they are both simultaneously operational and can, therefore, be pointed in different locations to provide varied coverage. It is crucial to remember what antenna diversity requires and how it behaves when deploying APs with external antennas. Antenna diversity is an important feature to have on an access point but not an absolutely crucial one. In environments where there is a shortage of antennas or less of a desire for antenna diversity, some Cisco APs can be configured to support transmitting and receiving from only one antenna. This is usually seen in outdoor bridging or point-to-point scenarios but can be applied to indoor deployments as well.

RSSI (Received Signal Strength Indicator)

RSSI or *received signal strength indicator* tells you, in essence, the strength of the signal that you are receiving from an access point. This measurement is usually in dBm (decibels of the measured power referenced to one milliwatt or mW) and is a negative number where values approaching zero signify increasing signal strength.

Radio transmit power measurements are usually expressed in milliwatts (mW), and signal strength readings at the client are usually measured in dBm. The radio frequency mathematics involved in this conversion are relatively simple. The specific math is beyond the scope of this book, but the general rule-of-thumb is to add 3 dBm for every doubling of transmit power in mW and to add 10 dBm each time the transmit power in mW is multiplied by 10. The most commonly used mW to dBm conversions are shown in Table 1-1.

RSSI is often displayed in site survey and wireless sniffer tools as well as some wireless client devices. RSSI is an excellent way to determine the proximity of an AP from your current location and to ascertain the effective coverage of an access point.

TABLE 1-1 Common Milliwatt (mW) to Decibel-Milliwatt (dBm) Conversions

Milliwatts (mW)	Decibel-Milliwatts
1 mW	0 dBm
2 mW	3 dBm
5 mW	7 dBm
10 mW	10 dBm
20 mW	13 dBm
50 mW	17 dBm
100 mW	20 dBm
1 W	30 dBm

SNR (Signal-to-Noise Ratio)

SNR or *signal-to-noise Ratio* is the ratio of your RSSI to the measurement of noise (also in dBm). SNR is a single number that can immediately tell you how strong the wireless signal is in relationship to the ambient radio frequency noise, known as the *noise floor*. SNR is measured in dB, and it is desirable to have a higher SNR at the client/receiver end in order to achieve higher data rates and more reliable associations.

Gain

Gain is an increase in an RF signal's amplitude (or the height of the wave) and can be affected by the following factors:

- **Antenna type used** Certain antennas, known as high-gain antennas, focus the RF signal by concentrating the RF energy in a certain direction. This reallocation of RF energy toward a certain focused area increases the signal's amplitude.
- **Transmit power** As the signal strength is increased for a given signal by the access point, more power goes into the antenna and increases gain.
- **Amplifiers** These are discussed in the "Antenna Accessories" section.

It is important to note that gain does not always mean an increase in transmit power. Gain is a result of increased transmit power but can be achieved by other means.

Loss

Loss is a decrease in transmitted signal strength and can be introduced by one of several factors. Unlike gain, loss means a decrease in transmit power and not just amplitude of the signal:

- **Lengths of antenna cable** A longer cable from the access point to the antenna causes signal loss.
- **Connectors** Loosely connected or faulty connectors may let signal leak and result in RF loss.
- **Distortions of the RF signal** All of the factors listed in the previous section introduce loss to an RF signal.
- **Attenuators** Devices that are purposely and discussed in the antenna accessories section of this chapter.

EIRP (Effective Isotropic Radiated Power)

EIRP (Effective Isotropic Radiated Power) is the amount of power that is actually radiated from an antenna. EIRP takes into account all gain and loss factors up to the point of antenna transmission, as well as the power being output by the access point. Because of the unpredictable nature of loss beyond the antenna by distortion or external factors, EIRP does not include loss beyond the cable and connectors between the transmitter (access point) and the antenna. EIRP can be broken down to the following equation:

EIRP = transmitter power + antenna gain – cable and connector loss

Line of Sight

Line of sight refers to the straight line between a transmitter and a receiver at which they can see each other. In visual line of sight, there is a straight line path between your eye and any specific object your eye can see. RF line of sight works in a very similar manner. Many outdoor wireless site surveys for bridged point-to-point wireless links are often performed using visual line of sight to determine the placement of antennas and bridges. Although line of sight is a relatively simple concept, there is another concept that sets RF line of sight apart from visual line of sight: the Fresnel Zone.

Fresnel Zone

The *Fresnel Zone* (Figure 1-12) refers to an elliptical area immediately around an RF signal's beam from transmitter to receiver. This elliptical area, if blocked, may only constitute a partial disruption of line of sight in the visual sense but can actually mean a break in a connection in an RF sense. In an outdoor environment, things such as trees, buildings, mountains, bodies of water, and even the curvature of the earth over a long enough distance can cause a blockage of the Fresnel Zone and create problems for a wireless link. The best ways to adjust to these issues are to either raise the transmitting and receiving antennas or eliminate the obstacles.

Antenna Types

A common misconception about antennas is that they can somehow amplify or increase coverage or transmit power on an access point (AP). Although an antenna is absolutely necessary for providing proper coverage from an AP (without one, coverage is very poor and extremely inconsistent), an antenna plays no part in adding or increasing the amount of RF energy that is emitted by an access point. Instead, antennas are used to shape the energy that is provided to it through the antenna connector from the AP's radio. When you replace an antenna with a different model to improve coverage, you are simply reshaping the RF signal coming from the access point in order to better suit the environment that you are providing coverage for. Antennas come in many shapes and sizes but can typically be broken down into two or three different categories. Simply put, the two *major* categories

FIGURE 1-12 The Fresnel Zone

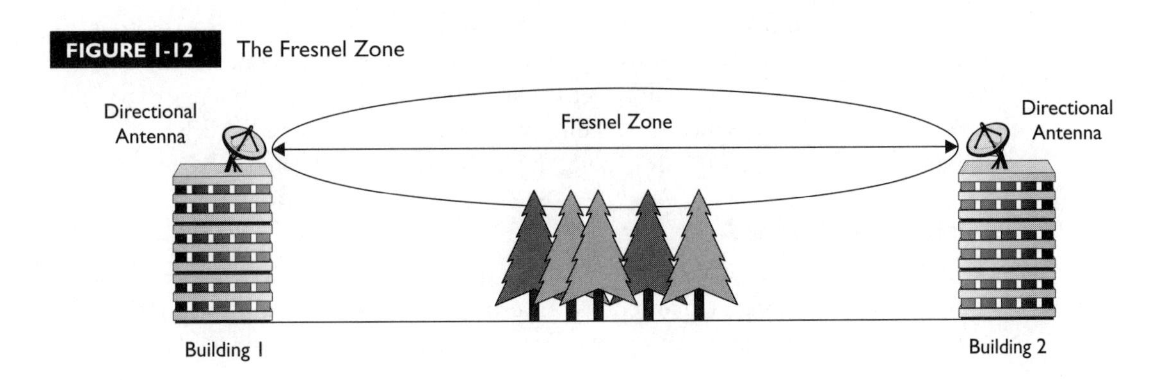

of antennas are omnidirectional antennas and directional antennas. There are also antennas, sometimes referred to as semidirectional, that have coverage characteristics of both omnidirectional and directional antennas.

When we talk about antennas, we refer to their coverage pattern—the shape of their coverage—in terms of the following two planes:

- **H-plane** Also known as the *horizontal* or *azimuth plane*, this is the coverage pattern that is parallel to the ground. This is normally the more important coverage pattern because it determines the range that most RF engineers take into account when performing a site survey. It is important to note that the H-plane applies only if the antenna is mounted and oriented properly.
- **E-plane** Also known as the *elevation plane*, this is the coverage pattern that is perpendicular to the ground or, more accurately, parallel to the antenna itself. This coverage pattern allows you to determine vertical coverage of an antenna and estimate things such as interfloor coverage in multifloor wireless deployments.

In addition to understanding the shape of antenna coverage, it is important to understand what the *dBi* or *decibel-isotropic* measurement for different types of antennas refers to. dBi is a unit that allows you to measure gain on a specific antenna. Earlier in this chapter, we discussed dBm (decibel-milliwatt), which referred to the RSSI measurement of a transmitting access point. The dBi measurement, in contrast, is not based on one milliwatt, but on a concept known as an *isotropic radiator*. An isotropic radiator is best thought of as a "perfect" antenna. More accurately, the isotropic radiator is a theoretical antenna, which transmits in all directions at full power and with complete 100 percent efficiency. All other antennas are measured against this perfect antenna. Because antennas are always used to introduce gain, you will never see an antenna measured with a negative value for dBi. Throughout this section, you will see antennas with a dBi gain designation. The higher the dBi value, the higher the overall gain of the antenna.

Omnidirectional Antennas

Omnidirectional antennas (Figure 1-13) are the most widely used antennas and are often used as the integrated antennas on certain access points. Omnidirectional antennas are characterized by a generally round H-plane pattern and a relatively

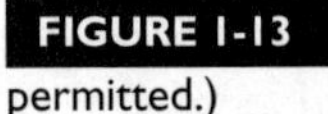

FIGURE 1-13 Typical omnidirectional antennas (Courtesy of Cisco Systems, Inc. Unauthorized use not permitted.)

narrow E-plane. A properly oriented omnidirectional antenna will provide coverage around the antenna element that is similar to a doughnut, where the antenna is in the middle of the doughnut's hole. As you increase the gain on an omnidirectional antenna, the H-plane will extend while the E-plane contracts. This will be visually similar to the doughnut of RF coverage around the antenna flattening and widening shown in Figure 1-14.

Omnidirectional antennas are most often used in office and indoor environments due to their lack of specific directionality and because of their relative ease of placement compared to more directional antennas which, you will see, have more specific placement restrictions. Omnidirectional antennas are also used in outdoor applications for point-to-multipoint bridge setups, which we will explore in Chapter 2.

FIGURE 1-14 Effect of increasing gain on omnidirectional antenna coverage pattern

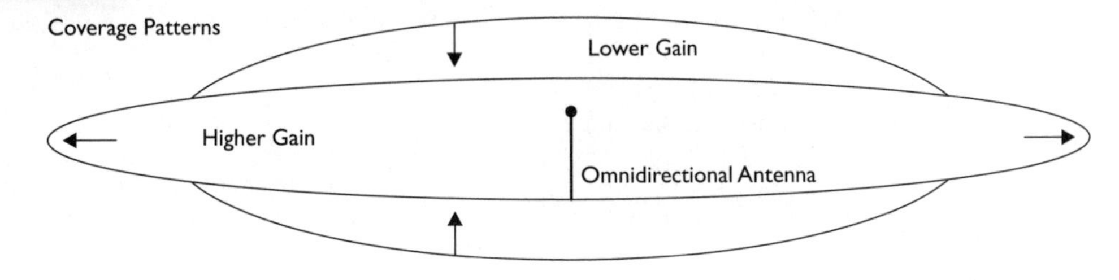

Cisco indoor access points such as the Cisco Aironet AP1130 series and Cisco Aironet AP1140 series feature integrated omnidirectional antennas and are not able to accept an external antenna of any kind. The most common omnidirectional antennas found on the Cisco Aironet 1230, 1240, and 1250 series antennas are the 2.2 dBi omnidirectional dipole (commonly known as a "rubber ducky" antenna) and the ceiling mounted AIR-ANT5959 on the 1230 and 1240 series access points. The typical H-plane and E-plane coverage patterns for a standard 2.2 dBi omnidirectional dipole (Figure 1-15) are very similar to the coverage patterns you will see for most omnidirectional antennas.

Semidirectional Antennas

Semidirectional antennas (Figure 1-16) encompass the more widely known *patch* or *yagi* antennas. These antennas are characterized by an H-plane that generally travels in one direction from the antenna (as opposed to the omnidirectional antennas where coverage radiates from all directions) and a normally round and E-plane pattern. As you increase the gain on a semidirectional antenna, you will decrease

FIGURE 1-15

Coverage pattern for 2.2 dBi dipole AIR-ANT4941 (omnidirectional)

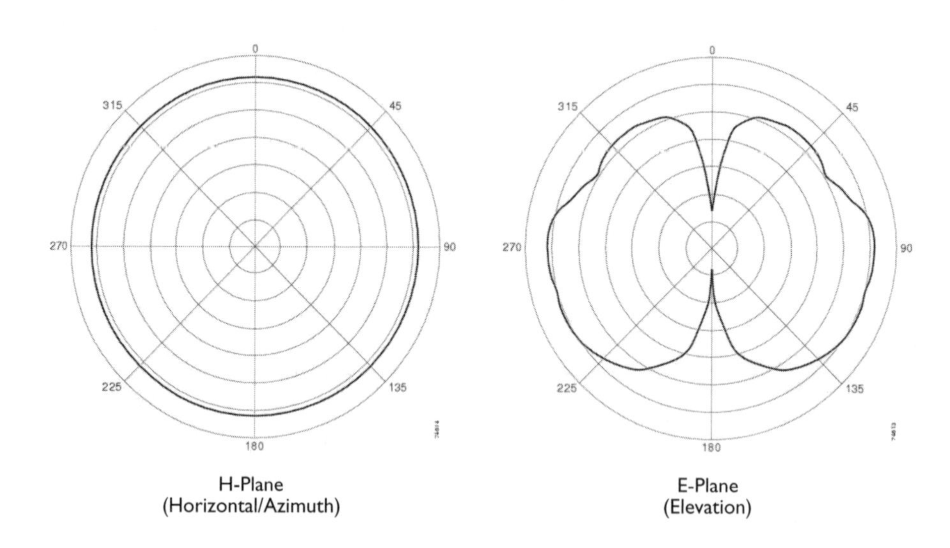

FIGURE 1-16

Typical patch antenna and yagi antenna (Courtesy of Cisco Systems, Inc. Unauthorized use not permitted.)

Patch Antenna

Yagi Antenna

the width of the E-plane while increasing the distance in which the H-plane radiates in the specific direction specified by the antenna.

Semidirectional antennas can be used in both indoor environments for providing coverage to hallways or pointing coverage into a specific direction of the building. Semidirectional antennas have seen a great deal of use in warehouse and distribution center environments where coverage is often directed down inventory aisles or between large pieces of machinery and equipment. Semidirectional antennas are also utilized in outdoor environments for providing coverage to open areas from a building or for short-distance point-to-point or point-to-multipoint bridging applications.

The Cisco Aironet 1230, 1240, and 1250 series antennas can all support various flavors of patch and yagi antennas, where the Cisco Aironet BR 1310 and 1410 series bridges have integrated patch antennas and the ability to connect external patch and yagi antennas. An example of a widely used 6.5 dBi diversity patch antenna exhibits an H-plane and E-plane coverage (Figure 1-17) that is typical of semidirectional patch and yagi antennas.

FIGURE 1-17

Coverage pattern for 8.5 dBi wall mount AIR-ANT2485P-R (semidirectional patch)

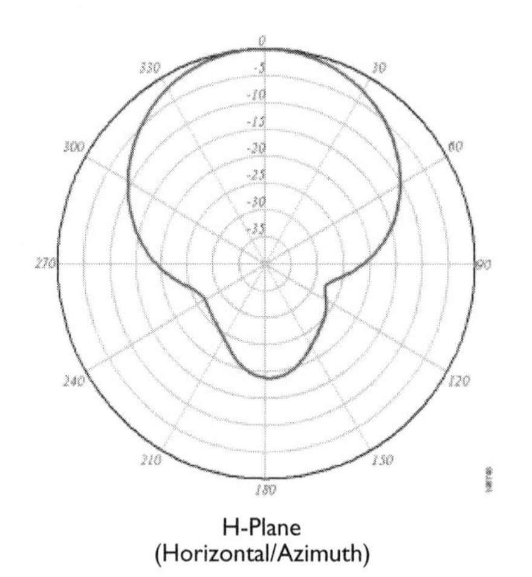

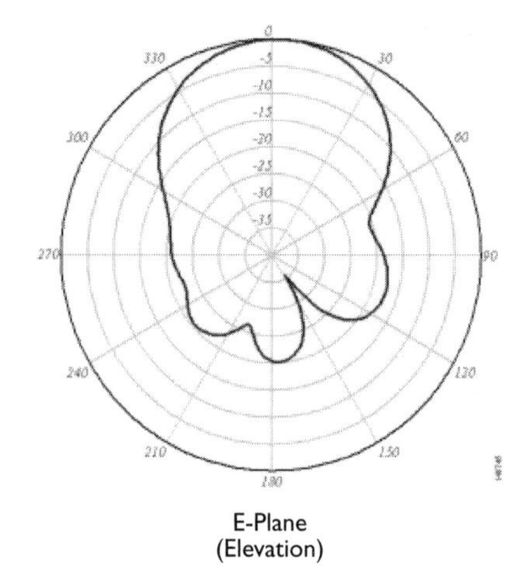

Directional Antennas

Directional antennas (Figure 1-18), sometimes known as *highly directional* antennas, are just as their name describes. Directional antennas are extremely high-gain devices that provide a very narrow, focused beam of coverage that can go a very long distance. When *parabolic dish* or *grid* antennas are mentioned, they refer to directional antennas.

FIGURE 1-18

Typical dish antenna and grid antenna (Courtesy of Cisco Systems, Inc. Unauthorized use not permitted.)

Parabolic Dish Antenna

Grid Antenna

Directional antennas have no practical application in indoor environments or for typical access point applications. This is because the highly focused coverage makes it very difficult for clients to share the wireless transmitter. Directional antennas are almost exclusively used in long distance point-to-point wireless bridge links.

Directional antennas are exclusively deployed on Cisco Aironet BR 1310 and 1410 series bridges for links that can go to up to several miles, depending on the transmit power and the specific gain of the antenna used.

Antenna Accessories

In addition to antennas, the often forgotten elements of a wireless system (and key elements in calculation of EIRP) are the various accessories that are found between an external antenna and its access point. In this section, we will cover the different categories of antenna accessories that are seen in various indoor and outdoor wireless installations.

Connectors

Connectors are the method by which an antenna is attached to a transmitting access point. There are several different kinds of connectors that are found on typical access points and bridges, but the ones that are used on Cisco equipment are RP-TNC or N connectors. Antenna connectors, by nature, are a source of some signal loss and this is exacerbated if an antenna or antenna cable is not properly connected or mismatched for the connector type on the access point or bridge.

Cables

In certain situations, an AP or bridge cannot be placed in the location where coverage is needed. In these cases, cables can be used to connect the access point connector to the external antenna. Antenna cables are manufactured and sold by a variety of vendors and can be used to adapt antennas to various different connectors as well. The primary drawback of antenna cables is that they introduce unintended and often undesirable signal loss. The longer an antenna cable is, the more severe the signal loss. Cisco does offer low loss and ultra-low loss cables to mitigate the effects of cable loss.

Splitters

Splitters are used to split a radio signal from a transmitter into several different directions. Conceptually, a splitter allows you to take a single wireless signal and have it transmit to antennas in two different locations. Although this has some practical applications, it severely degrades the original transmission and is generally considered an undesirable element to introduce into a wireless deployment, especially in cases where it is relatively easy to add another access point.

Amplifiers

Amplifiers, which we touched on in the "Gain" section of this chapter, are simply devices that can be placed between the access point and the antenna to reduce the effects of cable and/or connector loss. Most amplifiers are powered by an electrical DC injector, which enables the signal amplification to occur.

Attenuators

Attenuators do the opposite of what amplifiers do. Attenuators *attenuate*, or introduce loss, into a transmitted signal. Like amplifiers, attenuators are also placed between the access point and the antenna to lower the transmit power of a signal. In certain environments where restricted or lowered coverage is desired, attenuators can be a very useful way to control a signal's coverage pattern.

Lighting Arrestors

Lightning arrestors are used in outdoor access points and bridges to mitigate the effects of lighting strikes on antennas that may be located on building rooftops or high locations. A wireless antenna, which is effectively a specialized cable, can become a lightning rod under the right circumstances and cause damage to the cabling and wireless equipment that it is connected to. Because of this, it is absolutely necessary to have lightning arrestors installed on wireless components, antennas, and equipment where there is a likelihood of lightning strikes.

INSIDE THE EXAM

Spread Spectrum Technology

For the exam, you should be able to understand the advantages of spread spectrum technology. Know that spread spectrum is characterized by low peak power and wider bandwidth and that it resists interference while enabling higher data rates. Understanding the concept of modulation is important. Know that modulation is a process by which data is placed onto a signal. Be familiar with the differences between the types of modulation and know that DSSS is typically associated with 802.11b and OFDM with 802.11a and 802.11g and that MIMO is specifically used in 802.11n.

Have an understanding of the unlicensed frequency bands that wireless networking runs on. The 900 MHz band provides good range but does not allow for high enough data rates to be used by IEEE 802.11 wireless technologies. 2.4 GHz is the most highly utilized band and faces the most interference but is also the easiest to support. 5.0 GHz is a quieter band that is seeing more utilization. Understand the difference between U-NII 1, U-NII 2, and U-NII 3. U-NII 1 is used only in indoor environments, while U-NII 2 is used in indoor and outdoor environments and U-NII 3 is used exclusively for outdoor environments.

Be familiar with the concepts of channel overlap and channel reuse. Remember that there is only one combination of three nonoverlapping channels in the 2.4 GHz band. Those channels are always 1, 6, and 11. Understand that channel reuse involves placement of access points next to other access points with nonoverlapping channels so that channels can be reused within the same facility. Have a basic understanding of how CSMA/CA works. CSMA/CA is simply a method by which transmitting devices send out communications to other devices on the wireless network and selectively hold back or delay their transmissions so that others can proceed.

Wireless LAN RF Principles

You need to remember the different sources of RF signal distortion. In addition to interference, there is reflection, multipath, scattering, refraction, diffraction, and absorption. Have a detailed understanding of multipath interference and how diversity antennas can mitigate its effects.

Know antenna-related concepts such as polarity, diversity, and elements of RSSI and SNR. RSSI and SNR are important measures for wireless site surveys and are an excellent way to determine whether a given wireless client is or will receive adequate coverage from a transmitting access point. Be familiar with RF gain and loss and understand what can affect both. Gain is not necessarily an introduction of more transmit power. Understand the equation for EIRP (EIRP= transmitter power + antenna gain – cable and connector loss). Know the basic types of antennas. Semidirectional and directional antennas can be placed into the same effective group, but they are separated by the coverage distance and gain on them. Be familiar with the accessories associated with wireless access point antennas and how they can affect RF gain and loss.

CERTIFICATION SUMMARY

Spread spectrum technology is a method by which data is placed on a signal medium, such as radio frequency waves. Modulation is the process by which the date is placed on the medium and can be performed with different methods of modulation. In wireless networking, these methods are DSSS, OFDM, and MIMO. DSSS is seen in 802.11b, OFDM is seen in 802.11a and 802.11g, and MIMO is seen in 802.11n

There are three primary unlicensed frequency bands in use today: 900 MHz, 2.4 GHz, and 5 GHz. 2.4 GHz and 5 GHz are used extensively by IEEE 802.11 technology because they allow for higher data rates due to their higher bandwidth and ability to utilize modulation. 802.11b/g/n uses 2.4 GHz, and 802.11a/n uses 5 GHz. Within each of these bands are channels that are utilized. In the 2.4 GHz band there is one combination of three nonoverlapping channels of 22 MHz in width: 1, 6, and 11. In the 5 GHz band, there is a combination of eight nonoverlapping channels in the U-NII 1 and U-NII 2 band for indoor use. More channels in 802.11a can be used if U-NII 2 is extended or U-NII 3 is implemented. Dynamic rate shifting also enables an access point to provide coverage at greater distance by allowing a less complex modulation and using a lower specific data rate at a given distance and SNR.

Radio frequency signals can be distorted in a number of ways by objects between the receiver and transmitter. Reflection bounces a signal into a different direction. Multipath consists of a number of reflections leading to delayed and repeated signals and/or signal degradation and dispersion. Scattering severely weakens a signal as it breaks off into several weaker signals streams. Refraction and diffraction are ways that signals are bent and degraded by specific mediums. Absorption represents the total dissipation of a signal in certain mediums.

Access point antennas need to be properly polarized, use diversity, and be adjusted to achieve the highest possible EIRP by managing loss and providing the correct amount of gain to provide coverage over a desired area. In outdoor wireless bridging, it is important to consider line of sight and the Fresnel Zone, which states that a bridged link needs more than visual line of sight to achieve a successful wireless connection.

Antenna types can be broken down into omnidirectionals, patch/yagi (or semidirectional) and dish/grid (or directional). Each group of antennas has its own application with specific coverage patterns. Antennas often utilize connectors or cables and can use splitters, amplifiers, and attenuators, depending on the environment and administrator preferences.

TWO-MINUTE DRILL

Spread Spectrum Technology

- Spread spectrum technology is characterized by high bandwidth and low peak power and is naturally resistant to narrowband interference.
- The goals of radio transmissions are to send as much data as possible, as far as possible, as fast as possible.
- The three laws of radio dynamics are that higher data rate = shorter transmission range; higher power output = increased range and lower battery life on devices; and higher frequency radios = higher data rates and shorter ranges.
- Modulation is the process by which data is placed onto a signal medium. There are three major types of modulation used in IEEE 802.11: DSSS, OFDM, and MIMO.
- Unlicensed frequency bands include 900 MHz, 2.4 GHz, and 5 GHz. 2.4 GHz is used by IEEE 802.11b/g/n, and 5 GHz is used by IEEE 802.11a/n.
- The goal in channel assignment is to provide for as many access points as possible while having as little channel overlap as possible. This is accomplished by channel reuse.

Wireless LAN RF Principles

- RF signals can be distorted by reflection, multipath, scattering, refraction, diffraction, absorption, and interference.
- Antenna polarization and diversity are physical methods of improving antenna effectiveness.
- RSSI and SNR are methods for monitoring signal strength from an access point and its antenna.
- Gain and loss represent signal strength from a transmission device.
- EIRP (Effective Isotropic Radiated Power) = transmitter power + antenna gain – cable and connector loss.
- The Fresnel Zone represents the area above and below the visual line of sight between two outdoor bridges connected by a point-to-point link. If an obstacle blocks any part of the Fresnel Zone, wireless transmissions are negatively affected.

- Omnidirectional antennas include dipoles and are used indoors or in point-to-multipoint outdoor links.
- Semidirectional or patch and yagi antennas are used both indoors and outdoors for a variety of applications. They have more directionality than omnidirectional antennas.
- Directional or dish and grid antennas are used primarily outdoors for point-to-point bridging applications. They are highly directional due to high gain and transmit powers allowed.
- A variety of devices can introduce gain or loss into a wireless system. Connectors, cables, and attenuators introduce loss while amplifiers, higher transmit power, and higher gain antennas introduce gain.

SELF TEST

The following self test questions will help you measure your understanding of the material presented in this chapter. Read all the choices carefully, as there may be more than one correct answer. Choose all correct answers for each question.

Spread Spectrum Technology

1. Which of the following is not an advantage of spread spectrum technology?

A. Higher security

B. Less complex modulation required

C. Greater resistance to interference

D. Higher data rates

2. Which of the following is not a true modulation type used in IEEE 802.11 wireless networks?

A. CSMA

B. OFDM

C. MIMO

D. DSSS

3. In the United States, all but which of the following nonoverlapping channels in the 2.4 GHz range is used for IEEE 802.11b/g?

A. 6

B. 1

C. 14

D. 11

4. Which of the following unlicensed bands is utilized by IEEE 802.11a?

A. 900 MHz

B. 2.4 GHz

C. 5.0 GHz

D. None of the above

5. Which of the following is not true of modulation?

A. More complex modulation equals lower data rates.

B. The process of modulation alters a waveform.

C. Modulation produces a carrier signal.

D. None of the above.

Wireless LAN RF Principles

6. An RF signal can be negatively distorted by all but which of the following?
   A. Refraction
   B. Interference
   C. Raising the elevation of the Fresnel Zone
   D. Multipath
7. Which of the following is not a way to introduce loss into a wireless system?
   A. Add an attenuator.
   B. Utilize a shorter antenna cable.
   C. Place the wireless access point in a shielded room.
   D. Loosen antenna connectors.
8. When installing an access point for an indoor wireless system, which of the following antennas would you not use?
   A. 2.2 dBi omnidirectional dipole
   B. Parabolic dish
   C. Diversity patch
   D. All of the above can be used effectively in an indoor wireless system.
9. When a signal is refracted, which of the following cannot occur?
   A. The signal reaches the receiver in a shorter amount of time.
   B. A part of the signal is reflected.
   C. The strength of the signal is degraded.
   D. The direction of the signal is altered.
10. Which of the following is not true of multipath interference?
    A. Dead spots of coverage are created.
    B. Signals arrive at the receiver distorted.
    C. Diversity antennas can mitigate the effects of multipath.
    D. All of the above are true of multipath interference.

SELF TEST ANSWERS

Spread Spectrum Technology

1. Which of the following is not an advantage of spread spectrum technology?

A. Higher security

B. Less complex modulation required

C. Greater resistance to interference

D. Higher data rates

☑ **B.** Spread spectrum technology depends on more complex modulation to accommodate higher data rates without utilizing more frequency bandwidth.

☒ **A, C,** and **D** are incorrect. All of these are valid advantages and positive characteristics of spread spectrum technology.

2. Which of the following is not a true modulation type used in IEEE 802.11 wireless networks?

A. CSMA

B. OFDM

C. MIMO

D. DSSS

☑ **A.** CSMA is not a modulation type, it is a method of avoiding collisions.

☒ **B, C,** and **D** are valid modulation types used in IEEE 802.11 wireless networks.

3. In the United States, all but which of the following nonoverlapping channels in the 2.4 GHz range is used for IEEE 802.11b/g?

A. 6

B. 1

C. 14

D. 11

☑ C. Although Channel 14 is nonoverlapping in the 2.4 GHz range, it is not used in the United States.

☒ **B, C,** and **D** are valid nonoverlapping channels used in the United States.

4. Which of the following unlicensed bands is utilized by IEEE 802.11a?

A. 900 MHz

B. 2.4 GHz

C. 5.0 GHz

D. None of the above

☑ **C.** 802.11a utilizes some or all of the three U-NII bands of the 5 GHz range.

☒ **A, C,** and **D** are incorrect. 900 MHz is not used by any 802.11 wireless networks and 2.4 GHz is used by the 802.11b and 802.11g standards.

5. Which of the following is not true of modulation?

A. More complex modulation equals lower data rates.

B. The process of modulation alters a waveform.

C. Modulation produces a carrier signal.

D. None of the above.

☑ **A.** More complex modulation increases data rates in a given bandwidth.

☒ **B, C,** and **D** are incorrect. The process of modulation does alter waveforms and produces a resulting carrier signal.

Wireless LAN RF Principles

6. An RF signal can be negatively distorted by all but which of the following?

A. Refraction

B. Interference

C. Raising the elevation of the Fresnel Zone

D. Multipath

☑ **C.** Raising the elevation of the Fresnel Zone is an effective way to avoid objects that can interfere with radio transmissions.

☒ **A, B,** and **D** are incorrect. Refraction, interference, and multipath are all negative forms of distortion that can affect an RF signal.

7. Which of the following is not a way to introduce loss into a wireless system?

A. Add an attenuator.

B. Utilize a shorter antenna cable.

C. Place the wireless access point in a shielded room.

D. Loosen antenna connectors.

☑ **B.** The longer the antenna cable, the more loss introduced into the wireless system. Shortening an antenna cable decreases loss.

☒ **A, C,** and **D** are methods of introducing or increasing RF signal loss and are incorrect.

8. When installing an access point for an indoor wireless system, which of the following antennas would you not use?

A. 2.2 dBi omnidirectional dipole

B. Parabolic dish

C. Diversity patch

D. All of the above can be used effectively in an indoor wireless system.

☑ **B.** A parabolic dish antenna, because of its coverage characteristics, has no practical application in indoor environments.

☒ **A, C,** and **D** are incorrect. In various indoor wireless environments you are very likely to see 2.2 dBi omnidirectional dipoles, and diversity patch antennas have been known to be used in select scenarios.

9. When a signal is refracted, which of the following cannot occur?

A. The signal reaches the receiver in a shorter amount of time.

B. A part of the signal is reflected.

C. The strength of the signal is degraded.

D. The direction of the signal is altered.

☑ **A.** Refraction involves the passing and reflecting of a signal off a medium and therefore increases the amount of time it takes that signal to reach its receiver.

☒ **B, C,** and **D** are incorrect. During refraction, an RF signal can be partly reflected, have its direction altered, and almost always be degraded.

10. Which of the following is not true of multipath interference?

A. Dead spots of coverage are created.

B. Signals arrive at the receiver distorted.

C. Diversity antennas can mitigate the effects of multipath.

D. All of the above are true of multipath interference.

☑ **D.** All of the choices are characteristics of multipath interference.

☒ **A, B,** and **C** are incorrect. Although each of these individual answers is true of multipath interference, the correct answer is all of the above.

2

Wireless LAN Standards and Topologies

CERTIFICATION OBJECTIVES

2.01 Wireless Regulatory Bodies

2.02 Wireless LAN Standards and Certifications

2.03 Wireless Topologies and Frame Types

✓ Two-Minute Drill

Q&A Self Test

Maturing standards and protocols have aided the acceleration of wireless LAN adoption in the past 10 years. Clear regulatory rules and definitions have also provided a healthy environment for the industry to continue to evolve. Depending on where you are in the world, wireless LAN may have slightly different meanings to you based on your local regulatory rules and standards. Let's first understand who is making these rules and standards for wireless LAN. In this chapter, we will discuss briefly the roles the FCC, ETSI, IEEE, IETF, and Wi-Fi Alliance play in creating standards that allow computers and devices to communicate with one another wirelessly. We will also discuss the use of frequencies and the 802.11 family of protocols that provide the foundation of wireless LAN communications. Furthermore, we will discuss the basic building blocks of the 802.11 standards and the enhancements introduced in the 802.11 extensions.

CERTIFICATION OBJECTIVE 2.01

Wireless Regulatory Bodies

In the wireless LAN world, there are two influential regulatory bodies that help define rules and standards for the rest of the world: the FCC and ETSI. Regulatory authorities from many countries adopt and reference rules and regulations set by the FCC and ETSI as foundations for their own countries. Furthermore, the regulatory bodies from different countries usually work and collaborate closely to ensure that their regulations work well with neighbor countries.

FCC

The Federal Communications Commission (FCC) is the regulatory authority in the United States and its territories for all forms of telecommunications, including radio, Internet, telephone, television, satellite, and cable. Established by the Communications Act of 1934, the FCC is an independent agency similar to FTC and SEC and is directly accountable to Congress. FCC sets the rules and regulations within which wireless LAN and devices operate and follow. FCC also regulates the radio frequencies, power settings, transmission techniques, and hardware, such as radios and antennas used in wireless LAN. FCC rules are often adopted and followed by authorities from other countries and regulatory domains to set countries' local rules and regulation on wireless LAN communications.

ETSI

The European Telecommunications Standards Institute (ETSI) is somewhat equivalent to the FCC in United States for the European Union countries. ETSI is legally and officially recognized by the European Commission as a European Standards Organization that sets standards for Information and Communications Technologies (ICT), including mobile, radio, broadcast, and Internet technologies. ETSI does not enforce the rules and regulations in each European Union (EU) country, but rather it defines the standards to be used by EU countries.

Knowing the regulatory domain in which a wireless LAN will be deployed is one of the most important steps in planning a wireless LAN deployment. In different countries, government regulations may direct APs and clients to use different radios and antennas, which in term will greatly influence how a wireless LAN is surveyed and designed. Furthermore, if a wireless user is expected to travel across different regulatory domains (usually traveling between countries), the wireless adapter he/she uses should also be certified to support local rules and regulations set by local authorities.

CERTIFICATION OBJECTIVE 2.02

Wireless LAN Standards and Certifications

Apart from government regulations and regulatory rules, technical standards also contribute largely to the success of wireless LAN and modern day information technology. The standards bodies set standards and protocols for the industry that wishes to build technology using a common set of languages. The standards and protocols are designed to ensure that communications equipment from different vendors can interoperate. Industry organizations such as the Wi-Fi Alliance offer interoperability testing and certifications to ensure the proper implementations of standards and protocols.

IETF

The Internet Engineering Task Force (IETF) is another influential standards body that creates standards and protocols for the Internet, including the core protocols of the Internet Protocol Suite, TCP/IP. Another important standard published by IETF is the Extensible Authentication Protocol (EAP) defined in RFC 3748 and later updated in RFC 5247. EAP provides a framework for modern wireless LAN security standard, working in conjunction with IEEE 802.1X, to negotiate authentication mechanism between client and infrastructure, but it does not specify the actual authentication methods. The Wi-Fi Alliance has adopted five different EAP authentication types for its WPA and WPA2 standards. The five EAP types are

- EAP-TLS
- EAP-TTLS/MSCHAPv2
- PEAPv0/EAP-MSCHAPv2
- PEAPv1/EAP-GTC
- EAP-SIM

IEEE

The Institute of Electrical and Electronics Engineers (IEEE) is one of the most influential standards bodies in the world for information technology. IEEE has created many widely adopted standards, such as port-based Network Access Control (IEEE 802.1X), Ethernet (IEEE 802.3), Firewire (IEEE 1394), and wireless LAN (IEEE 802.11). The 802.11 standards and its variant wireless LAN standards are now the foundation used by virtually all wireless LAN vendors and products in the world. The following sections provide an overview of 802.11 and some key variant standards that are widely adopted for wireless LAN.

Wi-Fi Alliance

The Wi-Fi Alliance is a nonprofit industry organization and was formed in 1999 with the goals of driving adoption of IEEE 802.11 wireless LAN standards, promoting proper implementation of standards, and ensuring interoperability through its rigorous certification process. Wi-Fi Alliance launched the Wi-Fi CERTIFIED program in March 2000, and the program provides a widely recognized designation of interoperability. The program also helps to ensure that Wi-Fi enabled products deliver the best user experience. To date, Wi-Fi Alliance has completed more than 5,000 product certifications. The program consists of mandatory and optional programs. The mandatory program provides the following certifications:

- 802.11a
- 802.11b
- 802.11g
- 802.11 a/b/g in single, dual mode (802.11b and 802.11g)
- 802.11 a/b/g in multiband (2.4GHz and 5GHz)
- WPA (Wi-Fi Protected Access)—Enterprise, Personal
- WPA2 (Wi-Fi Protected Access 2)—Enterprise, Personal
- EAP Types

The optional program provides the following certifications:

- 802.11n DRAFT 2
- Wi-Fi Protected Setup
- WMM—QoS
- WMM Power Save

Figure 2-1 shows the Wi-Fi certificate of a Cisco 1242 AG AP and the certification designations this AP bears. The certificate also shows the based IEEE technologies (802.11 a/b/g/d/h), security options (WPA/WPA2), EAP types, and multimedia standards (WMM and WMM Power Save) this AP supports.

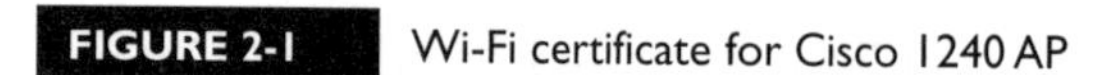

FIGURE 2-1 Wi-Fi certificate for Cisco 1240 AP

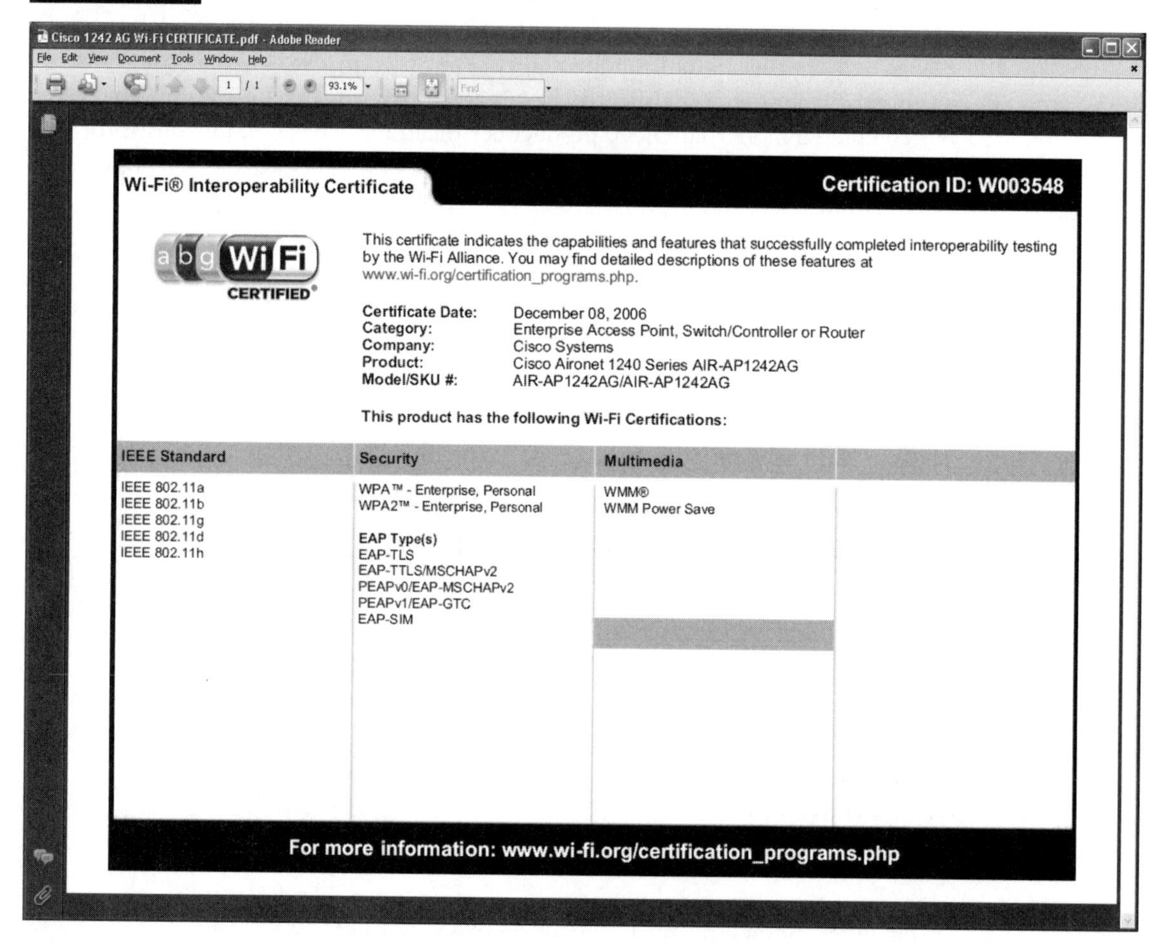

Cisco 1242 AG Wi-Fi CERTIFICATE.pdf - Adobe Reader

Wi-Fi® Interoperability Certificate **Certification ID: W003548**

This certificate indicates the capabilities and features that successfully completed interoperability testing by the Wi-Fi Alliance. You may find detailed descriptions of these features at www.wi-fi.org/certification_programs.php.

Certificate Date: December 08, 2006
Category: Enterprise Access Point, Switch/Controller or Router
Company: Cisco Systems
Product: Cisco Aironet 1240 Series AIR-AP1242AG
Model/SKU #: AIR-AP1242AG/AIR-AP1242AG

This product has the following Wi-Fi Certifications:

IEEE Standard	Security	Multimedia	
IEEE 802.11a IEEE 802.11b IEEE 802.11g IEEE 802.11d IEEE 802.11h	WPA™ - Enterprise, Personal WPA2™ - Enterprise, Personal **EAP Type(s)** EAP-TLS EAP-TTLS/MSCHAPv2 PEAPv0/EAP-MSCHAPv2 PEAPv1/EAP-GTC EAP-SIM	WMM® WMM Power Save	

For more information: www.wi-fi.org/certification_programs.php

802.11

The wireless LAN networking is a very crowded space that consists of many different technologies and protocols competing for acceptance and popularity. Similar to the evolution of wired LAN, several different technologies and standards have come and gone. Today, there is one standard that is, almost if not already, synonymous with the term "wireless networking" for consumer and enterprise users: IEEE 802.11 and its variants. For convenience, unless otherwise stated, wireless LAN or WLAN will refer to a wireless LAN network based on 802.11 and its variant standards.

802.11 is primarily a physical (PHY) and media access control (MAC) layer specification for wireless LAN connectivity developed by IEEE, and the two layers correspond to physical and data link layers in the OSI reference model. The 802.11 variant protocols address either the PHY or MAC or both while assuming the upper layer protocols work properly. In 1997, IEEE published the first WLAN standard, IEEE std 802.11, which was revised in 1999. The original 802.11 describes three PHY, which includes infrared (IR) PHY, frequency hopping spread spectrum (FHSS) PHY, and direct sequence spread spectrum (DSSS) PHY, all of which support 1 Mbps and 2 Mbps data rates. Although three PHY layers are defined in the standard, IR and FHSSS never gained wide acceptance and DSSS became the dominant PHY for WLAN in homes, enterprises, and other vertical segments. In 1999, two additional PHY types were added, one described in IEEE Std 802.11a using orthogonal frequency division multiplexing (OFDM) operating in the Unlicensed National Information Infrastructure (U-NII or UNII) bands supporting up to 54 Mbps and the second as an extension to original DSSS PHY described in IEEE Std 802.11b supporting up to 11 Mbps in the 2.4 GHz spectrum. In 2002, the 802.11 working group published IEEE Std 802.11g adding OFDM to the 2.4 GHz band.

exam watch

The CCNA Wireless exam will focus on 802.11 and a subset of 802.11 protocols and amendments, including 802.11a, 802.11b, 802.11g, and 802.11n. All of these protocols are based on the original 802.11 protocol that was published in 1997. Each of the variants and amendments enhanced the original protocol and provided additional benefits to the end users. Although the 802.11 standard includes IR and FHSS, the exam will focus on DSSS as base technology for 802.11.

The original 802.11 devices operate in the 2.4 GHz spectrum and within the Industrial, Scientific, and Medical (ISM) bands. The 2.4 GHz spectrum was divided into fourteen 22 MHz–wide channels, and the center frequency of each channel is 5 MHz apart. Table 2-1 shows the frequency to channel mapping.

TABLE 2-1 2.4 GHz Frequency and Channel Mapping

Channel	Lower Frequency (GHz)	Center Frequency (GHz)	Upper Frequency (GHz)
1	2.401	2.412	2.423
2	2.404	2.417	2.428
3	2.411	2.422	2.433
4	2.416	2.427	2.438
5	2.421	2.432	2.443
6	2.426	2.437	2.448
7	2.431	2.442	2.453
8	2.436	2.447	2.458
9	2.441	2.452	2.463
10	2.446	2.457	2.468
11	2.451	2.462	2.473
12	2.456	2.467	2.478
13	2.461	2.472	2.483
14	2.473	2.484	2.495

Depending on the country or regulatory domain you're in, you may use all or a subset of the 14 channels. For example, in Japan, all 14 channels are allowed to be used in the 2.4 GHz spectrum, whereas only 11 of the first 14 channels are allowed to be used in the United States. In some monitoring tools, you will see that all 14 channels are available, but use of channels 12–14 should be in passive mode (listening) only, and no active transmission should be allowed in channels 12–14 in the United States. Figure 2-2 illustrates the channel to frequency mapping in the 2.4 GHz spectrum.

Because the channels are closely defined in the 2.4 GHz spectrum with as little as 5 MHz separation from an adjacent channel, interferences from an adjacent channel or adjacent channels are highly likely unless there is sufficient spacing between channels used in a WLAN. In the 802.11 specification, IEEE states:

> In a multiple cell network topology, overlapping and/or adjacent cells using different channels can operate simultaneously without interference if the distance between the center frequencies is at least 25 MHz.

FIGURE 2-2 2.4 GHz channel to frequency map

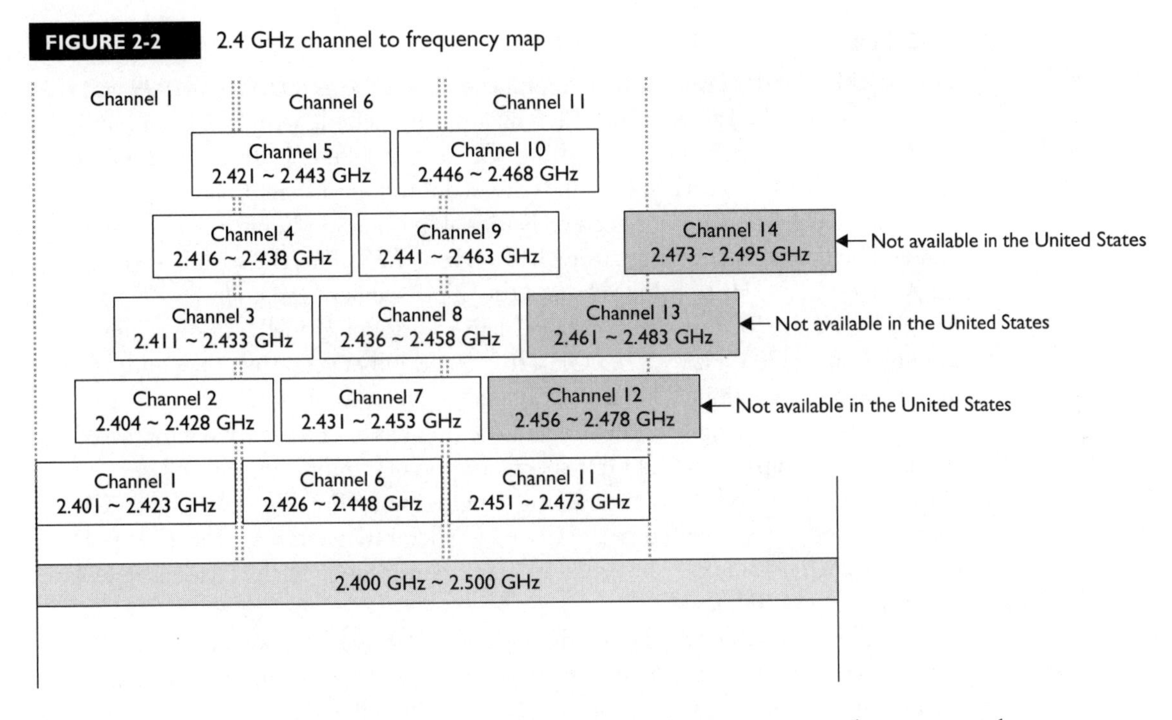

Therefore, in the 2.4 GHz spectrum, there are at most three nonoverlapping channels available to use with 5-channel separation between each. For example, channels 1 and 6 are considered nonoverlapping because the center frequencies of channel 1 (2412 MHz) and channel 6 (2437 MHz) are 25 MHz apart. You can use any two channels to avoid interferences as long as the center frequencies are separated by five channels or 25 MHz; nonetheless, in order to maximize the capacity of the WLAN, channels 1, 6, and 11 are widely regarded in the industry as "the nonoverlapping channels" in the 2.4 GHz frequency band.

For ETSI or Japan regulatory domains, although there are more channels available to choose from, there are still only three nonoverlapping channels available in the 2.4 GHz spectrum.

Some field engineers have promoted use of a 4-channel plan in the 2.4 GHz spectrum in the hopes of increasing overall capacity and performance of wireless LAN. Nonetheless, there simply aren't enough channels and spacing in the 2.4 GHz spectrum to allow this type of channel planning without artificially introducing adjacent channel interferences. In a dense deployment, the overall wireless LAN performance may actually degrade because of the adjacent interferences introduced by its own AP(s).

802.11a

The 802.11a standard was ratified around the same time as 802.11b in 1999 and operates in the 5 GHz spectrum, which makes it inoperable with 802.11 or any radio operating in the 2.4 GHz spectrum. One major difference from the 2.4 GHz spectrum is that the 5 GHz spectrum is divided into four different band segments, which are referred to as the Unlicensed National Information Infrastructure (U-NII) bands. The first segment ranges from 5.15 GHz to 5.25 GHz, the middle segment ranges from 5.25 GHz to 5.35 GHz, and the upper segment ranges from 5.725 GHz to 5.825 GHz. In 2003, the FCC proposed an additional 255 MHz of spectrum, ranging from 5.470 GHz to 5.725 GHz, to be used in between the upper and middle segments, making 23 channels available over a total of 555 MHz of radio frequencies. Even though bandwidth and the number of channels have increased significantly compared to 2.4 GHz spectrum, the maximum transmit power allowed for each U-NII band is different, ranging from 50mW to 1W. The FCC's specific ruling is documented in the Code of Federal Regulations, Title 47 Part 15 (47CFR15.407). Table 2-2 shows the transmit power range in the United States frequency band (GHz).

The ETSI has published slightly different rules for available frequencies and transit power for use of the 5 GHz spectrum in the EU countries; the rules are documented in Harmonized European Standard: Broadband Radio Access Networks (BRAN); 5 GHz High Performance Radio Local Area Networks (RLAN) (ETSI EN 309 893 v1.5.1 2008-08). Table 2-3 shows the transmit power range specified by the ETSI for frequency band (GHz).

TABLE 2-2 Maximum Transmit Power in the 5 GHz Spectrum in the United States

	Maximum Transmit Power	Maximum EIRP	Use
U-NII 1 5.125–5.250	50mW	200mW	Indoor only
U-NII 2 5.250–5.350	250mW	1W (+30 dBm)	Indoor and outdoor
U-NII 2E 5.470–5.725	250mW	1W (+30 dBm)	Indoor and outdoor
U-NII 3 5.725–5.850	1W	200W (+53 dBm)	Indoor and outdoor

TABLE 2-3 Maximum Transmit Power in 5 GHz Under ETSI Rules

Frequency Band (GHz)	Mean EIRP Limit
5.150–5.250	200mW
5.250–5.350	200mW
5.470–5.725	1W

Figure 2-3 illustrates the channel to frequency mapping in the 5.0 GHz spectrum.

exam watch

802.11a is not compatible with 802.11 and 802.11b and does not support any data rates that are supported in 802.11 and 802.11b. Furthermore, because 802.11a operates in 5 GHz, 802.11a does not interfere with 802.11 and 802.11b/g radios.

802.11a uses OFDM as default modulation and supports different data rates from 802.11; the 802.11a standard supports 6, 9, 12, 18, 24, 36, 48, and 54 Mbps data rates; 6, 12, 24 Mbps are the mandatory rates. The initial FCC regulations required the use of "integral" (permanently attached) antennas for the U-NII1 band. In 2004, the FCC updated the regulations and now permits the use of external antennas in the U-NII1 band.

FIGURE 2-3 5 GHz channel to frequency map

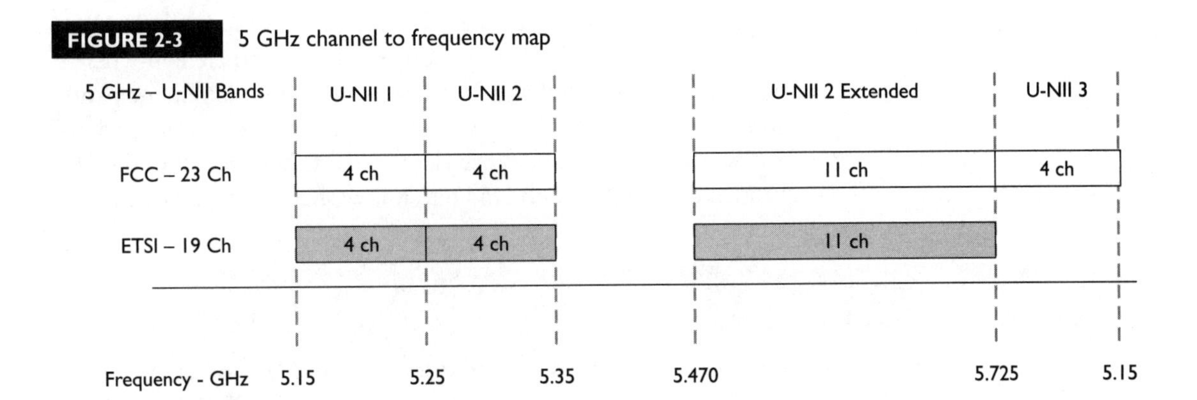

802.11b

The 802.11b standard is an enhancement to the original 802.11 standard. In 1999, IEEE standardized 802.11b using a higher rate PHY extension to provide faster connectivity to WLAN operating in the 2.4 GHz because technologies had quickly outgrown the original 802.11 standard. The 1 and 2 Mbps offered by the original 802.11 standard could no longer satisfy higher-speed WLAN needs. The 802.11b standard offers data rates of 5.5 and 11 Mbps and is backward compatible with 802.11 at 1 and 2 Mbps. The higher data rates and backward compatibility were the main reasons that 802.11b gained instant popularity and brought wireless LAN, once considered to be a very specialized area, into the mainstream. Although 5.5 and 11 Mbps may seem slow by today's standards, they were considered high speed back in 1999, when DSL ran at 256–512 Kbps and USB 1.0 ran between 1.5 Mbps and 12 Mbps were considered high speed broadband connections.

A different encoding scheme is used for 802.11b to operate in data rates higher than 1 and 2 Mbps. The 802.11b standard uses CCK to encode information for 5.5 and 11 Mbps with information modulated using QPSK at 2, 5.5, and 11 Mbps and BPSK at 1 Mbps; while 802.11 uses Barker Code for 1 and 2 Mbps. The change in modulation allows more information to be transmitted using the same time frame.

One problem associated with the 802.11 and 802.11b standards when the standards were published was that the standards were defined to operate in the United States, Canada, France, Spain, part of Europe under ETSI, and Japan only. Anything outside of these six regulatory domains could not claim to be IEEE 802.11 or 802.11b compliant. To address this issue, the IEEE 802.11 Task Group d (TGd) was formed. The solution chosen by TGd was to add an information element (IE) to the 802.11 management frames specifying location, regulatory requirements, and specific configurations for PHY (FH). The new IE announced the regulatory and location information, so that client devices can determine if they are allowed to operate in the location and to configure the radio to follow the local regulatory rules.

exam watch

802.11b is backward compatible with 802.11 using DSSS and is also called "High-Rate DSSS" (HR-DSSS). Some people use 802.11b and 802.11 loosely and interchangeably. However, 802.11b does not support FHSS PHY, while it does interoperate with 802.11 devices at 1 and 2 Mbps data rates.

802.11g

In 2002, IEEE ratified 802.11g as an extension to 802.11 and 802.11b in the 2.4GHz ISM spectrum to answer increasing demands for higher speeds than 802.11b provided. However, because 802.11g operates in the same radio frequencies as 802.11 and 802.11b, it also inherited the same shortcoming as 802.11 and 802.11b: limited bandwidth available in the 2.4 GHz spectrum. Although 802.11g has enjoyed great success and has been widely adopted by both small-office-home-office (SOHO) and enterprise users, it faces similar challenges and limitations in large, dense, or congested WLAN environments. The most non-overlapping channels 802.11g can support remain at three: channels 1, 6, and 11 in the 2.4 GHz spectrum.

The 802.11g also does not include specifications for FHSS, and a single 802.11g AP can serve both 802.11b and 802.11g devices. As 802.11g matured, another problem became apparent: it was designed to be compatible with 802.11b, but 802.11b does not understand 802.11g transmissions. As a result, when an AP tries to serve both 802.11b and 802.11g devices, an 802.11b device may inadvertently transmit over 802.11g transmissions because the 802.11b device does not understand 802.11g transmissions; thus, may regard 802.11g transmissions as non-802.11 or non-802.11b background noises and transmit over them. To avoid this, 802.11g can use the existing carrier sense mechanism called "Request-to-Send/Clear-to-Send" (RTS/CTS) to "reserve" the wireless medium for 802.11g transmission. In addition, 802.11g has also developed a protection mechanism called CTS-to-Self to notify 802.11b clients, so that 802.11b clients do not interfere with an 802.11g transmission in the same wireless LAN, which IEEE refers to as a basic service set (BSS). BSS will be discussed later in this chapter. The protection mechanisms avoid unnecessary interferences by 802.11b clients in a mixed environment, but they also reduce the throughput of the BSS when compared to a purely 802.11g BSS.

Similar to 802.11a, 802.11g adopted OFDM as its default modulation and supports the same data rates as 802.11a. The 802.11g was designed with backward compatibility in mind, supporting the same modulation and coding scheme as 802.11 and 802.11b for 1, 2, 5.5, and 11 Mbps data rates. For the extended data rates supported using OFDM, IEEE defined Extended Rate PHY (ERP), also referred to as ERP-OFDM to distinguish from 802.11a (OFDM). The 802.11g supports legacy rates using DSSS/CCK (sometimes referred to as ERP-DSSS/CCK), and it supports 6, 9, 12, 18, 24, 36, 48, and 54 Mbps extended data rates using ERP-OFDM.

802.11n

802.11n is the supplemental amendment to the 802.11 standards by the 802.11 Task Group n (TGn). The standard is currently expected to be completed by the end of 2009 or early 2010. Meanwhile, the Wi-Fi Alliance has created a prestandard certification based on Draft 2 of the pre-802.11n standard, which has accelerated the adoption of 802.11n in the SOHO and Enterprise marketplace. The IEEE 802.11n standard was ratified in September of 2009. Most 802.11n Draft 2 certified products currently available on the market are expected to support the official 802.11n through firmware and software upgrades. Wi-Fi Alliance 802.11n certified products can be expected in the market in early 2010.

There are many new enhancements built into the 802.11n standard, both at the PHY and MAC. One of the most noticeable changes is the addition of Multiple-Input Multiple-Output (MIMO) support. Two other enhancements added to the 802.11n are the addition of channel-bonding to the PHY and frame aggregation to the MAC. These new enhancements will drastically increase the speed of wireless LAN. On the other hand, new deployment strategies should be considered when deploying 802.11n.

With adoption of MIMO, wireless transmission using an 802.11n AP is more reliable than a normal 802.11 a/b/g AP where the AP only has one transmitter and receiver. As mentioned in Chapter 1, the addition of antennas (receivers) increases the signal quality at AP; hence, it increases the overall reliability even when clients are not 802.11n capable. Usually, the number of transmitters and receivers will describe the MIMO configuration. For example, Cisco AP 1250 is described as a 2×3 device because it has two transmitters and three receivers. The number of spatial streams will determine the performance and speed of the 802.11n network, and 3×3 will enjoy higher performance and throughput compared to a 2×2 configuration due to additional data stream. The 802.11n draft allows up to 4×4.

The 802.11n uses 20 MHz and 40 MHz channels, and the 40 MHz channels are just two 20 MHz channels bonded together. One thing to note is that the channel bonding feature should only combine two adjacent nonoverlapping channels. In the 2.4 GHz spectrum, there will only be one channel-bonded channel due to the limited nonoverlapping channels available in the spectrum. The 5 GHz is the preferred spectrum to implement channel bonding because all the channels are nonoverlapped. Hence, Cisco recommends that channel bonding should only be implemented in the 5 GHz spectrum.

Another enhancement mentioned is frame aggregation at MAC. At the 802.11 MAC layer, there are two types of data units: The MAC Service Data Unit (MSDU) and the MAC Protocol Data Unit (MPDU), which connect 802.2 LLC to 802.11 MAC and to PLCP at the physical layer, as shown in Figure 2-4.

FIGURE 2-4

802.11 MAC and PHY layers

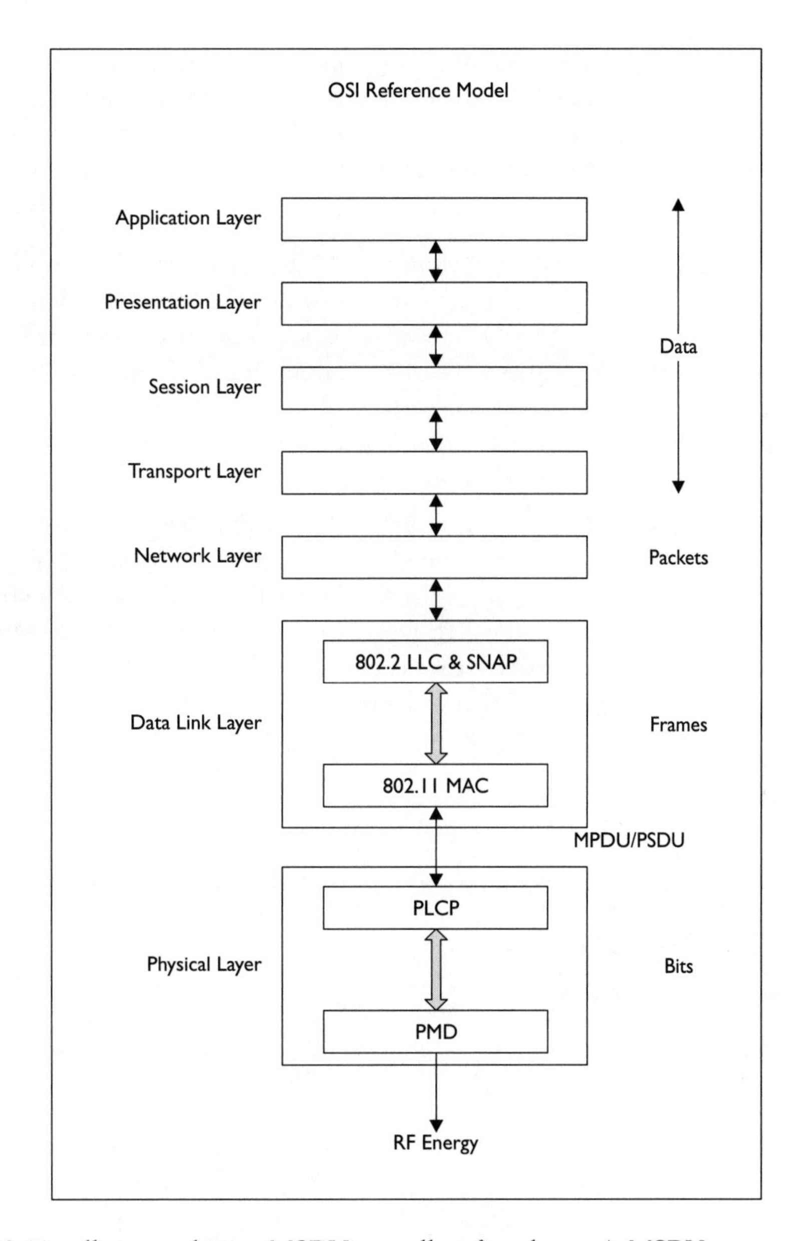

The 802.11n allows combining MSDUs, usually referred to as A-MSDU or MSDU aggregation, or combining MPDUs, usually referred to as A-MPDU or A-MPDU aggregation, to provide more efficient transmission. The aggregation will pack multiple MSDUs or MPDUs together to reduce the protocol overheads and increase overall performance. A-MPDU also allows ACK for a group of frames

instead of one ACK per frame. This is referred to as Block Acknowledgement or BlockAck, which was originally introduced in 802.11e and later optimized in 802.11n. Figure 2-5 illustrates this enhancement at the 802.11 MAC layer.

For backward compatibility, a number of protection mechanisms are built into 802.11n to protect legacy 802.11 a/b/g clients as follows:

- **802.11n running 20 MHz channels** To support legacy 802.11 modes, current implementation of 802.11n transmission use a mixed mode PHY. The 802.11n mixed mode PHY will encapsulate 802.11n HT PHY in the legacy PHY when transmitting at HT rates, so it is designed to be backward compatible with 802.11a and 802.11g. For 802.11b compatibility, 802.11n will take advantage in the protection mechanism built into the 802.11g specification.
- **802.11n running 40 MHz channels** A 40 MHz channel is basically two 20 MHz channels bonded together; therefore, it is possible that legacy 802.11a and 802.11g clients may exist on either 20 MHz channels. The 802.11n will use the CTS protection mechanism on both 20 MHz of the bonded channel, so that legacy clients understand when an 802.11n transmission occurs over the 40 MHz channel.
- **RTS/CTS protection mechanism** Similar to a mixed 802.11g environment, RTS/CTS is also an option available at 802.11n to protect subsequent 802.11n transmissions by 802.11b clients.

FIGURE 2-5 802.11n frame aggregation

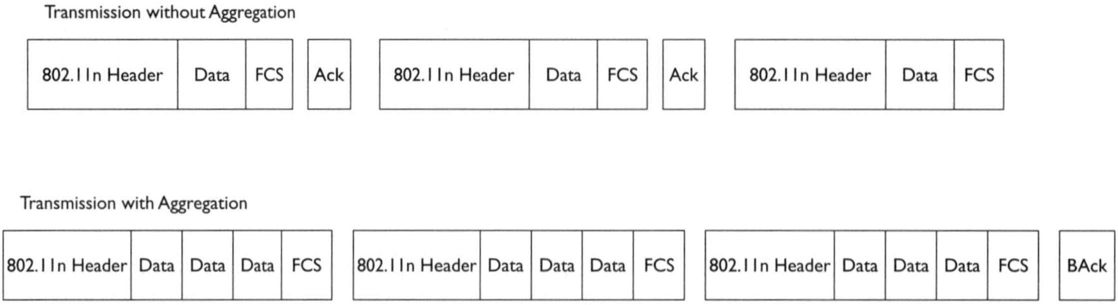

Transmission without Aggregation

802.11n Header | Data | FCS | Ack | 802.11n Header | Data | FCS | Ack | 802.11n Header | Data | FCS

Transmission with Aggregation

802.11n Header | Data | Data | Data | FCS | 802.11n Header | Data | Data | Data | FCS | 802.11n Header | Data | Data | Data | FCS | BAck

802.11e

802.11e was developed by IEEE in 2007 to address Quality of Service issues in the wireless LAN and to provide a mechanism to address traffic prioritization using a wireless medium. The Wi-Fi Alliance has created certification programs based on the IEEE 802.11e specifications for products intended for delay-sensitive devices and applications. CCNA—Wireless does not cover 802.11e.

802.11i

802.11i was developed by IEEE to address security issues in the wireless LAN. More details will be provided in Chapter 5.

CERTIFICATION OBJECTIVE 2.03

Wireless Topologies and Frame Types

To better understand how 802.11 and its variants work, it is important to understand the basic building blocks and terminologies used in the standards. These elements are used by the 802.11 family, and we will examine these elements in greater detail here by first looking at the 802.11 architecture, then reviewing different types of wireless topologies, and then discussing different frame types used in 802.11.

802.11 Architecture and Frame Types

The IEEE 802.11 architecture consists of these building blocks, and they are shared among all 802.11 family of protocols:

- STA and AP
- Basic service set (BSS)
- Extended service set (ESS)
- Distribution system (DS)
- DCF/PCF
- 802.11 frame types

STA and AP

There are two major components described by the 802.11 standard: mobile station (STA) and access points (AP). STA is the device that connects to the radio frequency (RF) or wireless medium and is usually referred to as the network adapter or network interface card (NIC) for those more familiar with the wired networks. In this book, STA and mobile client will be used interchangeably as STA is the official term used by the IEEE 802.11 standard. Because 802.11 consists of MAC and PHY, STA will support both MAC and PHY specified in the 802.11 standard. Each STA will support a well defined set of services, including authentication, deauthentication, privacy, and delivery of data.

Basic Service Set (BSS)

All 802.11 networks are built around a basic service set (BSS), which is simply a group of STAs trying to connect with one another, and a service set identifier (SSID), which is used to identify an ESS. When an AP is offering BSS services, it will send out a broadcast using its radio MAC address to advertise services it offers; this is called BSSID. There are two types of BSS, and each operates slightly differently from the other. When there is no AP in the BSS, it is referred to as an independent BSS (IBSS), more often an ad hoc network. STAs within the IBSS will communicate directly with one another using the same SSID, and a fully meshed relationship will be formed in order for an STA to communicate with all STAs within the IBSS. In addition, there is no concept of broadcast or multicast in an IBSS environment.

Because each STA must build a fully meshed relationship with all other STAs in the IBSS, it is not the most efficient way of sharing the wireless medium in a large WLAN deployment. IBSS, on the other hand, may be useful in a scenario where temporary information exchange among a small number of PCs is desired. In many client adapter configurations, IBSS mode is simply referred as ad hoc mode. While IBSS works well for temporary information exchange, it does not work well for a business network that requires high availability.

When there is an AP in the BSS, it is called an infrastructure BSS, or simply a BSS. In the client world, you will usually see the options of ad hoc mode and infrastructure mode, and that determines if client will join an IBSS or a BSS. Figure 2-6 illustrates an IBSS consists of four STAs.

FIGURE 2-6 STAs in an IBSS

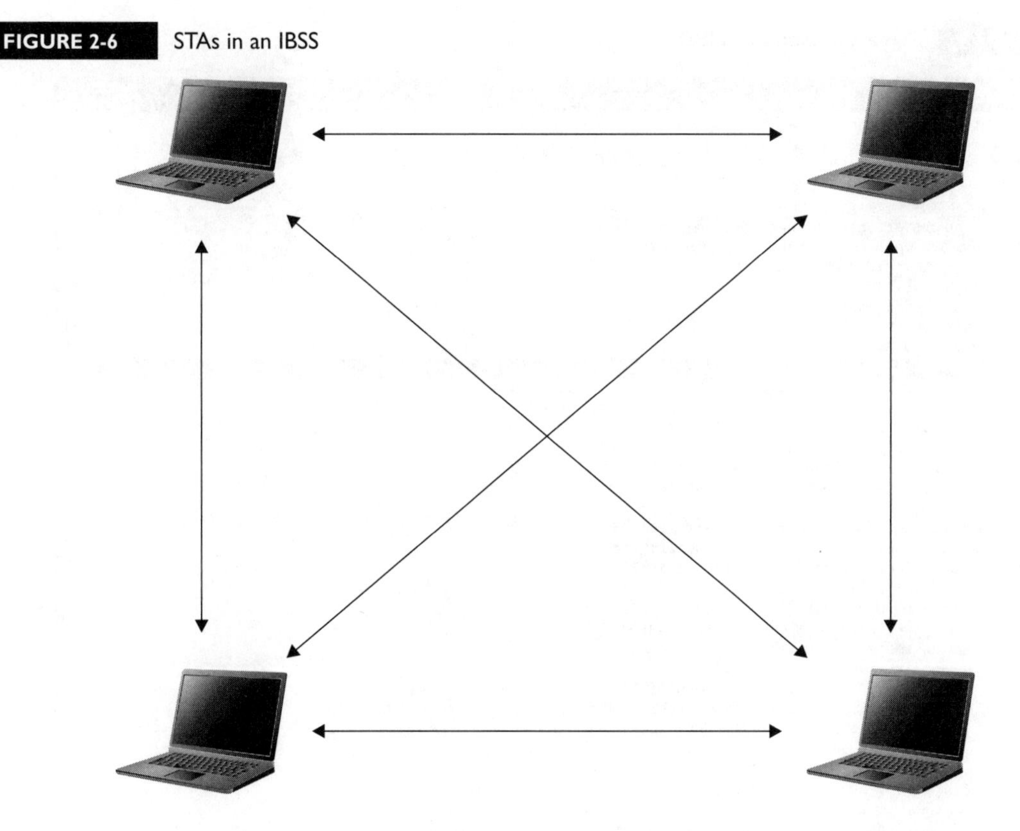

The packet trace decode displayed in Figure 2-7 shows an STA sending an association request to an IBSS.

In infrastructure BSS mode, all wireless and wired communications will go through an AP even when STA intends to send traffic to another STA wirelessly. For example, when one STA intends to send traffic to another STA in the BSS, it will first send the traffic to the AP, and then from the AP to the intended STA. As a result, the same traffic is sent over the air twice and consumes the available medium twice as opposed to a direct communication from STA to STA. While this may not seem to be an efficient use of bandwidth, the benefits outweigh the cost. Consider that the AP will then act as the hub in the wireless LAN and can provide

FIGURE 2-7 STA association to an IBSS

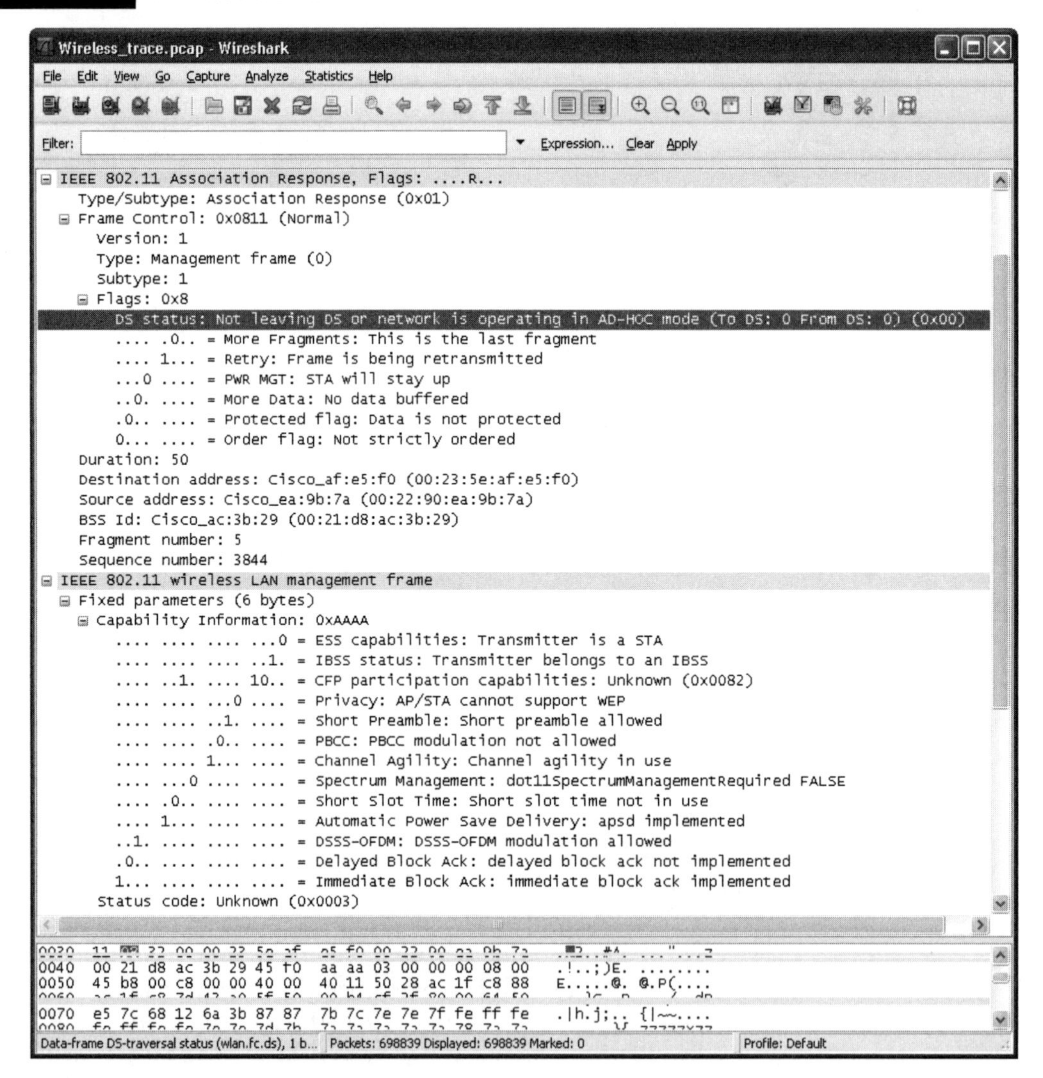

additional services to STA based service profiles, such as QoS, multicast, and client power save mode. STA, on the other hand, will no longer need to establish a fully meshed relationship with all the STAs in the BSS; it is a more efficient way of communications from a network architecture standpoint. Figure 2-8 illustrates an Infrastructure BSS consists of one AP and four STAs.

FIGURE 2-8 An infrastructure BSS

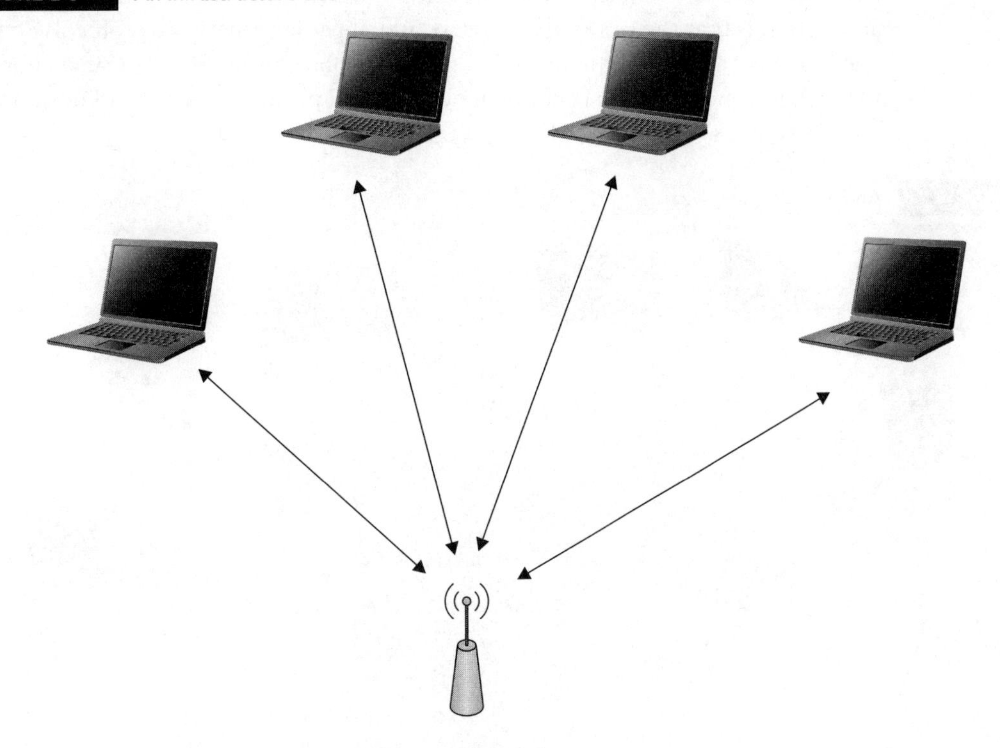

Extended Service Set (ESS)

As the name suggests, extended service set (ESS) is one level above BSS in the 802.11 architecture. When there are multiple AP and STAs (multiple BSSes) on the network to form a larger WLAN, it is called an extended service set (ESS). Within an ESS, STA can communicate with STA from another BSS through APs. The same SSID will be used. The medium that connects BSSes is called a distribution system (DS), which is usually, but not necessarily, the wired network infrastructure. AP will determine if the traffic received from the BSS should be forwarded back to the BSS for another STA, to another AP for another BSS, or to a destination on the wired network or the extended network. APs will also communicate among themselves to determine how to forward traffic originated from one BSS to another BSS or from an AP to another AP Furthermore, AP will also facilitate the roaming of STA from one BSS to another BSS.

From the network perspective, the ESS is a single MAC layer network, and all the STAs in the ESS are simply devices on the network bridged by the ESS. The ESS performs a similar function from a network perspective, with STA primarily living in a layer 2 network on the wireless LAN. The benefit of this network abstraction is that the rest of the network and upper layer network protocols do not need to know about the wireless LAN and the mobile nature of STAs since mobility is a foreign concept to the traditional network and protocols. Figure 2-8 illustrates an ESS consists of multiple APs and STAs.

FIGURE 2-9 An ESS

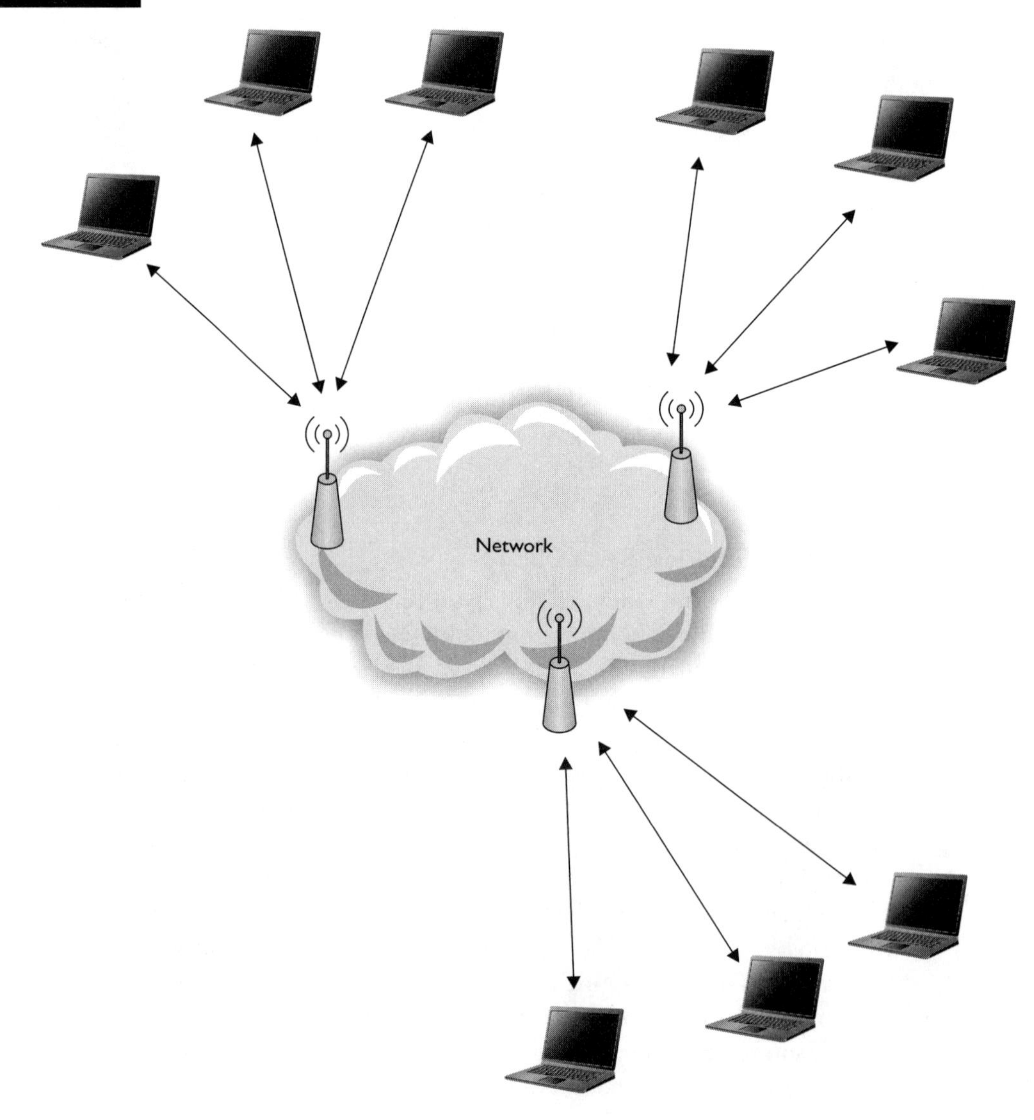

The communications between APs are beyond the scope of this book. The 802.11 working group has created a recommended guidelines on AP communications using Inter Access Point Protocol (IAPP) in 802.11F, but this protocol was never widely adopted by the industry and was later withdrawn from the standard track in 2006.

Distribution System (DS)

The distribution system (DS) is the mechanism specified by 802.11 to be used by AP to communicate with another AP to exchange traffic for STAs in BSSes. Although DS is not necessarily a network per the 802.11 standard, the 802.11 only specifies the services that a DS must provide and leaves the implementation DS open. Therefore, DS can be thought of as a network, such as Ethernet (IEEE 802.3), FastEthernet, or beyond. A DS can also be a wireless bridging system where APs are connected through a wireless bridge link.

Spread spectrum technology is characterized by wide bandwidth and low peak power. This allows it to resist interference and provides it with a certain level of security over traditional narrowband transmissions.

DCF/PCF

The 802.11 MAC defines two ways to coordinate use of the wireless medium: using the Distributed Coordination Function (DCF) or the Point Coordination Function (PCF).

DCF is most commonly seen in 802.11 implementations to share wireless medium between STAs and AP. DCF relies on CSMA/CA to avoid collision. CSMA/CA is described in detail in Chapter 1.

When MAC receives a request from STA to transmit, it verifies the medium is available. If the medium is not in use for a defined interval, the Contention Window, the transmission may begin. If the medium is busy, a back off interval will kick start, and the transmission will be put on hold. When the back off time expires, MAC will go through the same process until transmission is completed successfully. The transmission is considered incomplete when a collision occurs or an ACK is not received. In this scenario, CW will double and a back off time will be selected. The transmission will occur when the back off time expires.

There is also an optional 802.11 RTS/CTS mode to share the medium between stations. RTS/CTS will be discussed later in this chapter. DCF has several limitations:

- Collisions occur in spite of CSMA/CA (just as in Ethernet, which uses CSMA/CD). When collisions occur, the overall throughput of network decreases.
- 802.11 MAC has no concept of high or low priority traffic.
- 802.11 MAC has no concept of Quality of Service (QoS) guarantees.
- One station may monopolize access to the medium. For example, when a station has a low bit rate (such as 1 Mbps), it will take a long time to send its packet, and no other STA is allowed to transmit when an active transmission takes place.

PCF is an optional mode defined by 802.11; thus, it is not commonly adopted by vendors in 802.11 implementation. PCF is only available in infrastructure BSS where a "coordinator—AP" is needed to coordinate access to the wireless medium in BSS, and STAs connect to BSS through AP. PCF has two modes of operation: Contention Free Period (CFP) and Contention Period (CP). In CP, DCF is simply used. In CFP, AP sends Contention Free-Poll (CF-Poll) packets to STAs, one at a time, to control the right to access the wireless medium. PCF seems to provide a good mechanism for management of QoS. Unfortunately, it is not widely adopted and does have some limitations, such as lack of support for class of traffic.

802.11 Frame Types

To understand how 802.11 and supplement standards work, it is important to have a basic understanding of the different 802.11 frame types as a basis for design and administering a wireless LAN. We will provide an overview of 802.11 frame types to help you better understand wireless LAN operations using 802.11 and give you the knowledge needed for troubleshooting.

IEEE 802.11 has defined various frame types for STA and AP to use in the 802.11 MAC. The different types of frame will provide services needed for managing and controlling the wireless link. Every frame has a control field that depicts the 802.11 protocol version, frame type, and various indicators, such as security scheme, QoS, power management, and so on. All 802.11 frames contain a MAC, a frame sequence number, body, and frame check sequence (FCS/CRC). Figure 2-10 shows a high level view of the IEEE 802.11 MAC frame.

FIGURE 2-10 An 802.11 frame

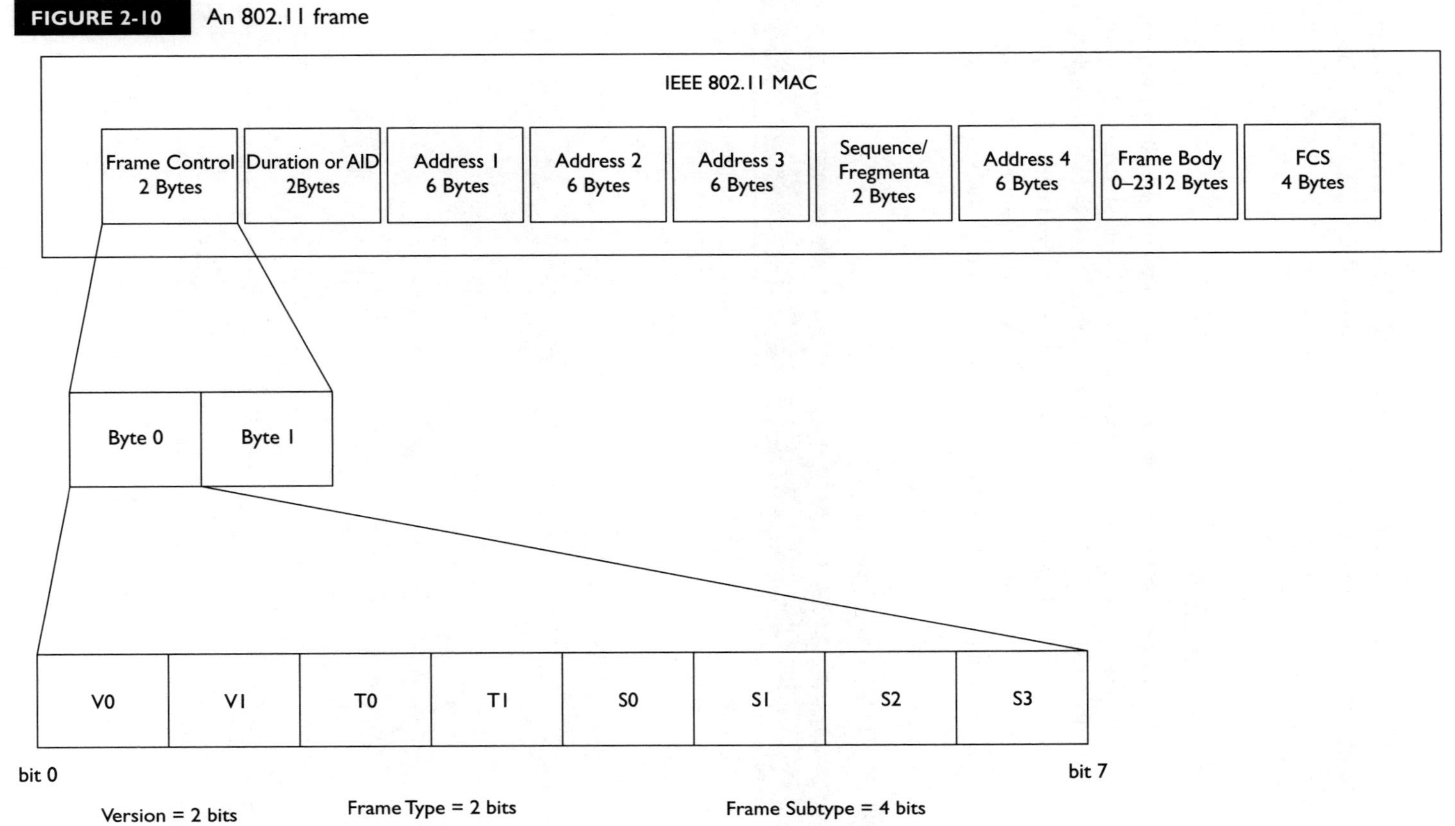

The 2-byte Frame Control field specifies the type of frame and subtypes. The 2-bit protocol version identifies the version of IEEE 802.11 MAC used and is always set to zero for the current version of standard. Table 2-4 shows the frame types and subtypes of 802.11 MAC:

Management Frames

The 802.11 management frames enable stations to establish and maintain communications. As mentioned earlier, 802.11 data frames live in Layer 2 of the OSI reference model and assume that protocols and data from higher layers, most commonly TCP/IP, work correctly. Management and control frames contain specific information regarding the wireless LAN. We will discuss Management frames first, and Control frames in the next section. For example, a beacon's frame body contains

TABLE 2-4 802.11 MAC Frame Types and Subtypes

Frame Type	Type Description	Frame Subtype	Frame Subtype
00	Management	0000 0001 0010 0011 0100 0101 1000 1001 1010 1011 1100 1101	Association request Association response Reassociation request Reassociation response Probe request Probe response Beacon Announcement traffic indication message Disassociation Authentication Deauthentication Action
01	Control	0000-0111 1010 1011 1100 1101	Reserved Power Save Poll (PS-Poll) RTS CTS ACK
10	Data	0000 0100	Data (payload) Data (null)

the service set identifier (SSID), timestamp, and other information regarding an AP. The following are common 802.11 management frame subtypes:

- **Beacon frame** Beacons are used to allow STA to locate and identify a BSS. The AP periodically sends a beacon frame to announce its presence and relay information, including SSID, support rates, timestamp, power management, and other parameters. For an 802.11d implementation, beacon frames will also include country specific information. For 802.11e support, additional IE may be included in beacon frames, such as QBSS load and EDCA IE. STA continually scans all 802.11 channels and listens to beacons as the basis for choosing the best AP to associate with or roam to. Figure 2-11 shows a beacon frame decode. It will help you understand better how information is assembled.

FIGURE 2-11 An 802.11 beacon frame

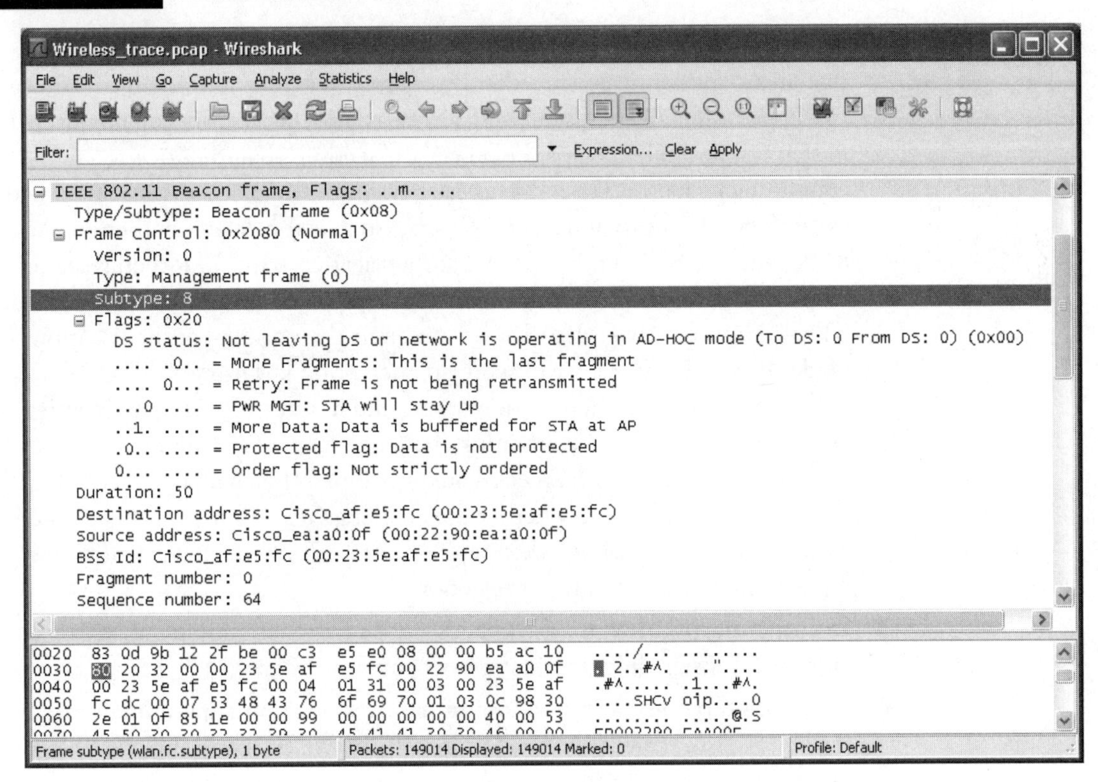

- **Probe request and response frame** STA sends a probe request frame when it needs to locate a BSS. For example, STA sends a probe request to determine whether an AP is within range. A probe request contains two information elements: SSID and supported rates. An AP or another STA will respond with a probe response frame after it receives a probe request frame.

on the job

It is a common misconception that SSID can be hidden when AP disables broadcasting SSID in beacon frames. SSID is a mandatory field in the beacon and probe response frames. When broadcasting SSID is disabled, it simply means beacon frames will contain SSID set to value of NULL. However, it does not prevent STA from finding SSID from AP. STA is allowed to send out a probe request with SSID set to NULL. When AP receives a NULL SSID in the probe request, it will respond with its SSID(s) that it supports. Also, do not confuse broadcasting SSID with BSSID advertised by AP.

- **Authentication frame** 802.11 authentication refers to an 802.11 process that an AP authenticates the identity of an STA before allowing it to associate to a BSS. Authentication can be used by two STAs, but it is not commonly used in that fashion. STA begins the process by sending an authentication frame containing its identity to AP. There are two modes of authentication defined by IEEE 802.11: open and shared. Open system is the default authentication mode, and the STA sends only one authentication frame. AP then responds with an authentication frame indicating acceptance (or rejection). If successful, STA will associate to AP. A shared system is the optional authentication mode. Both AP and STA will share an encryption key using WEP. STA will send an initial authentication frame, and in return, AP responds with an authentication frame containing challenge text. STA then must send a response back to AP with an encrypted version of the challenge text. AP will verify STA by examining whether the challenge text from STA is encrypted with the same key. The AP then replies to the STA with an authentication frame signifying acceptance or rejection.
- **Deauthentication frame** STA sends a deauthentication frame to AP or another STA signaling termination of communications.
- **Association request and response frame** Association request frames are used by STA to request an association with a BSS. STA will include SSID and supported rates in the association request frames to tell BSS of its

capability. Association response frames are used to signal success or failed association with the BSS with appropriate information, which includes status code, association ID (AID) and supported rates. An STA begins the process by sending an association request to an AP. If successful, the AP will reserve memory space and establishes an AID for the STA. If not successful, STA should respond appropriately based on status code. Nonetheless, 802.11 does not clearly state how status code should be used, so much of it is still up for interpretation by vendors.

- **Reassociation request and response frame** When an STA moves away from the associated BSS (AP) and attempts to associate with another AP using the same SSID, STA will send a reassociation request frame to the new AP. The new AP will then coordinate the forwarding that may be left in the buffer of the previous AP and establish association with the STA. The reassociation response frame contains an acceptance or rejection notice and is identical to the association response frame.
- **Disassociation frame** An STA sends a disassociation frame to BSS to signal termination of the association. AP will clear the memory and remove the STA from the association table.

Control Frames

The 802.11 control frames provide delivery of data frames between STAs. There are six types of control frames defined in 802.11, but we will only discuss the three most commonly used here:

- **Request-to-Send (RTS) frame** 802.11 defines an option mechanism using RTS/CTS to avoid frame collision. RTS is intended to inform STAs in BSS of a transmission so that STAs in the neighborhood will not transmit and cause collisions in the medium. The sending STA expects to receive a CTS from receiving STA of an intended transmission.
- **Clear-to-Send (CTS) frame** The intended receiving station, usually an AP, responds to an RTS with a CTS. The CTS signifies a clearance for the requesting STA to send a data frame. CTS includes a time value that halts transmission at all other STAs (including hidden nodes) for the requesting STA. This minimizes collisions among hidden nodes and can result in higher throughput.

- **Acknowledgement (ACK) frame** The acknowledgement (ACK) is sent immediately after receiving a data frame, management frame, or a Power Save Poll (PS-Poll) frame when the frame arrives without errors. This is the mechanism to prevent sending station from retransmit frames.

The other three types control frames that are not discussed here are Power Save Poll (PS-Poll), Contention-Free End (CF-End), and CF-End plus ACK (CF-End ACK).

Data Frames

The essence of having a wireless LAN is to transport data. The 802.11 defines a data frame to carry information from higher layers, such as TCP/IP type of traffic, in the body of the frame. The correct technical term is encapsulation. So, wireless LAN encapsulates the upper layer protocol packets and delivers them from STA to BSS. When viewing 802.11 data frames with a packet analyzer, you will find everything you know from TCP/IP exists in an 802.11 data frame. If the frame is not encrypted, you will be able to see the contents of the frame. There are several data frame types defined in 802.11: simple data, data with Contention-Free Acknowledgement (CF-ACK), data with Contention-Free Poll (CF-Poll), data with CF-ACK and CF-Poll.

Wireless Topologies

AP was originally designed only to connect clients to a local LAN segment. As technology matured and wireless LAN evolved, bridging and repeater functions were added to AP to connect LAN segments and clients with extended ranges. AP performing AP-only functionality is typically referred to as a root-mode AP. Two additional modes can be commonly seen on an AP:

- Wireless repeater
- Wireless bridge

The 802.11 and supplements do not specify bridge and repeater functionality in the standards; thus, most wireless bridging and repeater implementations remain proprietary, and vendors usually build wireless bridges and repeaters on top of existing AP functionality. In addition to wireless repeaters and wireless bridges, a new type of wireless solution called "mesh network" is emerging.

Wireless Repeater

As a wireless repeater, AP provides connectivity to AP (in root mode) and clients that are outside the reach of the root AP. A wireless repeater will connect to a client through association and connect to the root AP as a client to the root AP. Using a repeater is not usually encouraged as it reduces the overall bandwidth available to clients. Imagine that the repeater needs to communicate with both the root AP and the client. Thus, the same data frame will be transmitted twice to reach the root AP using a wireless medium that is half-duplex in nature. Figure 2-12 illustrates a typical repeater deployment connecting clients to the network.

AP deployed in repeater mode

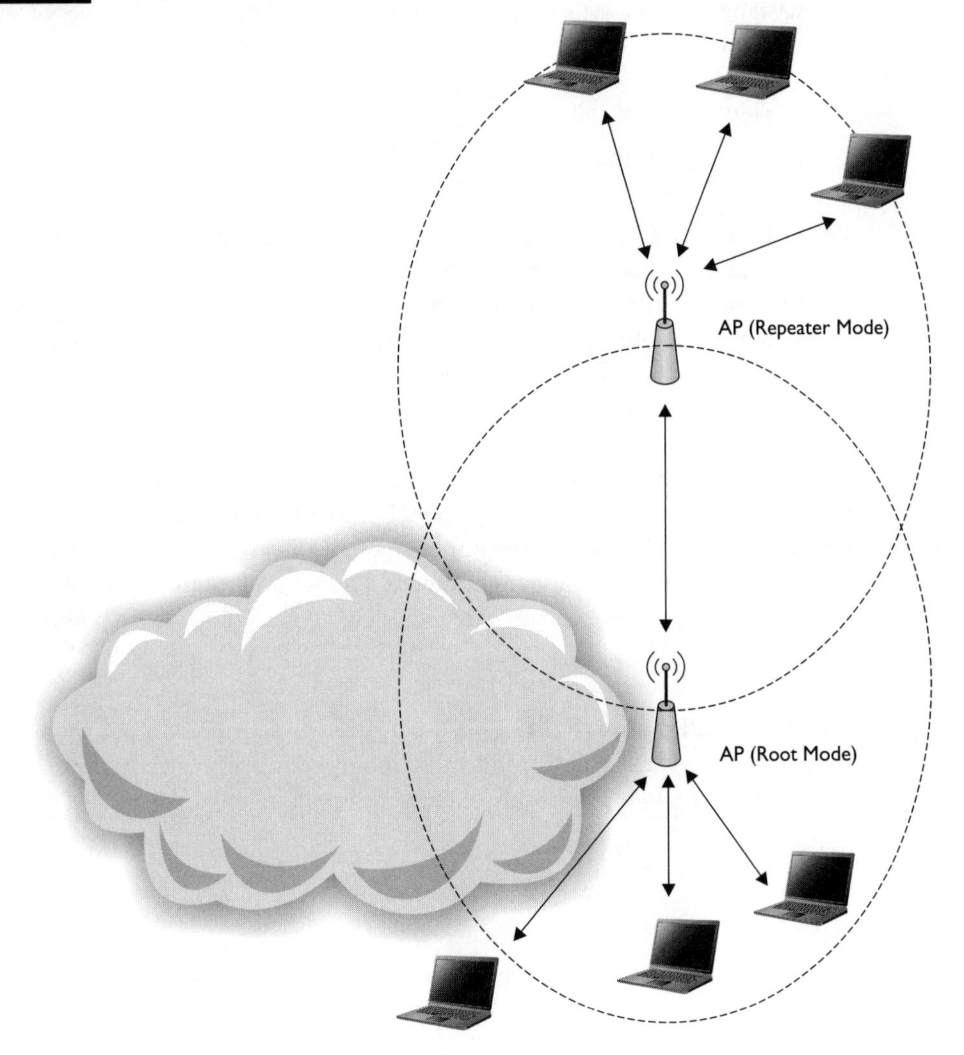

Understand the drawback of repeater mode and how it may impact the performance of WLAN.

Wireless Bridges

APs in bridge mode or dedicated wireless bridges create a wireless connection by associating with each other. Some vendors offer bridges that act as wireless bridges and APs simultaneously, so that AP can connect not only STAs, but also two LAN segments by wireless medium. Wireless bridges are usually deployed in outdoor environments and use semidirectional or directional antennas to make connections with greater ranges than a typical indoor wireless LAN. There are two types of wireless topologies:

- Point-to-point
- Point-to-multipoint

Point-to-Point (PtP) With a point-to-point topology, a direct wireless bridge link is established to connect two LAN segments using a pair of wireless bridges. Usually, directional or semidirectional antennas are used for creating a point-to-point connection although omni-directional antennas can be used to form a point-to-point connection. Typical connections include a building-to-building or site-to-site configuration when the bridge link is needed only between two locations. The distance can range from a few hundred yards to a few miles depending on the type of antennas used. Typically, higher gain antennas are used in a PtP bridge deployment. Semiomni-directional or highly directional antennas are most frequently selected for PtP links; however, omni-directional antennas will also work but may pick up unnecessary environmental noises and interferences, which may impact the performance of the bridge link. Figure 2-13 illustrates a point-to-point bridge link.

Point-to-Multipoint (PtMP) With a point-to-multipoint topology, a collection of direct wireless bridge links connection multiple sites with a central site tying remote sites together. The backhaul network (or main network) usually, but not

FIGURE 2-13 A point-to-point bridge link

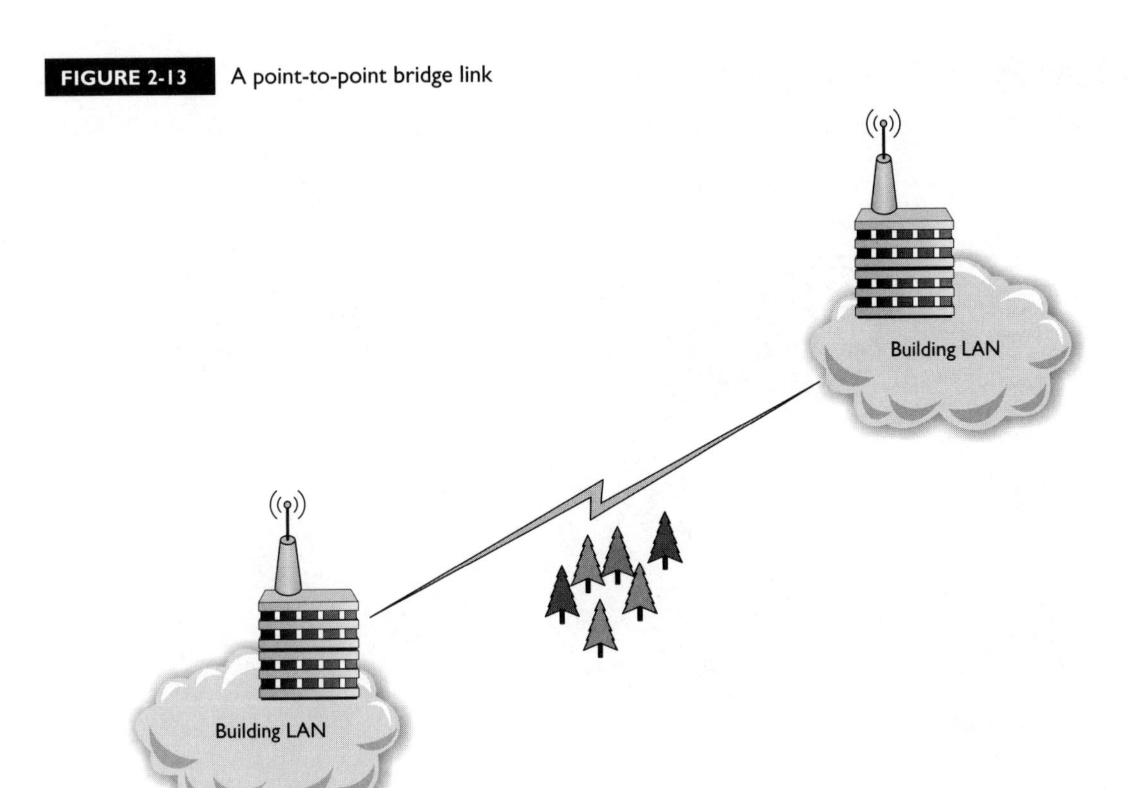

necessarily, resides at the central site. This topology is similar to the hub-and-spoke topology typically seen in a wide area network (WAN) deployment.

Similar to a PtP link, higher gain antennas are used in a PtMP bridge deployment. Semiomni-directional or highly directional antennas are most frequently selected by the remote sites, but omni-directional antennas will also work. The central site usually uses an omni-directional or semidirectional antenna, depending on where the central site is located in relation to the remote sites. Using omni-directional and semidirectional antennas may pick up unnecessary environmental noises and interferences, which may impact the performance of the bridge network. Figure 2-14 illustrates a point-to-multipoint bridge topology.

FIGURE 2-14 A point-to-multipoint bridge topology

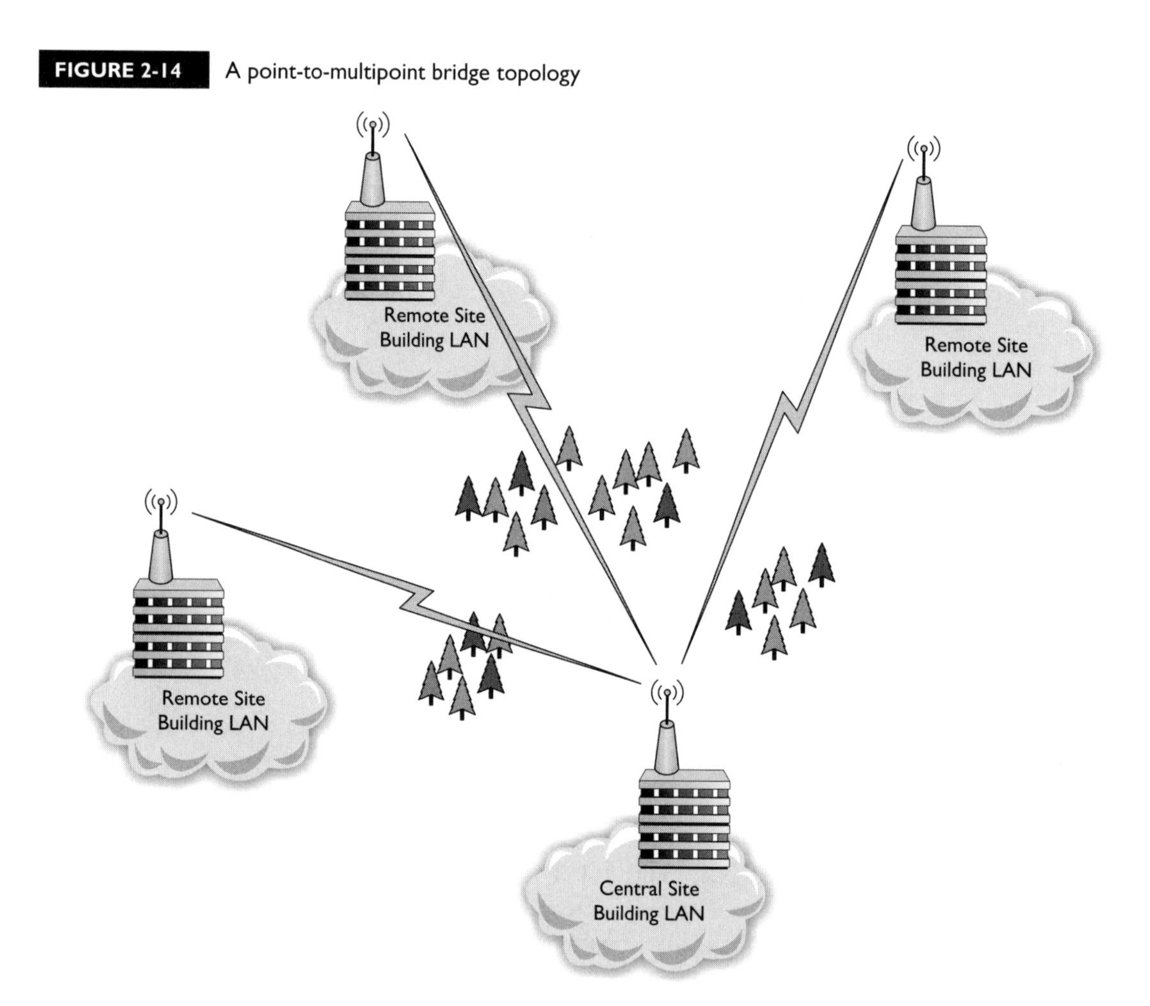

Mesh Networking

As the outdoor wireless LAN has matured, a new breed of outdoor bridging technology has been developed. Mesh networking has been built as enhancement to the traditional PtP and PtMP mode where there is usually one central AP or bridge that connects to one or more wireless bridges. The issue with the traditional PtP and PtMP model is that when the central bridge fails, remote sites also lose connectivity unless there is a backup bridge at the central site to resume connectivity, and the connection is usually not dynamically configured. Taking the example from the PtMP network we just discussed, when the bridge at the central site fails, all remote sites will lose connectivity to one another, as shown in Figure 2-15. To reconnect the remote sites may require reconfiguration of bridges.

FIGURE 2-15 A point-to-multipoint bridge topology with failed central site bridge

In mesh networking, all mesh APs can connect to all other mesh APs in a multipoint-to-multipoint fashion. Mesh AP is usually referred to as a node. The network will dynamically configure and reconfigure itself to find the most efficient way to transport traffic back to the backhaul WAN network. There may be multiple ways of connecting back to the main network, and the system will dynamically figure out the possible paths and choose the best path to use. Figure 2-16 illustrates a mesh network topology.

In the mesh network, APs are usually configured as either a mesh AP (MAP) or a root AP (RAP). RAPs have direct (usually wired) connections to the network, and MAPs usually connect to the network wirelessly through RAPs. Cisco has developed the Adaptive Wireless Path Protocol (AWPP) for the mesh network, where a

FIGURE 2-16 A mesh network topology

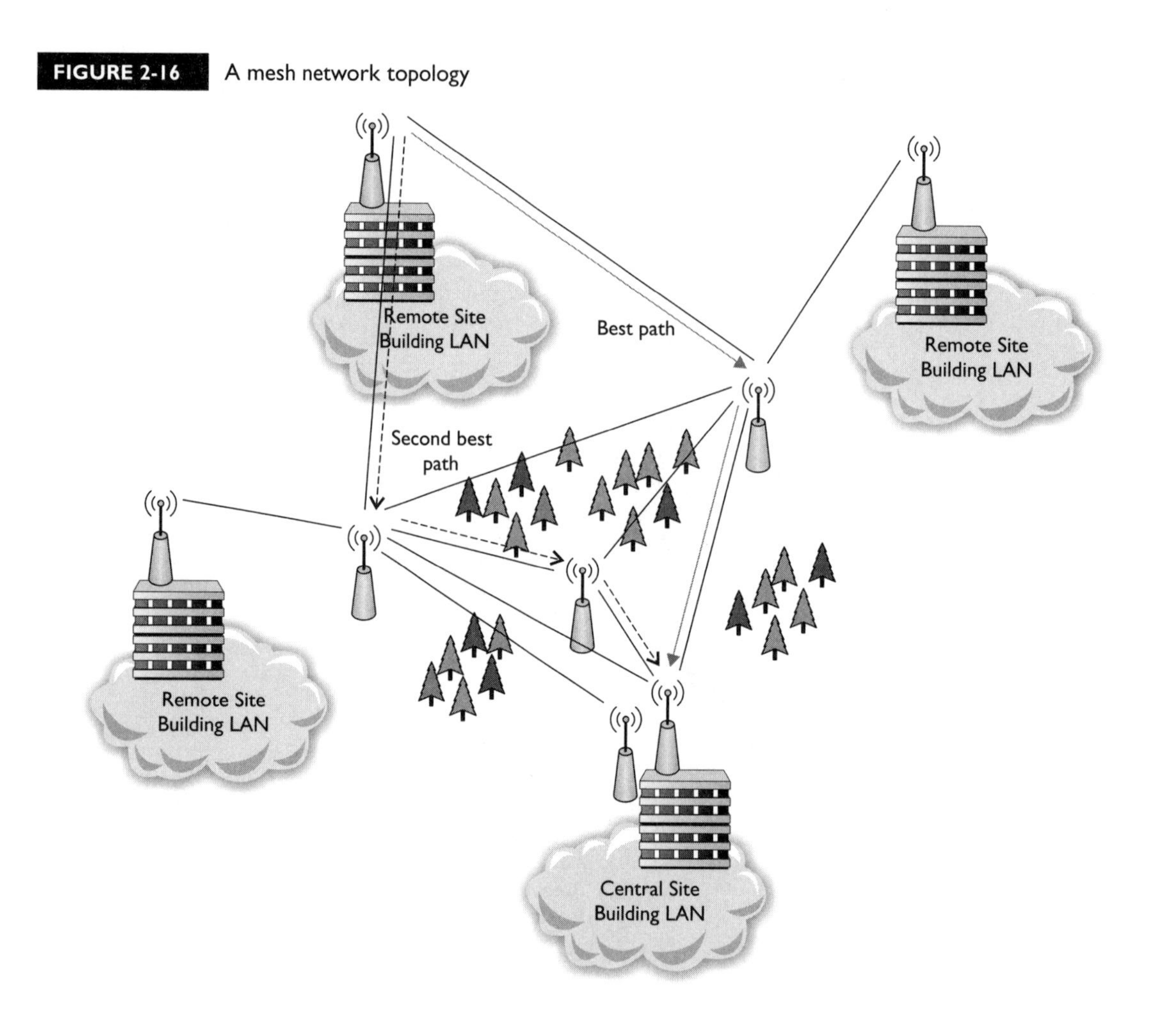

dynamic path decision is made based on link quality and the number of hops. A MAP actively solicits for neighbor MAPs and learns the available neighbors back to a RAP through AWPP. The AWPP will determine the best path to connect clients and MAPs to the network. In the event of a RAP failure, a backup RAP will be selected, and the system will converge when the best path is selected with the new RAP, as illustrated in Figure 2-17.

IEEE is currently developing a standard (802.11s) for the mesh network.

FIGURE 2-17 Mesh network recovery from a RAP failure

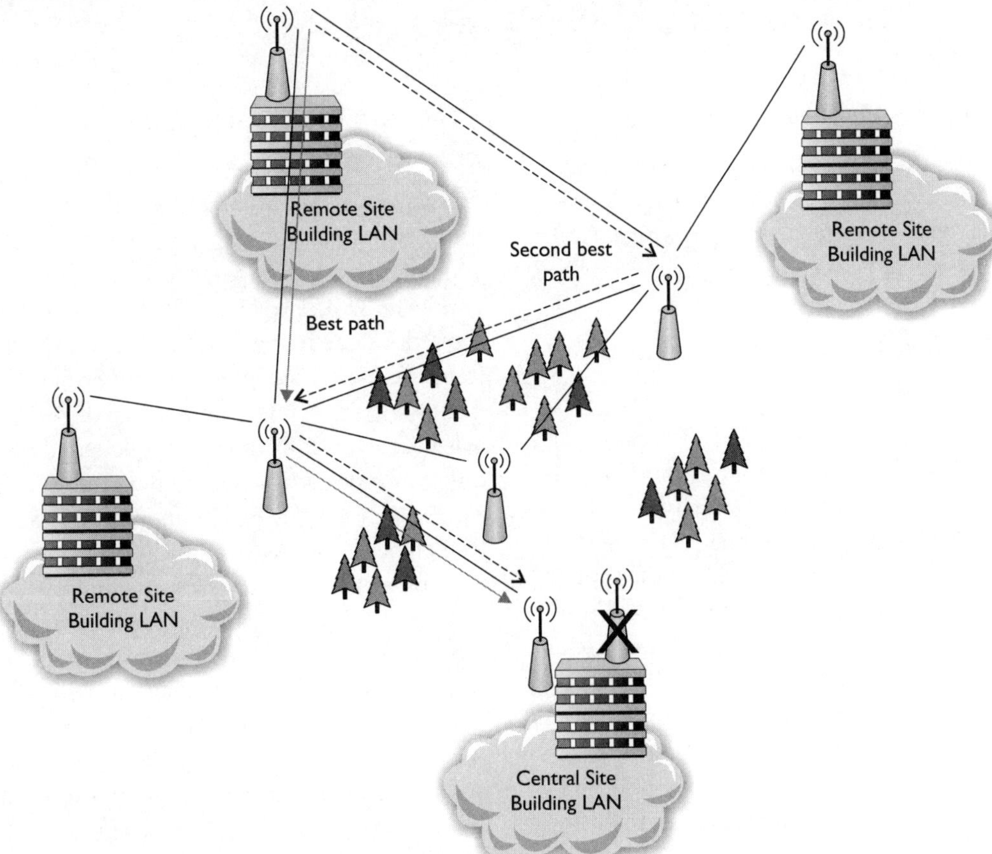

exam watch

Understand the advantage(s) of mesh over traditional point-to-point and point-to-multipoint bridging deployment. Mesh may also experience similar performance as each additional hop of MAP decreases available bandwidth by half.

INSIDE THE EXAM

Wireless Regulatory Bodies

For the exam, you should know how technologies work with rules and regulations set forth by country regulatory authorities and standards body, such as the FCC and ETSI.

Wireless LAN Standards and Certifications

For the exam, you should know the basic concepts of IEEE 802.11 and amendments. IEEE 802.11 is the base standard for the wireless LAN, and subsequent enhancements share the basic building blocks of 802.11 and the different frame types. You need to understand the main differences between 802.11, 802.11a, 802.11b, 802.11g, and 802.11n. It is important to understand the frequencies and power settings for the different standards. You may be asked about the compatibility between different standards, such as 802.11g and 802.11b, and specific enhancements added in each amendment.

Wireless Topologies and Frame Types

You should also be familiar with the different frame types defined in 802.11, as this helps you understand the operation of wireless LAN using 802.11. The 802.11a, b, g, and n use the same frame types despite different frequencies and modulations used at PHY. Because wireless medium is a shared medium, it is very important that you understand the different types of collision avoidance mechanisms, such as CSMA/CA and CTS/RTS. Remember how 802.11 uses DCF/PCF to control access to the wireless medium.

Wireless topology is another important subject. Understand the different roles an AP can play as you need to be familiar with AP in repeater mode and root mode. You will also need to understand the advantages and disadvantages of different types of bridge links and the types of antennas to use for each. Mesh networking is an emerging wireless technology that goes one step further than what traditional wireless bridges offer.

CERTIFICATION SUMMARY

Wireless LAN today is built primarily using the IEEE 802.11 standards suite; a regulatory agency in each country creates rules to regulate the use of frequencies and power settings in the regulatory domain. The FCC is the regulatory authority in the United States, and the ETSI governs the standardization in the European Union.

The Wi-Fi Alliance has created different certification programs based on the IEEE 802.11 and supplement standards to ensure interoperability.

The 802.11 and 802.11b operate primarily in the 2.4 GHz spectrum using DSSS, while 802.11a operates in the 5 GHz spectrum using OFDM. To provide high speed rates comparable to 802.11a in the 2.4 GHz spectrum, IEEE later created 802.11g. 802.11n is the up and coming standard currently in draft and provides wireless services in both 2.4 GHz and 5 GHz spectrum. All 802.11 standards use the same frame types: management, control, and data.

Wireless AP plays multiple roles as WLAN continues to evolve. The three commonly used modes are AP root mode, AP repeater mode, and AP bridge mode. In a bridging deployment, wireless LAN can be deployed in either point-to-point or point-to-multipoint fashion.

TWO-MINUTE DRILL

Wireless Regulatory Bodies

- ❑ The FCC is the regulatory authority in the United States that sets rules and regulations for all wireless related communications.
- ❑ The ETSI is the legal and official standards body that creates standards for European Union counties.
- ❑ The FCC and ETSI influence rules and regulations in countries outside of the United States and European Union countries.

Wireless LAN Standards and Certifications

- ❑ IETF is the main standards body that creates technical standards and protocols for Internet related communications, most notably the TCP/IP protocol suite.
- ❑ IETF standards are also being adopted for wireless LAN; EAP provides a framework for wireless LAN authentication under the IEEE 802.1x framework.
- ❑ IEEE is the main standards body behind the de facto wireless LAN standard suite 802.11 and has created many other popular technical standards, such as port-based Network Access Control (IEEE 802.1x), Ethernet (IEEE 802.3), and Firewire (IEEE 1394).
- ❑ IEEE 802.11 defines PHY and MAC for wireless LAN, including infrared (IR) PHY, frequency hopping spread spectrum (FHSS), and DSSS.
- ❑ IEEE 802.11 supports only 1 and 2 Mbps data rates and operates only in the 2.4 GHz ISM band.
- ❑ IEEE 802.11a is a supplement to the 802.11 standard and was created to support high speed data rates using OFDM in the 5 GHz spectrum. 802.11a support 6, 9, 12, 18, 24, 36, 48, and 54 Mbps data rates with 6, 12, and 24 Mbps being mandatory rates. 802.11a defines the MAC and PHY specifications in the 5 GHz band, and 802.11a is not compatible with IEEE 802.11 PHY in the 2.4 GHz band. The OFDM PHY is merged into the 802.11 standard 2003 revision.

- ❑ FCC and ETSI have different rules on how 802.11a should use the frequencies in 5 GHz and configure radio transmit powers.
- ❑ IEEE 802.11b was created to support high speed data rates using DSSS in the 2.4 GHz spectrum. The 802.11b supports 1, 2, 5.5, and 1 Mbps data rates and is backward compatible with IEEE 802.11 PHY.
- ❑ IEEE 802.11g was created to support the same high speed data rates using ERP-OFDM as 802.11a (OFDM), but only in the 2.4 GHz spectrum. 802.11g is backward compatible with the original IEEE 802.11 PHY and has a protection mechanism built into the standard to avoid collision with the 802.11b client.
- ❑ IEEE 802.11n is currently in draft status in IEEE TGn. MIMO, channel bonding (using the 40 MHz channel) and frame aggregation are new enhancements introduced in 802.11n to increase the performance and reliability of wireless transmission.
- ❑ The Wi-Fi Alliance is the industry body that promotes and certifies equipment from different vendors to ensure proper implementation of IEEE standards and interoperability. Important Wi-Fi certifications include 802.11a, 802.11b, 802.11g, 802.11n draft 2, WPA, WPA2, WMM, and WMM Power Save.

Wireless Topologies and Frame Types

- ❑ The main building blocks in IEEE 802.11 are STA, AP, BSS, ESS, and DS, which are also used in the supplemental 802.11 standards.
- ❑ DCF is the main mechanism used by 802.11 to ensure access to the wireless medium and avoid collisions from STAs. CSMA/CA is the main collision avoidance mechanism. PCF is an optional mechanism and is rarely implemented by equipment vendors.
- ❑ IEEE 802.11 is a layer 2 technology and has defined different frame types to ensure proper communications for the wireless LAN. Three frame types are defined: management frame, control frame, and data frame.
- ❑ The management frame type consists of the following subtypes: beacon, probe request, probe response, authentication, deauthentication, association request, association response, reassociation request, reassociation response, and disassociation.

- The control frame type consists of the following main subtypes: Request-to-Send (RTS), Clear-to-Send (CTS), acknowledgement (ACK)
- The data frame type simply carries information from upper layer protocols, such as TCP/IP, in the main body of the frame.
- AP is usually configured in one of three modes: root mode (AP mode), repeater mode, or bridge mode.
- APs configured in bridge mode or dedicated wireless bridges are usually deployed in either point-to-point (PtP) or point-to-multipoint (PtMP) topology.
- Mesh networking has evolved from PtMP, and the network will dynamically configure itself to find the best path to send traffic back to the main network.

SELF TEST

The following self test questions will help you measure your understanding of the material presented in this chapter. Read all the choices carefully, as there may be more than one correct answer. Choose all correct answers for each question.

Wireless Regulatory Bodies

1. Which organization determines the regulatory rules in the United States for using radio frequency and telecommunications?

A. ETSI
B. FCC
C. FTC
D. IEEE

2. Which organization determines the standards for wireless communications in the European Union, including use of radio frequency and transmission?

A. ETSI
B. FCC
C. FTC
D. IEEE

Wireless LAN Standards and Certifications

3. Which standards body published, maintained, and updated 802.11 and variant standards?

A. IEEE
B. FCC
C. ETSI
D. IETF
E. IANA

4. Which of the following IEEE standards operates in the 2.4 GHz spectrum? (Choose all that apply.)

A. 802.11n
B. 802.11g
C. 802.11
D. 802.1x

5. What modulation is selected by the IEEE and used by 802.11g to support high speed data rates beyond what is defined in 802.11b?
   A. OFDM
   B. DSSS
   C. FHSS
   D. CCK
6. What layer in the OSI model does 802.11 run on?
   A. Transport
   B. Network
   C. Data Link
   D. Physical
7. What is ad hoc network referred to in the IEEE 802.11 standard?
   A. Infrastructure BSS
   B. ESS
   C. Independent BSS
   D. BSS

Wireless Topologies and Frame Types

8. Which three types of frames are defined in the 802.11 standard? (Choose three.)
   A. Management
   B. Broadcast
   C. Control
   D. Data
9. What type of wireless network will dynamically reconfigure itself when connectivity to network is lost?
   A. Wireless LAN with AP in root mode
   B. Mesh networking using RAP and MAP
   C. Wireless bridges in point-to-multipoint topology
   D. Wireless LAN with AP in repeater mode

10. What is the most commonly used mechanism by 802.11 STA to secure access to the wireless medium?

A. CSMA/CA

B. PCF

C. RTS/CTP

D. DCF

11. What is the mechanism used by 802.11 STA to avoid collision to transmit a frame?

A. CSMA/CA

B. CSMA/CD

C. PCF

D. DCF

SELF TEST ANSWERS

Wireless Regulatory Bodies

1. Which organization determines the regulatory rules in the United States for using radio frequency and telecommunications?
   A. ETSI
   B. FCC
   C. FTC
   D. IEEE

☑ **B.** FCC is the regulatory agency in charge of the use of spectrum, power, and hardware.

☒ **A, C,** and **D** are incorrect. ETSI is the standard body for the EU. The FTC is in charge of trade and commerce in the United States. IEEE sets technical standards for technologies.

2. Which organization determines the standards for wireless communications in the European Union, including use of radio frequency and transmission?
   A. ETSI
   B. FCC
   C. FTC
   D. IEEE

☑ **A.** ETSI is the standards agency in Europe Union.

☒ **B, C,** and **D** are incorrect. The FCC is the regulatory agency in charge of the use of spectrum, power, and hardware in the United States. The FTC is in charge of trade and commerce in the United States. IEEE sets technical standards for technologies.

Wireless LAN Standards and Certifications

3. Which standards body published, maintained, and updated 802.11 and variant standards?
   A. IEEE
   B. FCC
   C. ETSI
   D. IETF
   E. IANA

☑ **A.** IEEE published, maintained, and updated 802.11 and variant standards.

☒ **B, C, D,** and **E** are incorrect. FCC is the regulatory agency in the United States. ETSI is the standards body in European Union. IETF published, maintained, and updated standards for the Internet. IANA assigns IP addressing.

4. Which of the following IEEE standards operates in the 2.4 GHz spectrum? (Choose all that apply.)

A. 802.11n

B. 802.11g

C. 802.11

D. 802.1x

☑ **A, B,** and **C.** 802.11, 802.11b, 802.11g, and 802.11n all operate in the 2.4 GHz spectrum.

☒ **D** is incorrect. 802.1x is a framework defined by IEEE for authentication.

5. What modulation is selected by the IEEE and used by 802.11g to support high speed data rates beyond what is defined in 802.11b?

A. OFDM

B. DSSS

C. FHSS

D. CCK

☑ **A.** OFDM is the modulation selected for speeds up to 54 Mbps for 802.11g.

☒ **B, C,** and **D** are incorrect. DSSS is supported in 802.11 and 802.11b, but 802.11g does not use DSSS as modulation for speeds above 11 Mbps. FHSS is only defined in 802.11 and supports 1 and 2 Mbps. CCK is an encoding scheme.

6. What layer in the OSI model does 802.11 run on?

A. Transport

B. Network

C. Data Link

D. Physical

☑ **C** and **D.** 802.11 defines PHY and MAC, which lives on the physical and data link layers, respectively.

☒ **A** and **B** are incorrect. TCP and UDP are typical protocols in the transport layer. IP is typical in the network layer.

7. What is ad hoc network referred to in the IEEE 802.11 standard?

A. Infrastructure BSS

B. ESS

C. Independent BSS

D. BSS

☑ **C.** Independent BSS is referred to as an ad hoc network, usually referring to BSS without an AP.

☒ **A, B,** and **D** are incorrect. Infrastructure BSS is BSS with an AP, sometimes also called BSS. ESS is a group of infrastructure BSS.

Wireless Topologies and Frame Types

8. Which three types of frames are defined in the 802.11 standard? (Choose three.)

A. Management

B. Broadcast

C. Control

D. Data

☑ **A, C,** and **D.** 802.11 defines three types of frames: management, control, and data.

☒ **B** is incorrect because broadcast is not a frame type defined by 802.11.

9. What type of wireless network will dynamically reconfigure itself when connectivity to network is lost?

A. Wireless LAN with AP in root mode

B. Mesh networking using RAP and MAP

C. Wireless bridges in point-to-multipoint topology

D. Wireless LAN with AP in repeater mode

☑ **B.** Mesh networking consists of RAP and MAP. When connectivity between RAP and MAP breaks, MAP will dynamically reconfigure an alternate best path to reach RAP and the network.

☒ **A, C,** and **D.** AP in root mode is normal wireless LAN operation and involves only clients and AP. Wireless bridges in point-to-multipoint topology consist of static links connecting LAN segments using wireless bridges. The topology is statically configured. AP in repeater mode extends the reach of an AP, but it does not configure wireless topology when it loses connectivity to the network.

10. What is the most commonly used mechanism by 802.11 STA to secure access to the wireless medium?

A. CSMA/CA

B. PCF

C. RTS/CTP

D. DCF

☑ **D.** DCF is the mechanism in 802.11 to control access to the wireless medium.

☒ **A, B,** and **C.** CSMA/CA is the mechanism for collision avoidance. PCF is an option mode in 802.11 for medium access control. RTS/CTS is another mechanism designed to avoid frame collision.

11. What is the mechanism used by 802.11 STA to avoid collision to transmit a frame?

A. CSMA/CA

B. CSMA/CD

C. PCF

D. DCF

☑ **A.** CSMA/CA is the mechanism for collision avoidance.

☒ **B, C,** and **D.** CSMA/CD is the collision avoidance mechanism used by 802.3 on Ethernet. It is not used by 802.11; PCF and DCF are mechanisms designed to control access to the wireless medium.

3

Other Wireless Technologies

CERTIFICATION OBJECTIVES

3.01 The Impact of Various Wireless Technologies

✓ Two-Minute Drill

Q&A Self Test

IEEE 802.11 is one of the many wireless technologies that utilize the unlicensed ISM 2.4 GHz and 5 GHz spectrum. There are many other wireless technologies operating in the same wireless spectrum and bringing convenience to people's everyday lives, and some may impact the 802.11 wireless LAN. This chapter will provide an overview of the various non-802.11 wireless technologies, how these different technologies operate, what levels of impact these technologies may have on an 802.11-based wireless LAN, and how to avoid or minimize impacts from outside interferences.

CERTIFICATION OBJECTIVE 3.01

The Impact of Various Wireless Technologies

Because the ISM band in 2.4 GHz and U-NII and ISM bands in 5 GHz spectrums are unlicensed, anyone who wishes to operate a radio frequency device can and is allowed to use the frequency as long as the power settings and EIRP stay within bound of government regulations. There are other industry standards supporting devices operating in the same spectrum and supporting everyday indoor applications, such as Bluetooth, IEEE 802.15.4 (ZigBee), 802.16 (WiMAX), DECT (typically seen in the home cordless telephones), and others. This chapter will provide a brief introduction to some of these devices and standards.

Bluetooth

Bluetooth is a wireless protocol designed for short range wireless communications for fixed or mobile devices in a Personal Area Network (PAN), which usually requires low power. It is an alternative technology to infrared (IrDA) and 802.11 technologies and well suited for personal mobile devices, such as cellular phone and PDA headsets, mobile accessories (such as cordless mouse for notebook or desktop PC), and low bandwidth devices (such as cameras and game console controllers for Wii and Sony's PlayStation 3).

The Bluetooth technology was originally developed by the Bluetooth Special Interest Group (SIG) in 1998. SIG published v1.0 specifications in 1999. As the standard evolved, IEEE adopted Bluetooth v1.1 Foundation Specifications to be the basis for the 802.15.1 Wireless Personal Area Network standard, and the official 802.15.1 was published in June 2002. In 2005, 802.15.1 was updated based on Bluetooth v1.2; nonetheless, IEEE has also determined to decouple Bluetooth with IEEE 802.15.1, so that future Bluetooth specifications will no longer be part of the IEEE standard. SIG continues to develop the Bluetooth standard and published v2.0 and v2.1 specifications in 2004 and 2007 respectively, supporting speeds up to 2 Mbps.

Bluetooth is a low power and frequency hopping technology operating in the 2.4–2.4835 GHz spectrum and divides the spectrum into 79-1MHz channels. Bluetooth devices can pair with other Bluetooth devices in ad hoc fashion and form piconet of up to eight devices. For example, a laptop PC with Bluetooth radio can pair up with seven different Bluetooth accessories. Although Bluetooth devices pose only minor impacts to wireless LAN in the 2.4 GHz due to low power transmitting radio and frequency hopping nature, Bluetooth may potentially interfere with 802.11 WLAN when many Bluetooth devices operate in close proximity of a wireless LAN in 2.4 GHz. Figure 3-1 shows raw energy trace of a Bluetooth device operating in the 2.4 GHz spectrum. The blue line shows the max RF energy observed in a given frequency, and the yellow line is real time RF energy plot. You can see the FHSS patterns of Bluetooth devices detected by the spectrum analyzer across the entire 2.4 GHz spectrum.

FIGURE 3-1 5 GHz channel to frequency map

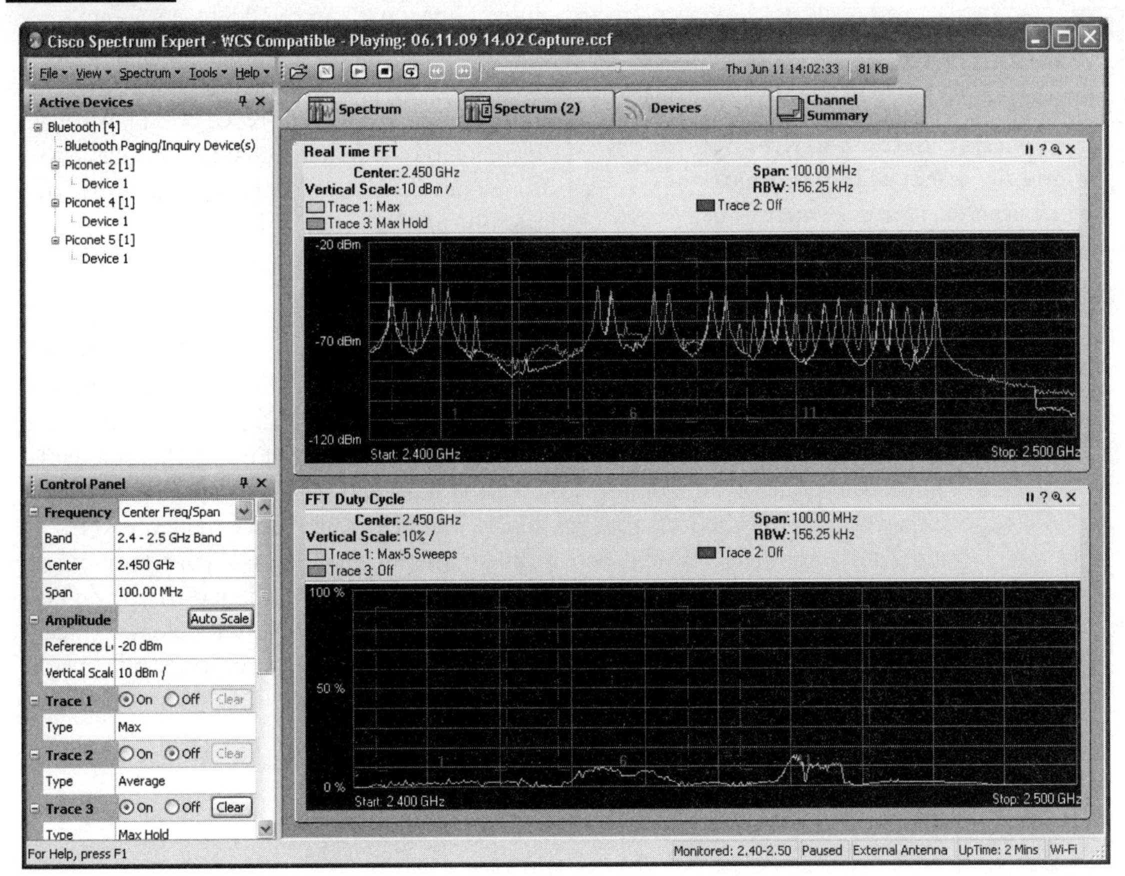

The Bluetooth and wireless LAN battle for the 2.4 GHz spectrum has been widely observed in the healthcare environment, where wireless LAN is being widely adopted to support patient care applications connectivity on mobile patient care carts on wheels (CoW) alongside Bluetooth peripheral/accessory devices, including mouse, bar code scanner, and voice headset. This is particularly true in patient care units where unused CoWs tend to congregate near the nursing station. To avoid contention between Bluetooth and 802.11-based WLAN, it is recommended that data and voice wireless LAN traffic be moved to the 5 GHz spectrum whenever possible and that congregation of large numbers of Bluetooth devices in one area be avoided.

Bluetooth v3.0 was released in April 2009. In v3.0, Bluetooth is allowed to use an alternate MAC and PHY (AMP), such as 802.11, to transport Bluetooth profile data. Bluetooth radio is still used for device discovery, initial connection, and profile configuration, but for high speed data transport, use 802.11 MAC and PHY as transport. SIG has created an 802.11 Protocol Adaptation Layer to interconnect with an 802.11-based system. Adaptive Frequency Hopping (AFH) has been introduced by SIG to avoid the impact of interference to 2.4 GHz–based 802.11 WLAN by reducing the number channels available to Bluetooth devices; the details are described in the Bluetooth Specification v3.0 + HS.

Understand the frequencies Bluetooth devices operate in and the effect of Bluetooth devices on active 802.11 communications.

ZigBee

ZigBee represents a different type of device that operates in the 2.4 GHz spectrum and is usually used for monitoring and controlling devices. The ZigBee standard is published by ZigBee Alliance, which promotes ZigBee as a standard for low cost and low power (long battery life) wireless devices based on the IEEE 802.15.4 standard with focus primarily in monitoring and control devices. ZigBee has been embedded in a wide range of products that are designed to be less expensive than WLAN (802.11) and other WPANs, such as Bluetooth. Another point worth mentioning about ZigBee is that the protocol is designed for mesh networking; thus, the ZigBee network is capable of dynamically changing the wireless network topology. Typically, ZigBee devices have been built into many industrial sensor monitoring and control devices, including the following:

- Smart meters
- Industrial and building automation

- Home automation
- Lighting controls
- HVAC controls
- Temperature sensors
- Smoke and CO_2 sensors

Mesh networking allows the ZigBee network to cover a large area without running costly cables. In addition, the self-healing nature of mesh increases the reliability of the network in case of a node failure. ZigBee operates in ISM; 868 MHz in Europe; 915 MHz in the United States and Australia; and 2.4 GHz in most of the regulatory domains worldwide. Because 802.11 b/g also operates in the ISM 2.4 GHz spectrum, the coexistence of 802.11 and ZigBee networks may interfere with each other when deployed in close proximity. ZigBee divides the 2.4 GHz spectrum (2.4–2.4835 GHz) into 16 3-MHz channels with 5 MHz bandwidth between adjacent channel center frequencies. The peak data rate is 250 kbps, and ZigBee too uses CSMA/CA for collision avoidance, allowing up to 256 nodes in a network. Because ZigBee devices are designed to be low power devices (that is, some ZigBee radios transmit a power range from –27 dBm to 4 dBm), compared to 802.11 devices (which typically transmit power up to 20 dBm), and because their channel bandwidth is narrower than 802.11 channels, when the two networks operate in the same frequency, it is more likely that the 802.11 devices will have greater impact on ZigBee devices than vice versa. The ZigBee Alliance plays a similar role as the Wi-Fi Alliance, which promotes interoperability of products based on the proper implementation of standards.

802.11 devices will regard ZigBee signal as noise and try to talk "over" any ZigBee communications. Site survey is critical when deploying 802.11 and ZigBee networks in the same environment using the 2.4 GHz spectrum. Cisco recommends careful channel selection and assignment, and impacts to the applications should be carefully evaluated.

WiMAX

Worldwide Interoperability for Microwave Access (WiMAX) is a broadband telecommunications technology offering fast wide area network (WAN), also referred to as wireless metropolitan area network (wireless MAN), wireless connectivity or the 4G mobile wireless technology. The technology is based on the IEEE 802.16 standard. There are several flavors of WiMAX standardized by IEEE and the most frequently mentioned are 802.16d—fixed WiMAX and 802.16e—mobile WiMAX.

802.16d is the standard for fixed WiMAX connections that utilize OFDM; it is not interoperable with 802.16e. The main intent of 802.16d was to provide high speed point-to-point or point-to-multipoint backbone connections, and typical 802.16d devices include indoor/outdoor customer premise equipment (CPE). Services providers can offer WiMAX as a "last mile" solution or an alternative to dedicated circuit connection, DSL, or cable modem of up to a few miles from the central office (CO).

802.16e is the mobile WiMAX standard; however, 802.16e is also applicable for fixed WiMAX connections and typically supports indoor/outdoor CPE, laptops, PDAs, cellular phones, and add-on PC cards. 802.16e also uses OFDM but is not compatible with 802.16d.

WiMAX operates in the licensed frequency spectrum; thus, the use of WiMAX in real life is regulated by FCC. WiMAX is also considered a mobile wireless backhaul technology, and some cellular services providers have regarded WiMAX as a 4G (fourth generation) mobile Internet technology. Other competing technologies include LTE Advanced. Since WiMAX only operates in the licensed spectrum, it does not interfere with wireless LAN or any of the 802.11 variants. WiMAX and wireless LAN are often regarded as complementary technologies: one is more focused on local area access, and the other provides an alternate solution to services provider backhaul. As technologies mature, there may be devices, such as laptop computers, equipped with multiple radios (WiMAX and 802.11) that will be able to roam between different types of networks without user intervention.

WiMAX Forum is the industry organization that promotes and certifies WiMAX devices interoperability through certification testing.

Cordless Telephones

Cordless telephones (sometimes referred to as DECT phones) use 900 MHz, 2.4 GHz, and 5.8 GHz of ISM bands in the United States. (This should not be confused with DECT developed by ETSI, which primarily operates in the 1.8–1.9 GHz spectrum.) The cordless phones operating in the 2.4 GHz spectrum are usually referred to as third generation cordless phones, and they operate between 2.4–2.4835 GHz with a channel spacing of 5 MHz. Because of the openness of 2.4 GHz spectrum, cordless phones operating in this spectrum do interfere with 802.11b/g based wireless LAN networks. Most of the phones use FHSS technology and will utilize the entire 2.4 GHz spectrum during operation. To ensure good voice quality, some cordless phones are equipped with powerful radios and often cause 802.11 b/g connections to drop. Figure 3-2 shows a spectrum analysis capture when a 2.4 GHz cordless phone was in use. The spectrum analyzer was placed 40 feet from

FIGURE 3-2 2.4 GHz RF energy reading when cordless phone in use

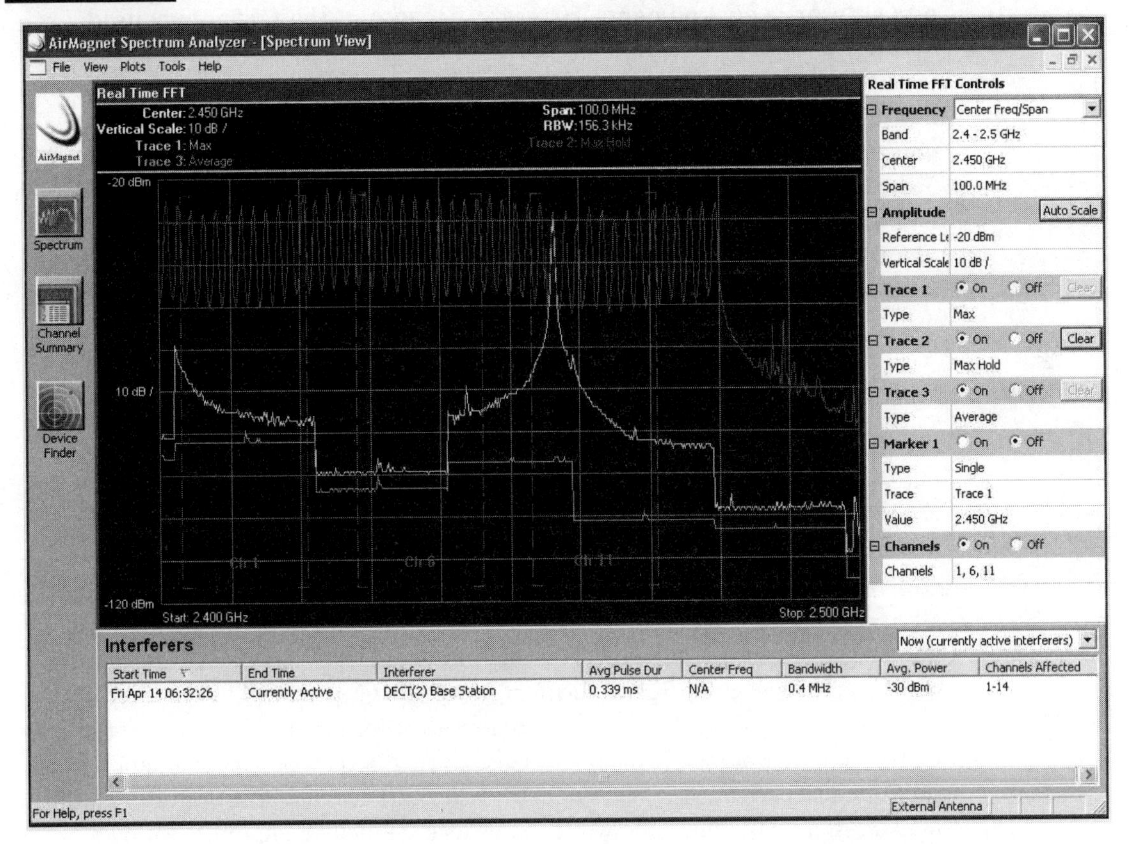

the cordless phone base station and the cordless phone handset. The yellow line shows the real-time RF energy utilizing the spectrum, while the pink line shows max energy reading hold. The pink line shows the FHSS pattern exhibited by the cordless phone operating in the 2.4 GHz spectrum.

The cordless phones operating in the 5.8 GHz spectrum are usually referred to as fourth generation cordless phones and operate between 5.725 and 5.850 GHz. Some cordless phones use 5.8 GHz frequencies for base-to-handset communications and 2.4 GHz frequencies for handset-to-base communications; hence, they utilize both 2.4 GHz and 5.8 GHz spectrum. The impact of 5.8 GHz cordless phones will likely have a similar effect on the 802.11a wireless LAN; however, the issue will be easier to mitigate as there are many more nonoverlapping channels available in the 5 GHz spectrum for wireless LAN to use.

on the Job

Cordless phones are very popular due to ease of use and low costs in home and enterprise environments. However, because of the technology used, cordless phones have also been found to be extremely disruptive to the wireless LAN. Migrating cordless phone devices to a different frequency has proved to be the most effective solution to avoid conflicts of airwave contention with a wireless LAN. The new DECT 6.0 standard in the United States has allocated 1.9 GHz to be used by the cordless phones. For field engineers, it is important to have tools such as a spectrum analyzer to help identify interfering devices; for example, a cordless phone that is interfering with wireless LAN when troubleshooting connectivity issues.

Microwaves

Microwave ovens are common household appliance that convert electricity to microwave energy to cook food. The magnetron used in microwave ovens typically produce microwave energy, usually in the 2.45 GHz range. Depending on the age of the microwave ovens, some produce microwave energy spanning across the entire 2.4 GHz spectrum. In an ideal situation, microwave ovens should prevent all the microwave energy from leaking outside of the microwave oven; however, all microwave ovens emit some levels of microwave energy. Even though the leakage may be very small compared to the microwave energy generated by the microwave oven, it may be enough energy to interfere with wireless LAN operation in the 2.4 GHz spectrum. Microwave energy will also disrupt other types of communications in the 2.4 GHz spectrum, but it may be to a lesser extent depending on the technology used. Possible causes for microwave energy leakage include improper sealing of microwave oven enclosures, normal wear and tear on microwave oven enclosures, improper use and abuse of the microwave oven by users, and defective material used for the microwave oven enclosures. Figure 3-3 shows a spectrum analysis capture in the 2.4 GHz spectrum when a microwave oven was in use in close proximity to the spectrum analyzer. The microwave oven occupied the entire 2.4 GHz spectrum and emitted strong energy.

The most effective workaround to mitigate the effect of microwave energy leakage from microwave ovens is to avoid using 802.11 b/g near microwave ovens. Microwave ovens do not interfere with devices using 802.11a in the 5 GHz spectrum. Site survey and RF planning in 2.4 GHz will help identify the extent of which microwave oven and microwave energy leakage will impact the 2.4 GHz spectrum. Periodic RF survey and validation is recommended to detect 2.4 GHz interference caused by microwave ovens.

FIGURE 3-3 Microwave energy in 2.4 GHz spectrum

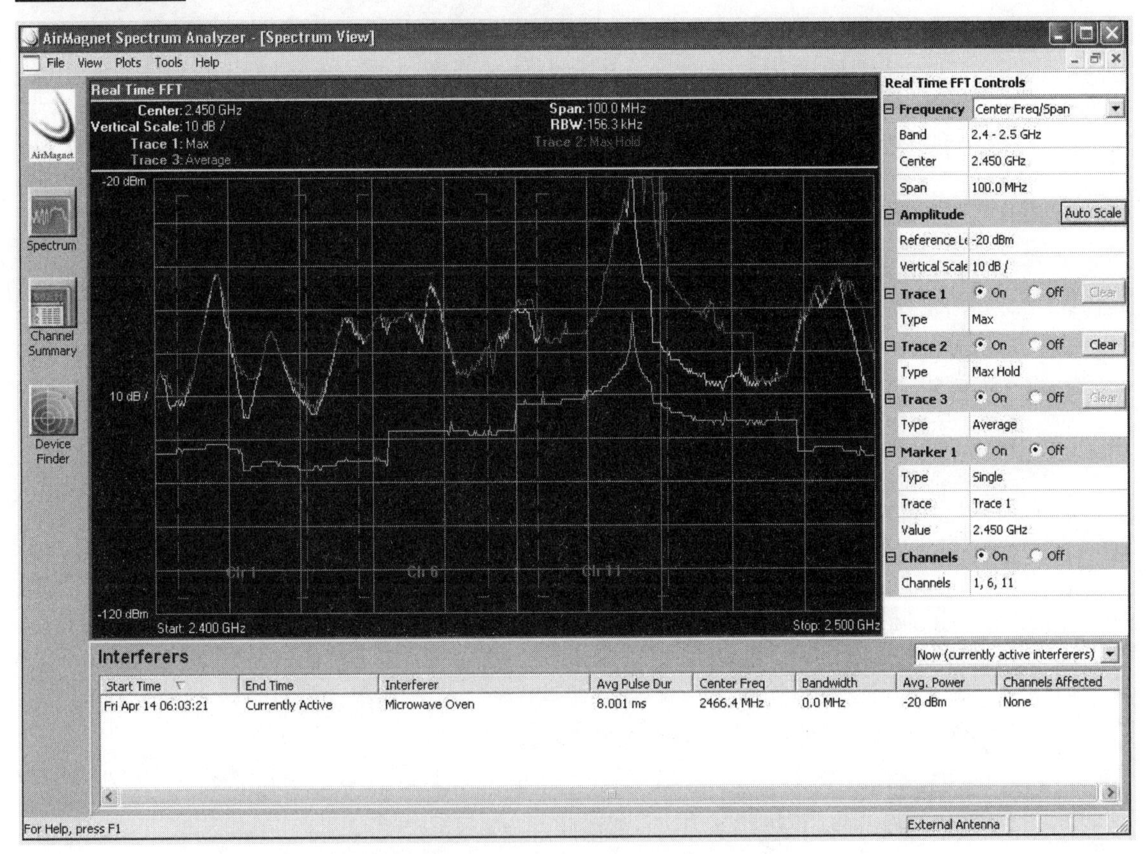

Most newer microwave ovens in the market today tend to operate in the higher range of the 2.4 GHz band, so, in some cases, setting the channel setting to 1 for AP near a microwave oven may help reduce the impact of the microwave oven to the WLAN, but you should verify the actual impact of microwave oven using a spectrum analyzer before making any configuration changes.

exam watch

Understand why cordless phones, microwave ovens, and Bluetooth devices are bad for 802.11 communications. Be familiar with these technologies and know how to mitigate them when you encounter interferences from these types of devices.

Radar

So far, most interference sources use ISM 2.4 GHz and impact 802.11 b/g devices. There is one type of outside interference that primarily impacts the 5 GHz spectrum and 802.11a devices. Radar operating in the 5GHz spectrum for military and civilian purposes may impact 802.11a devices using the same frequencies. Because both radar and 802.11a devices operate in a portion of the same 5 GHz spectrum, 802.11a devices may also impact radar operations if used in close proximity. To minimize impact, the FCC issued a Memorandum Opinion and Order (MO&O) in June 2006, requiring any devices operating in the 5.25–5.35 GHz and 5.47–5.725 GHz bands to employ Dynamic Frequency Selection (DFS) and Transmit Power Control (TPC) mechanisms so devices will avoid causing interferences with the radar system. DFS will continually monitor the RF for radar presence, and TPC will lower the radio transmitter and transmit power and operate at least 6 dB below the mean EIRP of 30 dBm. ITU-R Resolution 229 has referenced similar mechanisms for radar detection and has been adopted by regulatory authorities in many countries around the world. IEEE 802.11h has incorporated DFS and TPC into the standard and incorporated it into 802.11a based on IEEE 802.11—2007. Wi-Fi Alliance has also incorporated validation of proper implementation of standards into their certification testing. To avoid any issues with radar and to minimize impacts to wireless LAN, only products that are 802.11a and 802.11h certified by Wi-Fi Alliance should be considered.

Other Types of Outside Interference

There are other types of interferences operating in the 2.4 GHz and 5 GHz frequencies, including the following:

- **Game Controllers and Adapters** Many operate in the 2.4 GHz frequency range and use proprietary protocols.
- **Wireless Cameras** Many operate in the 2.45 GHz frequency range or use FHSS technology to connect to control and video storage or use proprietary protocols.
- **Motion Sensors and Alarm Systems** Many operate in the 2.4 GHz frequency, and some use infrared to connect to the alarm receiver or use proprietary protocols.
- **Baby Monitor** Many operate in the 2.40–2.483 GHz frequency range and some use FHSS technology or proprietary protocols.

To mitigate or reduce impacts of these devices on the wireless LAN, it is recommended that you use a standard-based solution utilizing 802.11 a/b/g/n or choose different frequencies for these devices. Site survey and continual and periodic RF validation are recommended to anyone running or planning to run a wireless LAN to avoid impact to other important applications utilizing the same spectrum. Furthermore, other RF interference may be produced by devices when there's an abrupt change in electric currents through some electrical switches, such as fluorescent lights or light dimmers. Use light fixtures with conductive shields to reduce RF interferences or avoid deploying wireless LAN in close proximity to the source of these interferences.

Knowing the enemy is usually half the battle when dealing with different types of RF interferences. Make sure you have proper tools that will help you identify and locate the sources of interferences. Having adequate information will also help you choose the right equipment and place the AP and client appropriately.

INSIDE THE EXAM

The Impact of Various Wireless Technologies

For the exam, you should know the different types of technologies operating in the 2.4 GHz spectrum. Some may be regarded as interferences by 802.11-based wireless LAN. It is not crucial to know all the nuts and bolts of each of the technologies, but be sure to know the frequencies they operate in and have basic understanding of each. Understand how these technologies may impact wireless LAN and how to coexist with them. Also, know what types of tool you will need to identify the different interferences and how to mitigate them if you're called upon.

CERTIFICATION SUMMARY

Although Wireless LAN today is built primarily using the IEEE 802.11 standards suite, there are many other technologies sharing the same frequencies that are designated for unlicensed use. Bluetooth is a popular technology to supplement many personal communications and electronics devices. With the proliferation of cellular phones and laptop computers, Bluetooth-based accessories have become more popular than ever. It is important to understand that Bluetooth shares the 2.4 GHz band with 802.11 b/g devices, and an FHSS-based Bluetooth device may impact normal wireless LAN operation in the same spectrum. Bluetooth specifications are maintained and published by Bluetooth SIG.

ZigBee is another technology operating in the 2.4 GHz spectrum. ZigBee is designed to be low cost and low power, so it may not impact 802.11 b/g–based wireless LAN as much. On the other hand, wireless LAN radio may impact ZigBee devices if transmitting in high power. ZigBee is defined by IEEE 802.15.4.

WiMAX is a complementary technology operating in the licensed spectrum. Similar to 802.11 devices, WiMAX may provide network access to end user devices, such as laptop PCs and cellular phones for data access, and fixed WiMAX provides last mile access connecting buildings and offices. WiMAX is defined by IEEE 802.16 and variants, and the most recent standard is 802.16e. WiMAX devices are tested and certified by WiMAX Forum.

FCC announced new regulations on the use of the 5 GHz spectrum. For devices operating in the U-NII 2 and U-NII 2 extended frequencies, TPC and DFS are required mechanisms to avoid interference with radar systems.

There are many other types of devices operating in the unlicensed 2.4 GHz, and it is important to know how to identify them using the tools available.

TWO-MINUTE DRILL

The Impact of Various Wireless Technologies

- ❑ Bluetooth is published by Bluetooth SIG.
- ❑ Bluetooth operates in the 2.4 GHz spectrum using FHSS technology.
- ❑ Bluetooth devices may impact 802.11 WLAN when many congregate in a concentrated area.
- ❑ ZigBee is defined by IEEE 802.15.4 and operates in the 2.4 GHz spectrum.
- ❑ ZigBee is designed as a low cost, low power technology and uses mesh networking technology to connect to different ZigBee nodes.
- ❑ WiMAX is defined by IEEE 802.16 and variants. The most recent standard is 802.16e, and it operates in the licensed spectrum.
- ❑ FCC has defined new regulations requiring the use of DFS and TCP for devices operating in the U-NII2 and U-NII2 extended frequencies to avoid interference with radar systems.
- ❑ IEEE 802.11h has incorporated DFS and TCP in the standard, and Wi-Fi Alliance requires support of DFS and TCP for 802.11a certification.
- ❑ The best way to avoid impacts on wireless LAN by other wireless devices is to avoid using the same operating frequencies or remove the sources of interferences.

SELF TEST

The following self test questions will help you measure your understanding of the material presented in this chapter. Read all the choices carefully, as there may be more than one correct answer. Choose all correct answers for each question.

The Impact of Various Wireless Technologies

1. Which IEEE standard defines the Bluetooth v 3.0?
   A. 802.15.1
   B. 802.15.4
   C. 802.16e
   D. None of above
2. Which IEEE standard defines the mobile WiMAX?
   A. 802.15.1
   B. 802.15.4
   C. 802.16e
   D. None of above
3. Which type(s) of outside interferences may impact the operation of 802.11b/g network? (Choose all that apply.)
   A. Bluetooth
   B. Radar
   C. WiMAX
   D. Microwave Ovens
4. Which IEEE standard impact ZigBee devices? (Choose all that apply.)
   A. 802.11a
   B. 802.11b
   C. 802.11g
   D. 802.11h
5. Which IEEE standard is frequently referred as "wireless MAN"?
   A. 802.11
   B. 802.15.1
   C. 802.15.4
   D. 802.16

6. What mechanisms are mandated by FCC for devices to operate in the 5 GHz spectrum? (Choose all that apply.)
   A. CSMA/CA
   B. DFS
   C. TPC
   D. TCP
   E. DSF
7. Which IEEE technology is designed to be used in the licensed spectrum?
   A. 802.15.1
   B. 802.11
   C. 802.15.4
   D. 802.16
8. Bluetooth is based on DSSS technology operating in the 2.4 GHz spectrum.
   A. True
   B. False
9. ZigBee is designed using mesh technology with a high power radio.
   A. True
   B. False
10. What frequency do microwave ovens operate in?
    A. 2.4 GHz
    B. 1.9 GHz
    C. 5 GHz
    D. 5.8 GHz

SELF TEST ANSWERS

The Impact of Various Wireless Technologies

1. Which IEEE standard defines the Bluetooth v 3.0?

A. 802.15.1

B. 802.15.4

C. 802.16e

D. None of above

☑ **D.** Bluetooth SIG publishes the latest Bluetooth standard and is not based on any IEEE standards.

☒ **A, B,** and **C** are incorrect. 802.15.1 is derived from Bluetooth v1.1 and v1.2 only. 802.15.4 is defined for ZigBee. 802.16e is defined for WiMAX.

2. Which IEEE standard defines the mobile WiMAX?

A. 802.15.1

B. 802.15.4

C. 802.16e

D. None of above

☑ **C.** 802.16e is defined for WiMAX.

☒ **A, B,** and **D** are incorrect. 802.15.1 is derived from Bluetooth v1.1 and v1.2 only. 802.15.4 is defined for ZigBee.

3. Which type(s) of outside interferences may impact the operation of 802.11b/g network? (Choose all that apply.)

A. Bluetooth

B. Radar

C. WiMAX

D. Microwave Ovens

☑ **A** and **D.** Bluetooth devices use FHSS and operate across the entire 2.4 spectrum. Microwave ovens operate in the 2.4 GHz spectrum.

☒ **B** and **C** are incorrect. Radar operates primarily in the 5 GHz range. WiMAX operates in the licensed frequency range.

4. Which IEEE standard impact ZigBee devices? (Choose all that apply.)

A. 802.11a

B. 802.11b

C. 802.11g

D. 802.11h

☑ **B** and **C**. 802.11b and 802.11g both operate in the 2.4 GHz spectrum. With a high transmit power setting, ZigBee may be impacted.

☒ **A** and **D** are incorrect. 802.11a operates in the 5 GHz spectrum; 802.11h defines TPC and DFS.

5. Which IEEE standard is frequently referred as "wireless MAN"?

A. 802.11

B. 802.15.1

C. 802.15.4

D. 802.16

☑ **D**. 802.16 defines WiMAX and is often referred to as wireless MAN.

☒ **A, B,** and **C** are incorrect. 802.11 is often referred to as WLAN. 802.15.1 and 802.15.4 are both considered WPAN.

6. What mechanisms are mandated by FCC for devices to operate in the 5 GHz spectrum? (Choose all that apply.)

A. CSMA/CA

B. DFS

C. TPC

D. TCP

E. DSF

☑ **B,** and **C**. The FCC mandates DFS and TPC for devices to operate in the 5 GHz spectrum.

☒ **A, D,** and **E** are incorrect. CSMA/CA is a collision avoidance mechanism defined by 802.11. TCP is a transport layer protocol.

7. Which IEEE technology is designed to be used in the licensed spectrum?

A. 802.15.1

B. 802.11

C. 802.15.4

D. 802.16

☑ **D**. 802.16 operates in the licensed spectrum.

☒ **A, B,** and **C** are incorrect. 802.15.1, 802.15.4, and 802.11 all operate in the unlicensed spectrum.

8. Bluetooth is based on DSSS technology operating in the 2.4 GHz spectrum.

A. True

B. False

☑ **A.** Bluetooth is based on FHSS technology.

☒ **B** is not correct because Bluetooth is not based on DSSS technology.

9. ZigBee is designed using mesh technology with a high power radio.

A. True

B. False

☑ **B.** ZigBee is designed to be low power.

☒ **A** is not correct because ZigBee is not designed to be high power.

10. What frequency do microwave ovens operate in?

A. 2.4 GHz

B. 1.9 GHz

C. 5 GHz

D. 5.8 GHz

☑ **A.** Microwave ovens operate in the 2.4 GHz spectrum.

☒ **B, C,** and **D** are incorrect. DECT 6.0 cordless phones operate in the 1.9 GHz spectrum, and 802.11a operates in the 5 GHz spectrum.

Part II

Wireless LAN Security

CHAPTERS

4

Wireless Security Framework

CERTIFICATION OBJECTIVES

4.01 The General Framework of Wireless Security and Security Components

✓ Two-Minute Drill

Q&A Self Test

Wireless LAN has been conceived as an insecure media for network connectivity. There has been a lot of publicity about wireless LAN security being compromised by tools that can be conveniently downloaded from the Internet and require little user knowledge to operate. As wireless LAN becomes widely adopted in homes and businesses, it is critical to implement proper security measures to ensure the fidelity of wireless LAN and integrity of personal and enterprise data. Government regulations, such as Sarbanes-Oxley (SOX), the Graham-Leach-Bliley Act (GLBA), and the Health Insurance and Portability and Accountability Act of 1996 (HIPAA), and industry standards, such as the Payment Card Industry Data Security Standard (PCI DSS, or PCI for short) provide sets of security requirements and guidelines for individuals and organizations to follow when deploying data networks, and these guidelines apply equally well for both wired and wireless LAN communications; hence, these guidelines and standards should only be treated as minimum security.

This chapter will provide an overview of wireless security framework with components that are commonly deployed to safeguard wireless LAN. Also in this chapter, we will discuss the various common security technologies and their applications from a high level, so you have a basic understanding of the role each device performs in a network. In the next chapter, we will explore the different types of encryption and authentication mechanisms available in the marketplace today. We'll then provide guidelines on how to properly select and configure a security solution with support for a variety of client types while maintaining application performance and high data throughput.

CERTIFICATION OBJECTIVE 4.01

The General Framework of Wireless Security and Security Components

You cannot rely on a single point solution to secure a wireless LAN; rather, it requires a holistic approach to secure the entire network and to address different types of security threats, including deploying purposely built security solutions, unified network management, and comprehensive policies. To find the appropriate security solutions, you will need to first understand the different types of security threats facing the wireless LAN. This chapter will provide an overview of common security threats, wireless LAN management and policy, and a more in-depth discussion of the following solutions and features to address specific wireless LAN security needs:

- Wireless intrusion prevention and detection system
- Firewall and access control lists
- Management frame protection

Common Security Threats

Before going into detail about wireless LAN security, it is important to know the security threats that security solutions are designed to protect against. This section will cover, at a high level, the primary security concerns of network administrators before and during a wireless LAN deployment while providing basic knowledge about the most prevalent threats to an enterprise wireless LAN network. It is important to note that the concept of physical security is often lost in this discussion. All of the following security threats require the attacker to be within the coverage range of the wireless network and can largely be prevented by physical security before the discussion of wireless LAN security even comes into play. (Note, however, that by using high gain antenna and RF amplifiers, it is possible that attacks can be carried out when attackers are physically outside of physical premise of your properties.)

Passive Wireless Sniffing and Active Scanning

Passive wireless sniffing attacks are a common and often undetectable breach of wireless security. People use different tools when conducting wireless sniffing, some of which may involve active scanning of RF channels. Pure passive wireless sniffing is virtually undetectable if the tool does not emit any RF transmission. Using a wireless sniffer is considered passive wireless sniffing when the client is in listening-only mode. Other techniques, such as eavesdropping, describe the same type of passive attack commonly seen against wireless LAN. The goal of wireless sniffing is to perform reconnaissance and gather information about a network by examining beacons that are sent by a wireless access point and a data that is transmitted between valid wireless transmitters and receivers. Information such as usernames, passwords, IP addresses, and various other information exchanges can be gathered by this manner and even used to perpetrate a more active attack on the wireless network as described in the next section. Because a passive attacker does not have to associate to the wireless network to get this information, passive wireless attacks are very difficult to detect and track. Tools for performing wireless sniffing are widely available and are easy to acquire and use, which means that an attacker does not have to be technically knowledgeable or spend much time to effectively sniff a wireless network.

Active wireless scanning attacks use tools that emit active RF transmissions to discover wireless networks in the surrounding area. Some tools will send out active probe requests and record the responses received. With correct security configurations, sometimes a tool may connect to a BSS network and perform basic performance tests. Figure 4-1 illustrate a snapshot of the wireless LAN available near a residential complex using NetStumbler 4, a freeware available on the Internet.

In addition to physical security, passive wireless sniffing can be mitigated by not broadcasting a wireless SSID and instituting strong encryption methods on data that is transmitted.

Active Wireless Access

Active wireless access attacks involve the actual penetration of a wireless network by an attacker who has either associated to an open wireless network or has breached the security measures on a wireless network to gain access. Active wireless attacks are used to acquire sensitive corporate information, to "spoof" illegal activity (spamming, Internet attacks) using network resources, and to make changes to the configuration of the network. An active attack can be used to set up or launch additional attacks, such as worms, viruses, and additional access holes within a network. As discussed before, the information need to get active access into a

FIGURE 4-1

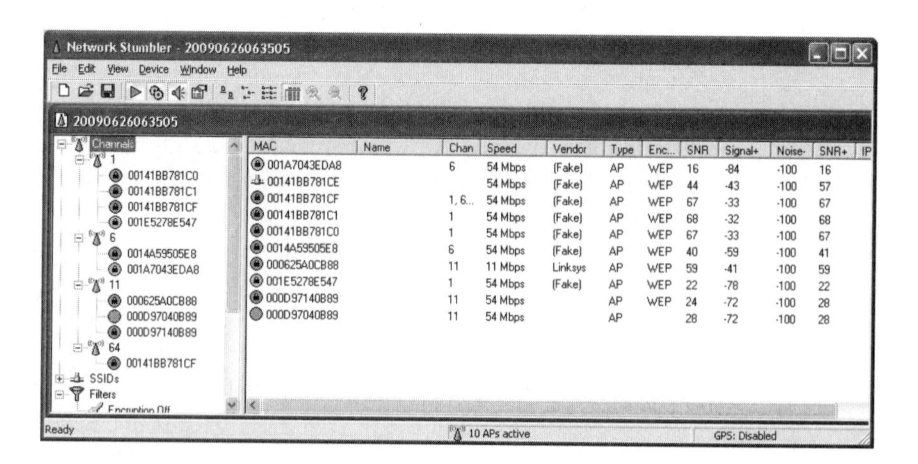

A snapshot of available wireless LAN

wireless network can be gained by using passive sniffing methods or gaining access credentials through social engineering tactics. Although an active attacker associates to the wireless network being compromised (increasing the likelihood of detection), the attacker can use this access to delete traces of the intrusion and association.

A common example of active attacks includes *war driving*, in which individuals find and associate to broadcasting wireless networks using a wireless laptop or PDA from a moving car; unsecured broadcasting wireless networks are most susceptible to this. The best methods for mitigating active wireless attacks include strong security policies combined with the application of robust authentication and encryption methods.

Jamming/Denial of Service

Jamming is a form of *Denial of Service (DoS)* attack that is used with the intention to disable or effectively neutralize the coverage and usability of a wireless LAN to legitimate users and clients. This is accomplished by having physical access and a device that can jam the legitimate radio frequency signal with a stronger disruptive signal. Due to the unlicensed nature of the 2.4 GHz and 5 GHz bands, devices to intentionally shut down a wireless network are relatively easy to acquire or build. Having strong physical security is one way to prevent jamming attacks. That said, the majority of RF jamming is caused by unintentional interference by existing devices such as neighboring wireless networks and many of the devices covered in Chapter 3. The best methods for preventing jamming or DoS attacks are judicious RF design through site surveys and implementation of wireless intrusion detection/protection systems and Management Frame Protection (MFP), which are both covered in later in this chapter.

exam watch

The three primary attacks on a wireless network can be categorized as passive (sniffing and reconnaissance), active (unauthorized access), and jamming (neutralization of the wireless network). These attacks can all be mitigated or prevented with a combination of strong encryption, authentication, and proactive security measures.

Unauthorized Access Points

In addition to the attacks just discussed, the effectiveness of a wireless network and its security can be weakened by the installation of external and unauthorized access points of different types. Three common forms of attacks involving unauthorized access points are listed in this section. All of these attacks can be prevented by strong physical security, rogue access point detection, judicious RF monitoring and, to an extent, wireless intrusion detection/protection systems.

Man-in-the-Middle (MITM) attack A man-in-the-middle (MITM) attack is accomplished when an attacker intentionally places an access point in between a legitimate wireless access point and its client(s). By using an external access point to spoof the original AP, the attacker is able to intercept any information that these clients send to the AP. Because the client and user believe that it is communicating with a legitimate access point, sensitive information such as usernames, passwords, and account numbers can be transmitted directly to the attacker's device. It is important to know that, in order to successfully accomplish a MITM attack, the attacker needs to know most, if not all, of the details of the legitimate access point (SSID, security type, security keys) to successfully hijack the client. An MITM attack is often combined with a limited DoS attack to knock clients off the legitimate AP and force them to reassociate to the attacker's access point, which also must have a substantially stronger signal than the legitimate AP.

Rogue Access Points Rogue access points, also known as unauthorized access points, are distinguished from man-in-the-middle attacks by the fact that rogue APs are more passive (in that they do not try to spoof a legitimate access point) and are often not even malicious. Rogue APs are typically installed on the same network that legitimate APs are on but can also be seen from neighboring networks in close proximity to the corporate wireless LAN. The problem with internally installed rogue APs is that they undermine corporate security policies by providing access to the same network resources that a secured and managed enterprise access point is intended for. In addition to this, a rogue AP complicates the RF environment and can create undesirable co-channel RF interference by crowding the frequency band in a particular area or by being improperly configured to coexist with the existing wireless network. It is often up to the wireless network administrator to proactively detect, identify, and contain rogue access points. Rogue AP Detection on the Cisco Wireless LAN Controllers and Wireless Control System helps to simplify this process.

Ad Hoc Wireless Clients One of the concepts that we touched on in Chapter 2 was the independent basic service set (IBSS), which is colloquially known as an *ad hoc network*. As we discussed, an ad hoc network allows clients to share their wireless or wired connection with other clients by using the client radio to act as an access point. A common application for this is Internet sharing and wireless ad hoc data transfers between clients without using an actual access point. Ad hoc clients present many of the same issues that rogue access points do in that they also undermine corporate security policies and introduce interference into the wireless LAN.

It is a little-known fact that, in many cases, an ad hoc client can introduce as much or more interference than an actual access point using only the client radio of a typical laptop.

on the job

Ad hoc mode is sometimes enabled by default on older or unpatched Windows clients. Some departments in certain enterprises utilize ad hoc to transmit data between members of the team or share a single LAN connection without realizing the impact that it can have on a corporate wireless LAN. It is important to have a strong corporate policy that addresses the use of ad hoc mode and informs users of the availability of a legitimate, secured wireless LAN.

Wireless LAN Management and Policy

Network management can help an organization achieve availability, performance, and security goals and should be considered an integral wireless security component. Most network management tools provide features including network planning, device configuration, and security management. Administrators can use network management tools, usually through a graphic user interface (GUI) to design, control, and monitor enterprise wireless networks from a centralized location. Most network management tools will also allow network administrators to perform RF prediction, policy provisioning, network optimization, troubleshooting, user tracking, security monitoring, and wireless LAN systems management, including detailed trending and analysis reports. Some tools will also perform RF location tracking and monitoring and use an intrusion prevention system (IPS). We will discuss Cisco Wireless Control Systems (WCS), a comprehensive network management solution for Cisco Unified Wireless Networks, in Chapter 8.

Wireless security policy addresses the general behavior and usage of a wireless LAN within an organization. Wireless security policy should be an integral part of the overall information security policy of an organization and provide rules and definitions of what is acceptable and unacceptable use of wireless networks. It should also define the processes of reporting and handling incidents and mitigation plans against threats from inside and outside of the organization. Different organizations usually have different philosophies when it comes to the use of information technology and computer networks, so there is no one standard for creating a wireless security policy. However, typical wireless security policy includes the following elements:

- **Network definition and risk assessment** Identify the scope of the security policy and the scope of wireless networks, the network's coverage, wireless network assets, and security vulnerabilities and risks. Wireless security policy should provide a clear definition of roles and responsibilities from an operations and planning perspective for wireless networks within an organization. A clear line of commands should be established so there is no ambiguity about whether a wireless security policy should be enforced. A high level executive, such as the CIO or CTO, usually serves as the executive sponsor for a wireless security policy. Cross functional teams sometimes prove to be necessary, and people coming from different areas of expertise will work together to support the wireless networks. Some security policies may include considerations of costs for security solutions and budgets for user and staff education and staffing on related issues.
- **Acceptable use and threat prevention** Determine acceptable and prohibitive usage of wireless networks and security measures to address areas of weaknesses. Depending on the nature of business in the organization, proper protection of organization intellectual properties, handling of sensitive data, patient healthcare information, the customer transaction database, and privacy information should be discussed. The security policy should provide technical, legal, and ethical guidelines for the organization for use of technologies. The organization should perform a periodic threats assessment, such as a security audit, to properly identify areas for improvement and assess the direction of the organization's security solutions and future projects. The threat assessment should also review design and implementation of the wireless networks.
- **Impact and incident response** Determine the necessary measures to assess the impact of incidents and create processes for appropriate actions when security incidents occur. There may be financial and legal consequences when a security incident takes place within the organization, in addition to many other issues such as damage of reputation, loss of customer sensitive information, loss of trade secrets, identity theft, and regulatory violations.

Proper procedures should be in place to minimize potential losses such as these, which can all occur as a result of a security incident.

- **Training** Increase awareness of the proper use of wireless networks and readiness of people to handle security incidents. This is one area that is often overlooked in many organizations, and no wireless security policy is useful if the users and administrators are ill prepared. The end users should be familiar with security tools that are provided for secure access of the network. Administrators and network staff should be trained on proper monitoring, maintenance, upgrading, and surveying of the wireless networks. Some organization may even specify the amount of training necessary for the network support staff to ensure the knowledge level, familiarity, and readiness of the wireless networks support team. The effectiveness of a wireless security policy is directly related to the awareness and readiness of people who use the wireless network.

For organizations without any wireless networks, it is still a good idea to have a wireless security policy so that users and network staff who wish to use wireless can follow predefined guidelines from the organization and avoid any unnecessary and undesired consequences.

Security Solutions and Features

There are many available commercial products designed to address different aspects of wireless security issues. We will discuss intrusion detection systems and intrusion prevention systems (IDS/IPS), firewalls and Access Control Lists (ACLs), and Management Frame Protection (MFP) in this chapter.

IDS/IPS

The wireless networks are usually seen as extensions to existing wired networks. Thus, any network based security threats, such as malicious Internet worms, denial of service (DoS) attacks, and business application attacks, also exist on the wireless networks. Network based IDS/IPS provides incremental protections from a network perspective to ensure that wireless networks remain an integral part of overall network security strategy. Usually, network based IDS/IPS inspect layer 2–7 traffic from policy violations, vulnerability exploitations, and anomalous activities. Some network based IDS/IPS solutions offer risk rating and event correlations to increase the confidence level of threats identified against false positives and lower the risks of dropping legitimate traffic.

With a shared medium, wireless networks face additional challenges that the traditional IDS/IPS solutions do not address, and this is where wireless IDS/IPS becomes critical in securing the wireless spectrum wireless networks depend on. This is particularly true even when wireless networks are not deployed on site. The wireless IDS/IPS provides wireless and spectrum specific detection and mitigation against attacks such as rogue APs; over-the-air, man-in-the-middle, zero-day attacks; authentication and encryption cracking; ad hoc and spoofing; and spectrum intelligence with threats from non-802.11 based devices. Some wireless IDS/IPS solutions require dedicated sensors for detection, and some can use the wireless LAN infrastructure for detection by setting the AP to monitor mode or using the AP's spare cycle to provide channel scanning. Cisco offers a comprehensive wireless IDS/IPS solution using the wireless infrastructure, WCS, and Mobility Services Engine (MSE), the analysis and policy engine for the Cisco adaptive wireless IPS solution.

Firewalls and ACLs

Firewalls and Access Control Lists (ACLs) are two of the most used solutions in a network to enforce policy and prevent unauthorized access. In a wireless network, a firewall and an ACL provide traffic segregation enforcement to ensure that only "authorized" traffic has access to critical business applications and infrastructure. Industry standards, such as the PCI DSS, have strict guidelines on how different types of wireless traffic, credit card payments in particular, should be segregated using a firewall or an ACL (in addition to modern encryption and authentication solutions) to protect business transactions. Furthermore, firewalls and ACLs provide extra layers of protection and prevent disruption of critical infrastructure in the events of network security breach.

As wireless networks become more widely deployed in the enterprise environment, it will become more common for network design to incorporate security and policy enforcement using firewalls and ACLs.

Management Frame Protection (MFP)

One important differentiator for 802.11 compared to other competing wireless technologies is the completeness of the management functions defined in the protocol stack that makes 802.11 robust and easy to deploy. In Chapter 2, we discussed the basic management functions provided by 802.11 management frames, such as authentication/deauthentication, association/disassociation, beacons, and probes, and how these frames play a critical role in 802.11. These frames are always transmitted unauthenticated and unencrypted as opposed to client data frames that are usually authenticated and encrypted in transmission. The openness of 802.11

management frames exposes 802.11 wireless networks to malicious attacks, such as Denial of Services (DoS), offline dictionary attacks, man-in-the-middle attack, and spoofing attacks. To address this pitfall in the 802.11 protocol design, Cisco introduced Management Frame Protection (MFP) with authenticated management frames to protect the integrity of 802.11 management traffic.

There are two modes of operation for MFP: Infrastructure MFP and Client MFP. The Infrastructure MFP enables the AP to add a message integrity check information element (MIC IE) to each management frame it transmits. Any attempt to tamper with the management frame will invalidate the MIC. In a Cisco Unified Wireless network, the wireless network will generate a unique key for each AP, and management frames from all AP within the wireless network will be protected. The keys will be shared with all APs within the network; hence, APs outside the wireless network can be easily identified.

In Client MFP, a similar mechanism is in place to protect the authenticated client from spoofed frames. Attacks such as deauthentication attacks do not usually take down a wireless network, but they degrade the overall performance of it. Client MFP allows the client to encrypt management frames sent between the AP and clients and prevents management frame attacks, such as deauthentication and disassociation. The client and AP will use the existing security mechanism defined in 802.11i and WPA2 (TKIP or AES-CCMP) to derive and distribute the Pairwise Master Key (PMK). Client MFP protects active client and AP sessions against common Denial-of-Service attacks. When a client or AP detects that the management frames have been tampered with, it drops the frame and the incident is reported back to the network. The current Cisco Client MFP requires clients to be CCX v. 5 compliant.

The current Cisco MFP implementation is a prestandard solution offered by Cisco to address the urgent needs of MFP while the standards body develops a standard based solution to address this issue. The IEEE 802.11w working group is in the process of developing Protected Management Frame (PMF) based on Cisco's MFP concept to protect 802.11 management frames.

Cisco Client MFP is a CCX v.5 feature; thus, when implementing a Cisco Client MFP solution, make sure the client and infrastructure are capable of supporting MFP. In addition, the standards-based 802.11w solution utilizes features from other 802.11 standards, including 802.11i, 802.11r, and 802.11u. Consult both equipment vendors to ensure the wireless network and client adapters can support MFP before implementing the solution on a production network.

Understand how different modes of MFP work with the existing infrastructure and how each protects the wireless network. Infrastructure MFP helps identify rogue BSSes, and the AP will discard management frames from unknown BSSIDs. Client MFP helps prevent attacks such as deauthentication attacks and disassociation attacks. Cisco client MFP requires the client adapter to be compliant with CCX v5.

Client Exclusion

Another security feature found on the Cisco wireless controller is called *client exclusion*. Client exclusion allows a wireless network to mitigate brute force attacks and is configured globally on a controller and applied on a per WLAN basis. When a wireless client fails a certain number of times while attempting to associate or authenticate to the wireless network, the wireless controller can exclude or ban the client from attempting to associate to the wireless network again. As you will see later in this section, the administrator can determine whether this exclusion is permanent or temporary. Figure 4-2 shows the global Client Exclusion Policies window, which allows you to enable or disable the following client behavior patterns or errors that will trigger a client to be excluded:

- **Excessive 802.11 Authentication Failures** An example of a client being excluded under this policy would be a device repeatedly failing open authentication due to an improperly configured or unsupported client/supplicant. The controller will allow a client five consecutive failures before it is excluded on the sixth one.
- **Excessive 802.11 Association Failures** An example of a client being excluded under this policy would be a device repeatedly failing to associate due to an improperly configured or unsupported client/supplicant or because of extremely weak RSSI at the client. The controller will allow a client five consecutive failures before it is excluded on the sixth one.
- **Excessive 802.1x Authentication Failures** An example of a client being excluded under this policy would be a device repeatedly failing Protected EAP (PEAP) authentication due to improperly entered credentials or an otherwise misconfigured client/supplicant. The controller will allow a client three consecutive attempts before it is excluded on the fourth one.

FIGURE 4-2 Client Exclusion Policies window

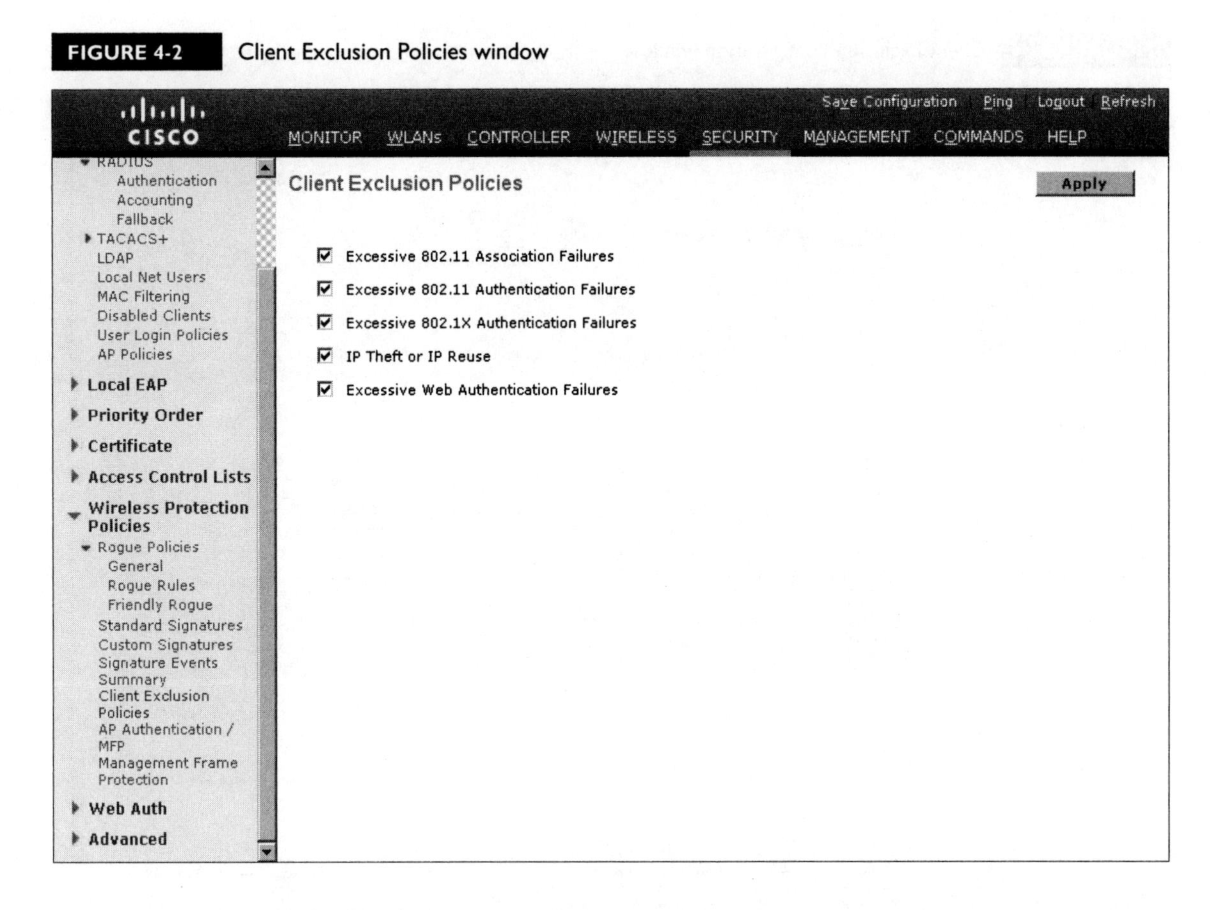

- **IP Theft or IP Reuse** An example of a client being excluded under this policy would be a device attempting to use an IP address that belongs to an already authenticated/legitimate client on the wireless system. The client is immediately excluded by the controller when this is detected.
- **Excessive Web Authentication Failures** In a guest access wireless network, a client will attempt and fail to authenticate through the web portal, most likely by entering invalid credentials. The controller will allow a client three consecutive attempts before it is excluded on the fourth one.

Although client exclusion policies can be enabled or disabled globally, there is the ability to select which specific WLANs will enforce client exclusion; you can also set the number of times that a client gets excluded. This is easily configured in the Advanced tab of any WLAN, as shown in Figure 4-3:

FIGURE 4-3 Client Exclusion configuration window

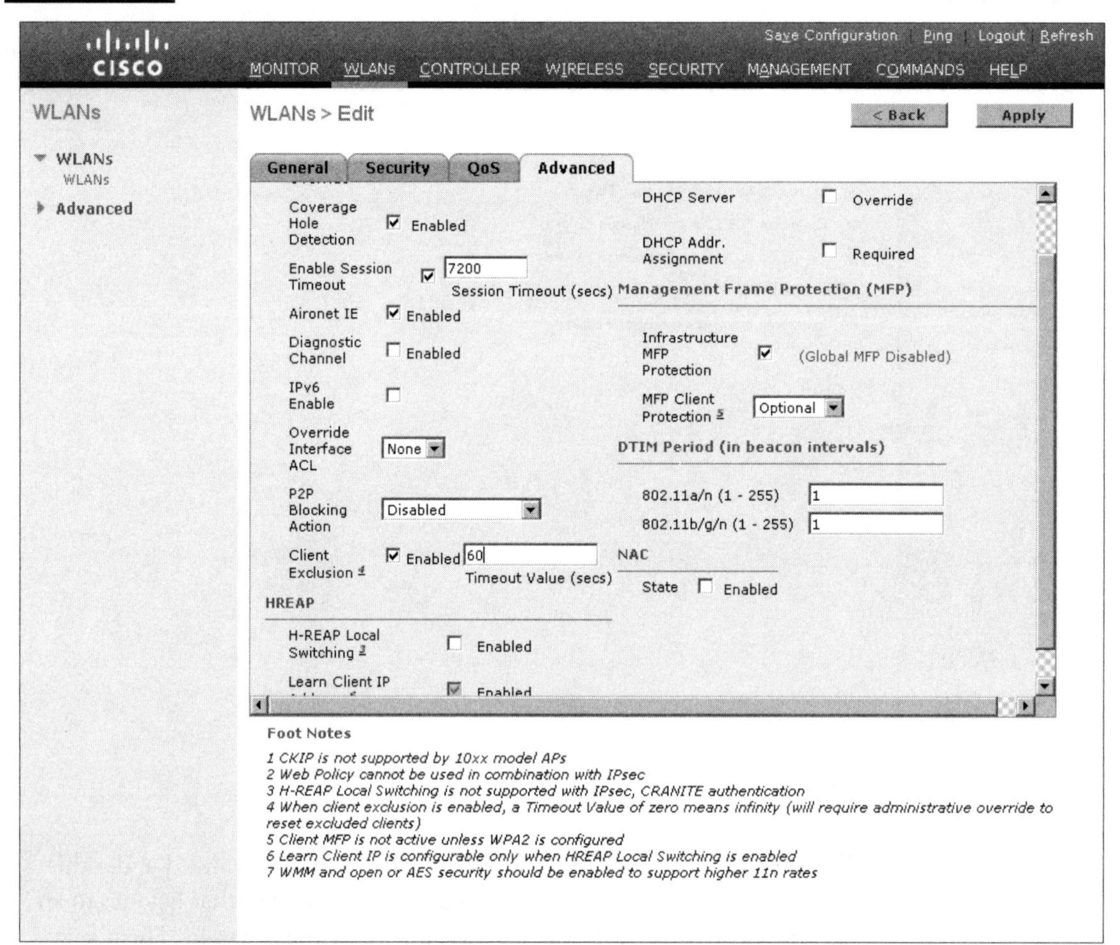

When configuring client exclusion on a per-WLAN basis, you have the option of disabling it altogether or enabling it with a specific timeout value. This timeout value will determine how long a client is excluded in seconds, with a value of 0 seconds indicating a permanent exclusion. A permanently excluded client will remain in this state until an administrator of the controller explicitly removes the client from the excluded state. After that is done, the client will be allowed to attempt association and authentication until it breaks policy and triggers another client exclusion event.

Other Security Solutions

There are other security solutions available in the marketplace that provide protections to wireless networks and clients by addressing different areas of security concerns. These security solutions include the following:

- Antivirus
- Virtual Private Network (VPN) client
- Host IDS/IPS, such as Cisco Security Agent (CSA)
- Network Access Control (NAC) solution, such as Cisco NAC Appliance

When deploying a wireless network, you can use a combination of security solutions discussed in this chapter, but keep in mind that no one can guarantee a network to be 100 percent secure. Every organization should continuously evaluate threats, improve security policy and processes, and implement security solutions that are appropriate for the organization. We will explore the different types of wireless security solutions, including encryption and authentication, in Chapter 5.

INSIDE THE EXAM

The General Framework of Wireless Security and Security Components

The goal of this exam is not to make you an expert in network security; thus, it is not critical to know all the ins and outs of an IDS/IPS solution or become an expert in firewalls. However, it is important to understand the different components of a security framework and the different types of threats, security policies, and security devices and solutions so that you are better prepared to deal with real world security related issues when deploying a wireless LAN network.

CERTIFICATION SUMMARY

There are four basic types of threats in a wireless network. Passive wireless sniffing attacks are usually hard to detect because the attacker may be passively listening for wireless traffic and gathering information for any weakness to be exploited. There are readily available tools that require little knowledge for users to operate and can perform passive wireless sniffing attack on a wireless network. Having a secured physical environment and strong encryption will help mitigate most passive wireless sniffing attacks. Active wireless access attacks can be easily detected as the attacker will need to perform active penetration of a wireless network. A strong security policy with robust authentication and encryption will provide protection against active wireless sniffing attacks. Jamming and DoS are the most difficult to mitigate, and often unintentional jamming attacks are caused by neighboring network networks and devices, such as microwave ovens and cordless phones. Securing the physical environment, conducting proper site surveys, removing sources of interfering RF signals, and implementing wireless IDS/IPS and MFP mitigate this type of attack. Unauthorized AP or ad hoc networks are also threats to the network. Tools such as wireless IDS/IPS, WCS, and MFP can help mitigate these threats.

Security policy provides a framework for the organization to handle different types of security issues and incidents. Typical security policy includes four different elements: network definition and assessment, acceptable use and threat prevention, impact and incident response, and training.

Three basic security components are discussed in this chapter. Network-based IDS/IPS inspects layer 2–7 traffic and detects any policy violations and anomalies. Wireless IDS/IPS provides additional insights into layer 1 and layer 2 threat detections and mitigates 802.11 threats and attacks in the medium. Firewalls and ACLs provide policy enforcement and filter capability so that different types of traffic can be properly segregated. MFP is a developing standard that provides additional protection to the management frames of the 802.11 protocol stack against rogue access points and mitigates deauthentication and disassociation attacks.

TWO-MINUTE DRILL

The General Framework of Wireless Security and Security Components

- Passive wireless sniffing attacks are easy to implement and often difficult to detect as attackers can stay in silent listening mode when carrying out such an attack.
- The goal of passive wireless sniffing is usually to gather information about a wireless network. Subsequent attacks may be carried out using information gathered through passive wireless sniffing. If not protected properly, valuable personal or commercial data may be obtained through passive wireless sniffing.
- Active wireless scanning attacks involve using tools that emit active RF transmissions to discover wireless networks in the surrounding.
- Active wireless access attacks involve the actual penetration of a wireless network by associating to the wireless network or breaching the security measures to gain access.
- To mitigate active wireless attacks include strong security policies with robust authentication and encryption methods.
- Jamming is a form of DoS attack but is often caused by unintentional interfering radio frequency energy.
- Having strong physical security, using wireless IDP/IPS, and implementing MFP can help detect and mitigate jamming attacks.
- Unauthorized access points attacks include man-in-the-middle, rogue AP, and ad hoc attacks.
- Wireless security policy is an integral part of any wireless security solution.
- Network IDS/IPS inspects layer 2–7 traffic and detects policy violations and network anomalies.
- Firewalls and ACLs provide policy enforcement and filtering to ensure only authorized users can access critical applications and information.
- An MFP protects 802.11 management frames from unauthorized tampering and helps wireless network detect rogue AP and mitigate wireless management frame attacks.

SELF TEST

The following self test questions will help you measure your understanding of the material presented in this chapter. Read all the choices carefully, as there may be more than one correct answer. Choose all correct answers for each question.

The General Framework of Wireless Security and Security Components

1. Which IEEE working group is developing a management frame protection mechanism to protect 802.11 management frames?
   A. 802.11k
   B. 802.11r
   C. 802.11u
   D. 802.11w
2. Which version of CCX will support client MFP?
   A. v4
   B. v3
   C. v5
   D. v6
3. Wireless IDS/IPS will monitor layer 2–7 traffic.
   A. False
   B. True
4. What type of attack is hard to detect?
   A. Man-in-the-middle (MITM)
   B. Active wireless access
   C. Ad hoc
   D. Passive wireless sniffing
5. Which security solution provides policy enforcement so only authorized users have access to appropriate network resources?
   A. Firewalls and ACLs
   B. Infrastructure MFP
   C. Antivirus
   D. Client MFP

6. A security policy provides what types of information for an organization? (Choose all that apply.)
   A. Processes and procedures on how to handle security incidents
   B. A particular type of security solution for the organization
   C. Training to the end users and staff on how to handle security threats
   D. A strategy on business application disaster recovery plan
7. A security policy would not include which of the following statements?
   A. End users should not bring personal devices and connect to the company network.
   B. End users should not share usernames and passwords with anyone, either internal or external.
   C. End users should not loan company assigned laptops to other people to download company information from a central depository.
   D. End users should not bring lunch to work and share it with other employees in case of food poisoning.
8. What type of attack is considered an ad hoc wireless network attack?
   A. Passive wireless sniffing
   B. Rogue wireless network
   C. Active wireless access
   D. Man-in-the-middle attack
9. What types of countermeasures are most effective against active wireless access? (Choose all that apply.)
   A. A spectrum analyzer
   B. Strong encryption algorithms
   C. Strong authentication methods
   D. A comprehensive security policy
10. Passive wireless sniffing is not considered a wireless threat because it does not actively attempt to connect to the wireless network.
    A. True
    B. False

SELF TEST ANSWERS

The General Framework of Wireless Security and Security Components

1. Which IEEE working group is developing a management frame protection mechanism to protect 802.11 management frames?

   A. 802.11k

   B. 802.11r

   C. 802.11u

   D. 802.11w

   ☑ **D.** IEEE 802.11w is the current working group developing the Protected Management Frame (PMF) based on Cisco's MFP concept as a solution to mitigate threats on unsecured 802.11 management frames.

   ☒ **A, B,** and **C** are incorrect. 802.11k is derived for radio resource management. 802.11r is defined for fast secure client roaming. 802.11u is defined for adding services to support external networks.

2. Which version of CCX will support client MFP?

   A. v4

   B. v3

   C. v5

   D. v6

   ☑ **C.** CCX v5 is defined to support Client MFP.

   ☒ **A, B,** and **D** are incorrect. CCX v4 focuses on support for voice over wireless LAN. There is no CCX v6 defined yet.

3. Wireless IDS/IPS will monitor layer 2–7 traffic.

   A. False

   B. True

   ☑ **A.** Wireless IDS/IPS monitors and detects threats and anomalies on the wireless medium, not layer 2–7.

   ☒ **B** is incorrect because network based IDS/IPS monitors layer 2–7 traffic.

4. What type of attack is hard to detect?

   A. Man-in-the-middle (MITM)

   B. Active wireless access

   C. Ad hoc

   D. Passive wireless sniffing

☑ **D.** Passive wireless sniffing is hard to detect because attackers only silently listen to wireless traffic and don't actively transmit any traffic.

☒ **A, B,** and **C** are incorrect. MITM can be detected as rogue AP transmit traffic in wireless. Active wireless access can be detected through use of identity based security solution. Ad hoc can be detected as it advertises IBSS to anyone who wishes to join the ad hoc network.

5. Which security solution provides policy enforcement so only authorized users have access to appropriate network resources?

A. CLs

B. Infrastructure MFP

C. Antivirus

D. Client MFP

☑ **A.** Firewalls and ACLs provides policy enforcement so only authorized users have access to appropriate network resources.

☒ **B, C, D** are incorrect. Infrastructure MFP detects unauthorized access points on the network. Antivirus doesn't administer user authorization. Client MFP ensures authenticity and integrity of wireless LAN connectivity, but does not enforce user authorization.

6. A security policy provides what types of information for an organization? (Choose all that apply.)

A. Processes and procedures on how to handle security incidents

B. A particular type of security solution for the organization

C. Training to the end users and staff on how to handle security threats

D. A strategy on business application disaster recovery plan

☑ **A, C.** Security policies usually outline actions on incident response and provide necessary training for end users and support staff.

☒ **B** and **D** are incorrect. A security policy usually does not cover a disaster recovery plan and seldom talks about a specific security solution to be selected.

7. A security policy would not include which of the following statements?

A. End users should not bring personal devices and connect to the company network.

B. End users should not share usernames and passwords with anyone, either internal or external.

C. End users should not loan company assigned laptops to other people to download company information from a central depository.

D. End users should not bring lunch to work and share it with other employees in case of food poisoning.

☑ **D.** A security policy does not regulate meal plans or food sharing among employees.

☒ **A, B,** and **C** are incorrect. **A** security policy regulates roles and responsibilities as well as processes and procedures related to the use of network, equipment, and credentials.

8. What type of attack is considered an ad hoc wireless network attack?

A. Passive wireless sniffing

B. Rogue wireless network

C. Active wireless access

D. Man-in-the-middle attack

☑ **B.** Ad hoc is considered a type of rogue wireless network.

☒ **A, C,** and **D** are incorrect. Passive wireless sniffing is listening mode only. An ad hoc network does not actively connect to company sanctioned wireless network. Man-in-the-middle is one type of rogue wireless network.

9. What types of countermeasures are most effective against active wireless access? (Choose all that apply.)

A. A spectrum analyzer

B. Strong encryption algorithms

C. Strong authentication methods

D. A comprehensive security policy

☑ **B, C,** and **D.** Strong encryption algorithms, strong authentication methods, and a comprehensive security policy are all effective measures against active wireless access attacks.

☒ **A** is incorrect. A spectrum analyzer can detect RF energy and activities but does not prevent active wireless access attacks.

10. Passive wireless sniffing is not considered a wireless threat because it does not actively attempt to connect to the wireless network.

A. True

B. False

☑ **B.** Passive wireless sniffing is considered a threat because it is usually the prelude to actually attacks. Information gathered during the sniffing can be used to gain unauthorized access to a wireless network.

☒ **A** is incorrect. Passive wireless sniffing is considered a threat because it is usually the prelude to actually attacks. With appropriate encryptions, firewall, ACL, and policy in place, it makes passive wireless sniffing less effective; nonetheless, information gathered during the sniffing can still be used to gain unauthorized access to a wireless network.

5

Wireless Authentication and Encryption

CERTIFICATION OBJECTIVES

5.01 Wireless Authentication

5.02 Wireless Encryption

5.03 Authentication Sources

✓ Two-Minute Drill

Q&A Self Test

In its infancy, the largest obstacle facing widespread wireless LAN adoption was its inherent lack of security when compared to similar wired LAN technologies. Because of this, the most complex part of deploying a wireless LAN, outside of RF design, has been and continues to be properly selecting and configuring a security solution that provides support for a variety of client types while maintaining application performance and high data throughput. As options for wireless security have evolved, the industry has seen a move from weak IEEE 802.11 standard security methods like WEP encryption to more robust technologies such as 802.1X standard authentication and AES encryption.

The most basic concept you need to understand when reading this chapter is that wireless LAN security of any kind is made up of two key elements: *authentication* and *encryption*:

- **Authentication** Authentication asks the questions, "Who are you?" and, when deployed with authorization, "Are you allowed inside this network?" In other words, authentication is the method by which a client device or user's identity is validated for access to the network. As you will see later in the chapter, there are several different ways to authenticate wireless devices or users utilizing specialized keys (WEP and PSK), username/password combinations, or digital certificates.
- **Encryption** Encryption refers to the method by which data is protected during a wireless transmission. Cleartext data is transmitted in an encrypted format using any of several different *ciphers*—algorithms for encryption—and decrypted at the receiving end. In this chapter, we will discuss the different kinds of encryption available for wireless LAN transmissions.

We will also describe the differences between the different wireless security methods currently available with IEEE 802.11 wireless technologies and examine the processes that are involved in their operation. Figure 5-1 shows how authentication and encryption fit in a wireless security framework, with delineations for legacy and enterprise class technologies.

FIGURE 5-1 Authentication and encryption in the wireless security framework

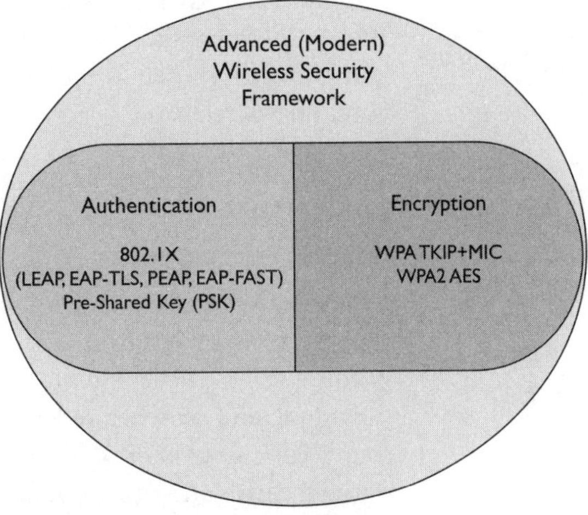

exam watch

Wireless LAN security can be broken down into the key processes of authentication and encryption. Both are independent elements of wireless security that work together to protect a wireless network from malicious attacks.

CERTIFICATION OBJECTIVE 5.01

Wireless Authentication

At the beginning of this chapter authentication was defined as the method by which a client device or user's identity is validated for access to the network. In this section we will describe the different forms of authentication that are used in today's wireless network deployments. These authentication methods can be broken down into basic (mostly IEEE 802.11 standard), advanced (802.1X standard), and guest/Web authentication. The majority of this chapter will focus on 802.1X standard authentication types, as they are the most secure and accepted forms used in modern enterprise wireless networks.

Basic Authentication

When the IEEE 802.11 wireless standard was ratified, wireless authentication was almost considered an afterthought. As a result of this, the authentication methods that were available in the early days of wireless networking are considered very primitive by today's standards. These authentication types are characterized by being very static and device-based and do not require any robust kind of authentication source (discussed later in this chapter). This section will give an overview of each of these early authentication methods.

Open Authentication

Open authentication is by far the most basic authentication available on IEEE 802.11 wireless networks. Open authentication simply uses a null authentication, meaning that all authentication requests are automatically accepted once they are received. In the early days of 802.11 wireless networks, open authentication was often paired with a Wired Equivalent Privacy (WEP) key that implicitly authenticated users (the client needed to have a specific WEP key that matched with the one configured on the access point) who wanted to gain access to the wireless network. Open authentication is still used today for passing users to a web/guest authentication network or to give users unsecured access to a wireless network. You will also see open authentication used in the early stages of the 802.1X authentication process.

Shared Authentication

Shared authentication, which was intended as a more secure form of 802.11 standard authentication, worked by using a concept of shared key exchange. In this process, the known key (text) could be encrypted, combining it with a 128-bit plaintext nonce (or random number). During the shared authentication process, the encrypted data was issued as the authentication request and response. If these matched, the client was then allowed to access the wireless network.

The fundamental problem with this process was that the known key could be easily discovered through passive sniffing by gathering the plaintext nonce and encrypted data and reversing the algorithm. A potential attacker would then be armed with the shared key and could not only authenticate into the network, but also decrypt any data that traveled over the wireless link. Paradoxically, shared authentication actually introduced more security threat than the relatively safe method of performing null authentication. As a result of this vulnerability, shared authentication has fallen heavily out of use and is considered a very undesirable method of authentication, even by early IEEE 802.11 standards.

MAC Address Authentication

MAC *address authentication* is also frequently referred to as MAC address filtering and is not covered by the IEEE 802.11 standard. MAC address authentication was a primitive way to provide device-based authentication for access to a wireless LAN. In MAC address authentication, a list of authorized MAC addresses is configured in the wireless access point or an external RADIUS server. When a client requested authentication into the wireless network, it used its MAC address as its credentials and the AP would either check for the MAC address in its local list or query the external RADIUS server. If the MAC address was found, the client was allowed to associate to the wireless network. MAC address authentication was often paired with open or shared authentication to provide additional security.

The obvious problem with MAC authentication was that if a user knew about an authorized MAC address, they could very easily spoof the MAC address on their own client device to match with a known address, so MAC authentication could be tricked into allowing the client to associate.

Advanced 802.1X Authentication

As you saw in the previous section, basic authentication methods are clearly deficient in providing the security necessary to secure an enterprise-class wireless network. To rectify this, the 802.1X standard (which was originally a standard for wired network port security) was adapted to wireless networks. The 802.1X standard is essentially a framework that specifies that access to a port or—in the case of wireless—a wireless network is allowed based on credentials that are authenticated on an AAA (Authentication, Authorization and Accounting) server running

RADIUS. Simply put, if a user is in the RADIUS server and provides the proper credentials, they are allowed access into the wireless network. The 802.1X wireless standard provides support for what are called *Extensible Authentication Protocols*, or EAP. EAP is an IETF standard that was later incorporated into the 802.1X standard. Although RADIUS is optional within 802.1X, it is expected that many 802.1X authenticators function as RADIUS clients, and most EAP types are based on RADIUS authentication. In this book, we assume RADIUS is used for all 802.1X authentication unless otherwise specified.

In order to understand how EAP works in the 802.1X framework, it is important to identify the elements of a basic 802.1X authentication process using EAP:

- **Supplicant** The *supplicant* is the device that is requesting access into the wireless network. This element is also known as the client and needs to have software capable of supporting and initiating the EAP authentication process. It is important to note that the term "supplicant" refers to the client itself in the 802.1X standard. In most normal use scenarios, a supplicant is recognized as the client software or adapter software and drivers on a wireless device such as a laptop, PDA, voice device, or other wireless-enabled equipment.
- **Server (RADIUS server)** Known as an *authentication server*, this is the device that contains the database of users and credentials (or has access to this database) and has knowledge of whether or not a supplicant's request can be validated or not. It is the responsibility of the server to send back an access "accept" or "reject" message based on these criteria. In many implementations, the authentication server further queries a back-end user database such as LDAP or Microsoft Active Directory (AD). The role of the authentication server as an authentication source is discussed in more detail later in this chapter.
- **Authenticator** The *authenticator* is the device that receives the request from the supplicant and passes it to the authentication server. In legacy wireless environments with autonomous access points, the authenticator is the access point itself. In modern centralized wireless deployments, the authenticator is the Cisco Wireless LAN Controller that manages the Lightweight Access Point that the client associates to. The best way to think of the role of an authenticator is to look at it as a RADIUS proxy device. In other words, the authenticator takes a RADIUS request (which comes from the supplicant), queries the authentication server, and grants or denies access to the wireless network based on the accept or reject response it receives from the authentication server. Figure 5-2 demonstrates how the supplicant, authenticator, and server interact with each other.

FIGURE 5-2 EAP within the 802.1X framework

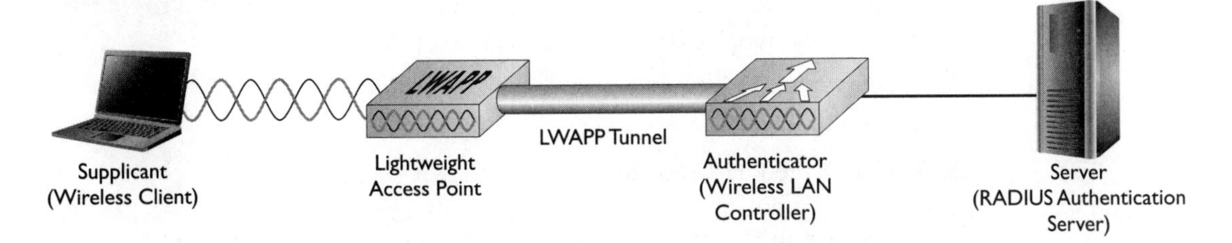

exam watch

The supplicant (wireless client) authenticates to the wireless network using the authenticator (access point or wireless controller) which proxies authentication requests to the server (authentication server such as RADIUS). The traffic exchange between these three components is the foundation for all wireless 802.1X authentication.

The generic EAP process uses these three elements to follow a very generic set of steps that can be applied to all the different EAP types. The EAP process also involves mutual authentication between the supplicant and the server. This means that there is a client authentication phase where the server validates the identity of the client and a separate server authentication phase where the client validates the server. In an 802.1X enabled secure wireless network, no traffic, including the exchange of encryption data, is passed until this process is complete. The step-by-step process is as follows:

1. A wireless client, or supplicant, associates to the wireless network.
2. The authenticator receives this request and responds with an EAP identity request.
3. The wireless client sends an EAP identity response to the authenticator, which is passed on to the authentication server.
4. The authentication server sends an authentication request to the authenticator, which is then passed to the supplicant.
5. The supplicant responds with an authentication response, which is sent to the authenticator and then passed to the authentication server.

6. The authentication server sends back an "accept" or "reject" to the authenticator. If the response is an "accept," the client is allowed to transmit and receive on the wireless network.
7. An encrypted key management session is started to protect transmitted data.
8. The encrypted data session is started.

As you will see in the next few sections, the only difference between the different 802.1X EAP authentication types is the credentials used in client and server authentication phases and how the phases are performed. The overall process remains the same regardless of the EAP type.

LEAP

Lightweight Encryption Protocol (LEAP), also known as *EAP-Cisco*, is a Cisco-proprietary authentication type that was originally in wide use when 802.11b was the prevailing wireless technology. LEAP is characterized by its use of username/password combinations as credentials for authentication and its use of dynamic WEP keys to encrypt data transmissions. Like other EAP types, LEAP supports mutual authentication but has seen far less use because of a dictionary attack vulnerability that was uncovered in 2003. This vulnerability raised the recommendation that a strong password policy be implemented along with any LEAP deployment and also accelerated Cisco's development of LEAP's successor, EAP-FAST. LEAP uses the following process (shown in Figure 5-3):

1. A wireless client, or supplicant, associates to the wireless network.
2. The authenticator receives this request and responds with an EAP identity request.
3. The wireless client sends an EAP identity response to the authenticator, which is passed on to the authentication server.
4. Client authentication of the username and password occurs and is verified by the RADIUS server.
5. Server authentication occurs and session keys are generated for use in the encryption process.

FIGURE 5-3 LEAP authentication process

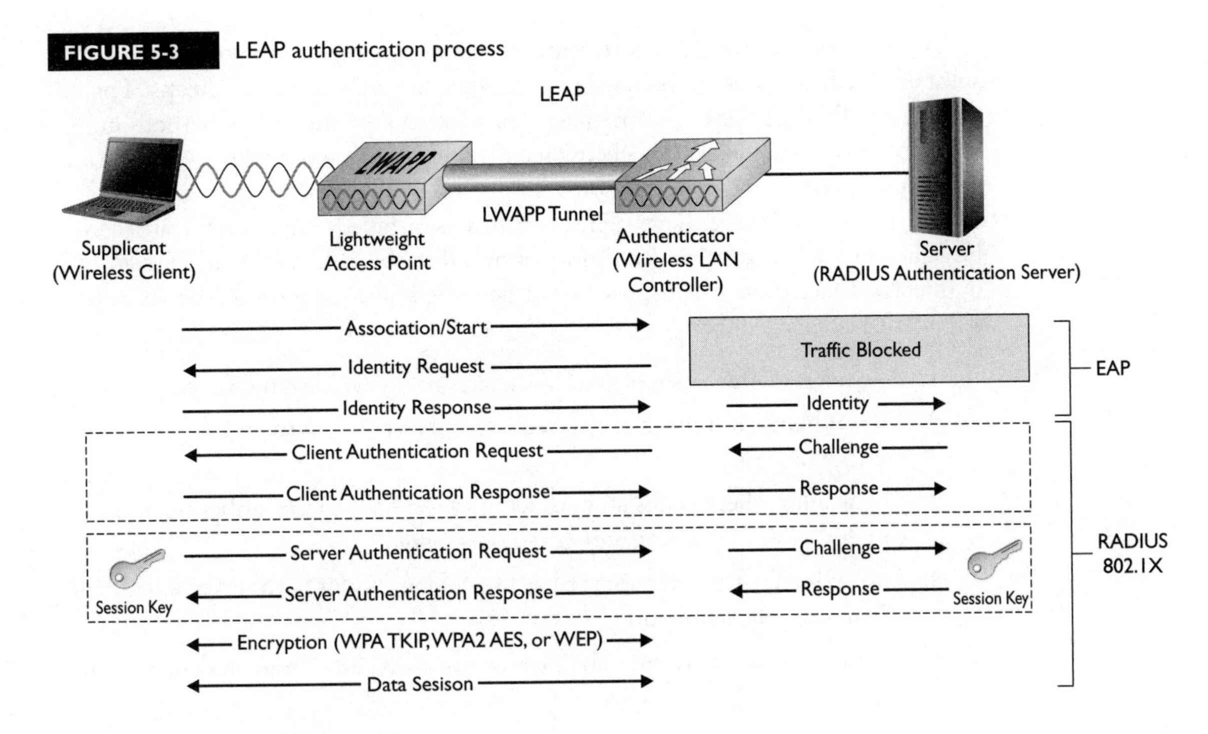

EAP-TLS

Extensible Authentication Protocol-Transport Layer Security (EAP-TLS) is a certificate-based authentication type that leverages an existing Public Key Infrastructure (PKI) utilizing a *certificate authority* (CA). EAP-TLS is the original EAP type and is still considered the most secure wireless authentication method because of the requirement for both client and server side certificates.

Digital certificates are documents that authenticate the identity of a subject and provide a public encryption key. The certificate authority is a third party that cryptographically signs (verifies) the certificate as valid. A digital certificate typically is tied into basic information about the subject such as:

- Name
- Serial number
- Public key
- Certificate issuing authority
- Valid dates

The drawback to the use of certificates is that deployment complexity is substantially higher due to the need to generate and verify client certificates. For this reason, EAP-TLS is not utilized as often as other authentication methods in most enterprises (the exceptions being security-sensitive environments such as government and financial institutions). EAP-TLS works by first requiring that both the client and server have signed certificates installed. These certificates can be signed by a local certificate authority or by a third-party CA such as Verisign or Thawte. Once these certificates are verified, the following process (shown in Figure 5-4) occurs:

1. A wireless client, or supplicant, associates to the wireless network.
2. The authenticator receives this request and responds with an EAP identity request.
3. The wireless client sends an EAP identity response to the authenticator, which is passed on to the authentication server.
4. The verified certificates are used as the credentials for both server and client side mutual authentication.
5. Session keys are then generated, which are also used to start the encryption process before data is passed.

PEAP

Protected Extensible Authentication Protocol (PEAP) represents a compromise between the ease of deployment seen with username/password based authentication (LEAP/ EAP-FAST) while still leveraging the security of digital certificates. Developed jointly by Cisco, Microsoft, and RSA Security, PEAP is the second major EAP type to use server-side only certificates (EAP-TTLS being the first) and has rapidly become the industry standard.

PEAP performs *two-phase authentication* utilizing a server-side only certificate in the authentication process to generate an SSL-like encrypted TLS tunnel within which client authentication is completed by an *inner-authentication* method such as username/password (PEAP-MS-CHAPv2) or generic token card (PEAP-GTC). PEAP follows this process (shown in Figure 5-5):

1. A wireless client, or supplicant, associates to the wireless network.
2. The authenticator receives this request and responds with an EAP identity request.

FIGURE 5-4 EAP-TLS authentication process

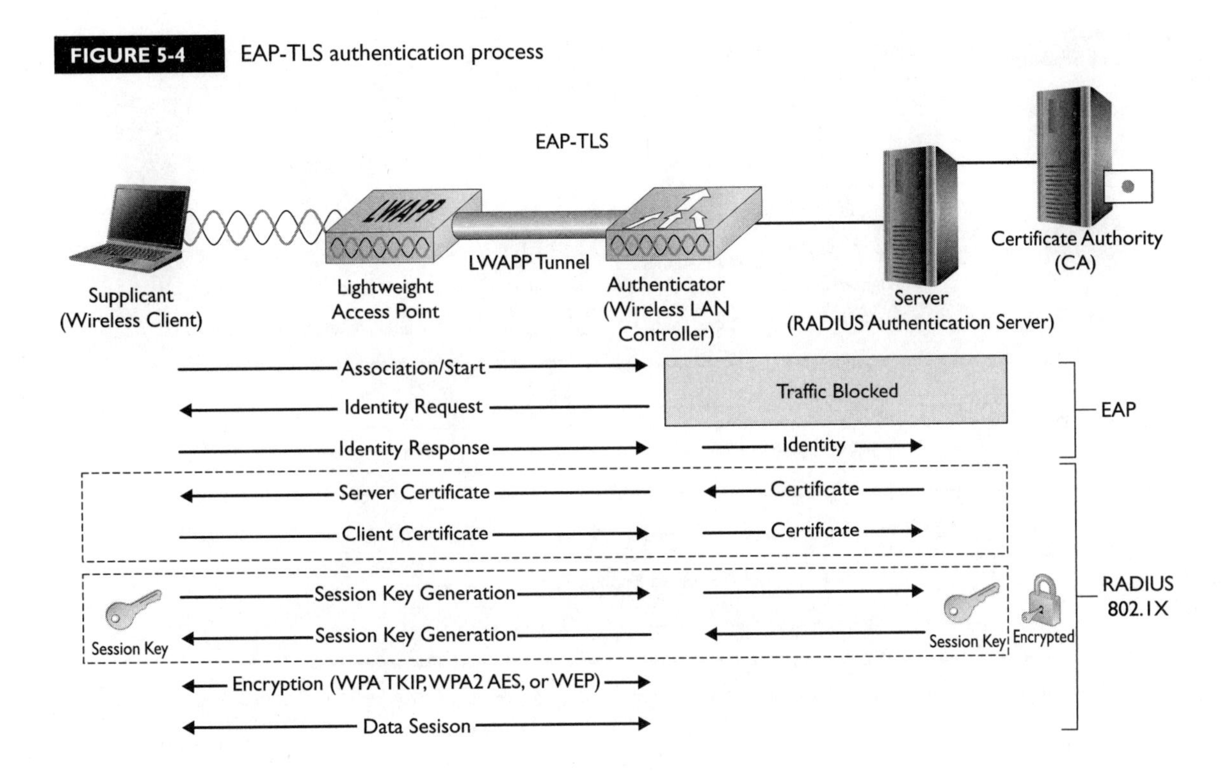

3. The wireless client sends an EAP identity response to the authenticator, which is passed on to the authentication server.
4. In the first phase, the server is validated with its digital certificate by the client.
5. The client then sends a *premaster secret*, which is used with the server certificate to generate an encrypted TLS (Transport Layer Security) tunnel where client authentication will occur.
6. In the second phase, the client (using either MS-CHAPv2 or GTC as inner authentication) is put through identity validation and then authentication against the RADIUS server before being accepted or rejected.
7. If the RADIUS server accepts the client, a session key is generated and sent to the client and encryption begins before data is passed.

FIGURE 5-5 PEAP authentication process

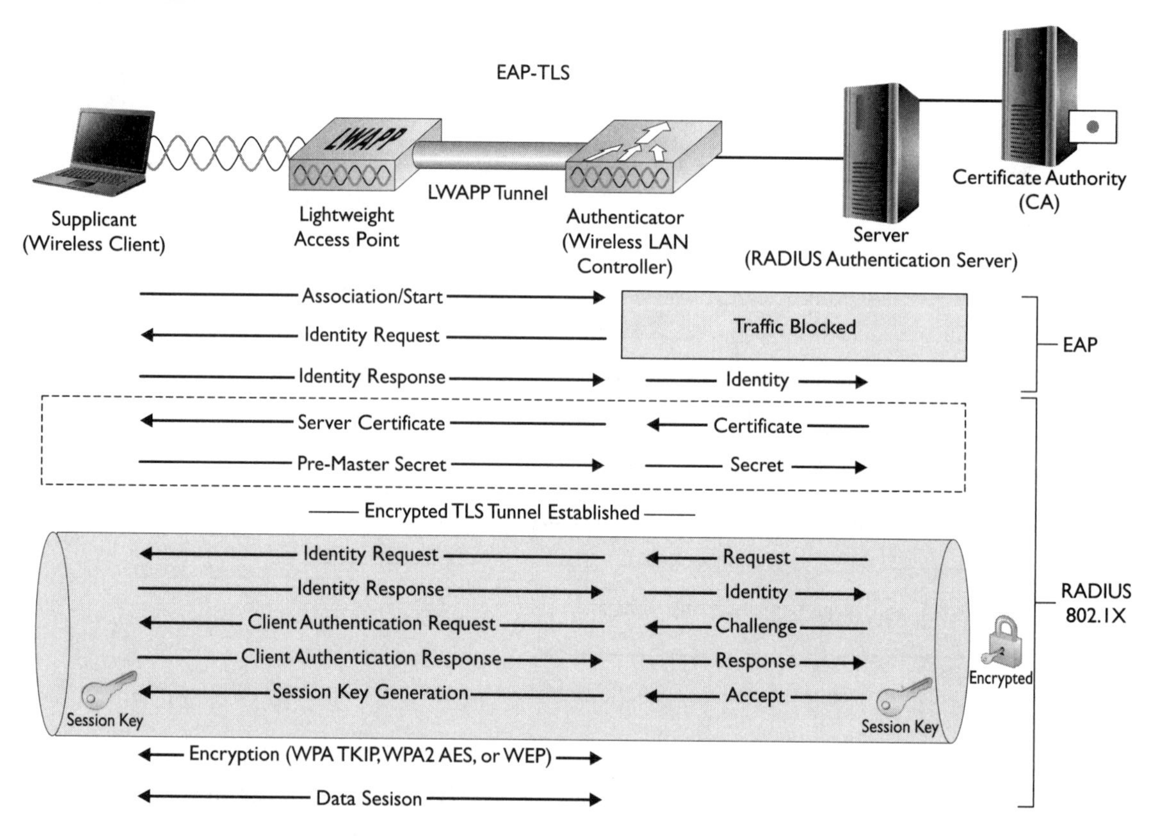

EAP-FAST

Extensible Authentication Protocol-Flexible Authentication via Secure Tunnel (EAP-FAST) is a Cisco-created authentication type specifically released to address vulnerabilities in LEAP. EAP-FAST is similar to PEAP except it does not utilize server-side certificates. Instead, EAP-FAST performs a process called *Protected Access Credential (PAC)* key provisioning. Like PEAP, EAP-FAST follows a multiple-phase authentication process with a "phase 0" devoted to PAC key provisioning followed by a similar two-phase process to the one seen in PEAP. The EAP-FAST authentication process, beginning at PAC provisioning is as follows (shown in Figure 5-6):

1. A wireless client, or supplicant, associates to the wireless network.
2. The authenticator receives this request and responds with an EAP identity request.
3. The wireless client sends an EAP identity response to the authenticator, which is passed on to the authentication server.

4. The RADIUS server initiates EAP-FAST and the PAC provisioning process by sending an *Authority-ID (A-ID)*, which is basically a master server key.
5. The client requests a PAC from the server by generating its own random PAC key as part of a *PAC-Opaque* reply. The PAC-Opaque field contains the PAC key, the client's *I-ID* (or user identity) and is sent along with the received A-ID to the server.
6. The server decodes the PAC-Opaque field using its A-ID and sends back a "final" PAC. This PAC is shared by both client and server and is used to create a TLS tunnel where client authentication then occurs.
7. The client (using an inner-authentication like MS-CHAPv2 or GTC) is put through identity validation and then authentication against the RADIUS server before being accepted or rejected.
8. If the RADIUS server accepts the client, a session key is generated and sent to the client and encryption begins before data is passed.

FIGURE 5-6 EAP-FAST authentication process

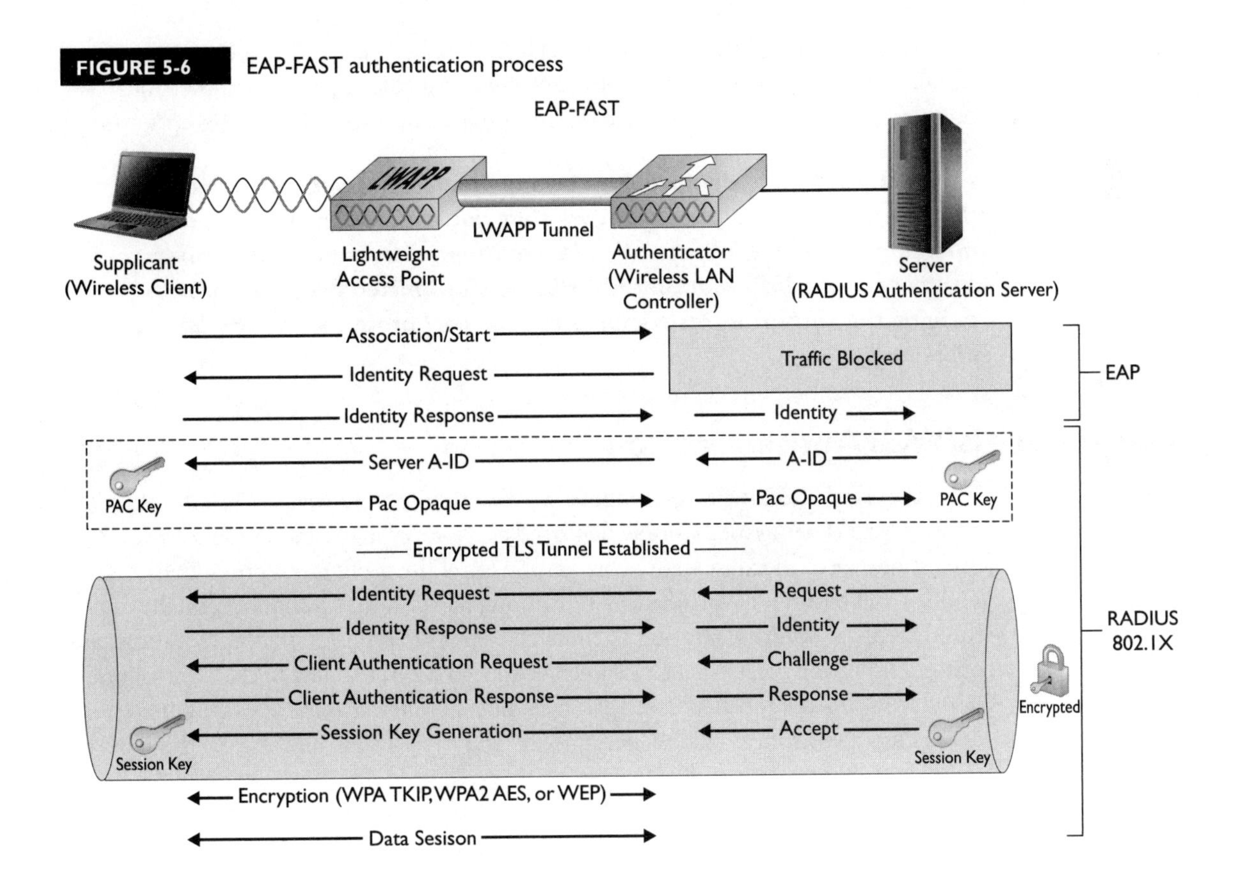

TABLE 5-1 802.1X EAP Type Comparison

	LEAP	EAP-TLS	PEAP	EAP-FAST
Single sign-on	Yes	Yes	Yes	Yes
Microsoft DB support	Yes	Yes	Yes	Yes
LDAP DB support	No	Yes	Yes	Yes
One-time password support	No	No	No	No
Offline dictionary attack vulnerability	Yes	No	No	No
Local authentication	Yes	No	No	Yes
WPA/WPA2 support	Yes	Yes	Yes	Yes
Server certificates required	No	Yes	Yes	No
Client certificates required	No	Yes	No	No
Deployment complexity	Low	High	Medium	Low
RADIUS server scalability impact	Low	High	High	Low/Medium

Table 5-1 compares the different 802.1X EAP types and their characteristics.

Some EAP types do support optional server certificates and/or client certificates. You should evaluate your needs and determine if you want to deploy the option(s).

In the early stages of deploying an enterprise wireless network, most administrators do not know what wireless authentication type to use. It is important to choose a proper wireless authentication type based on the sensitivity of the information that is being transmitted over the wireless network, the current infrastructure, and the corporate security policies and guidelines.

Guest Access/Web Authentication

Guest access has become a very popular driver for corporate wireless LAN adoption and is being increasingly deployed as a part of many modern enterprise wireless networks. Because many corporations find it beneficial to provide visitors, vendors, and customers with Internet or limited corporate network access, the requirement for security catering to these users has grown. Guest access solutions are continuously evolving to meet this need. On the Cisco Wireless LAN Controllers guest access is accomplished using the *Web authentication* feature. From a pure wireless technology perspective, guest access applications are agnostic to the type of authentication or encryption that is configured on the wireless LAN. The Web authentication feature for guest access enforces its security policies after a user has completed wireless association and, because of this, Web authentication is typically paired with open authentication. Web authentication can be tied into a back-end RADIUS database or a local database, or it can simply require a user to perform a basic action such as pressing a button or entering basic information. A typical user experience when associating to a guest wireless network using Web authentication is as follows:

1. A user opens their wireless device and authenticates and associates to the wireless network using the guest WLAN/SSID (open authentication process).
2. The device receives an IP address and is placed in the guest network.
3. The user is blocked from most network resources within the network until they open a web browser session.
4. When the web browser session is opened, the user is redirected to a *splash page* or default web page. This default web page typically includes login instructions, a company logo, legal disclaimers, and/or a login prompt or button.
5. The user enters their credentials to get access to the guest network. These credentials can be a name, an e-mail address, a username/password combination, or simply an acceptance of legal disclaimers.
6. If the user is accepted, they are allowed to access the guest network and transmit/receive data.

CERTIFICATION OBJECTIVE 5.02

Wireless Encryption

As discussed before, encryption refers to the method by which data is protected during a data transmission. Encryption has evolved over time to become stronger and more secure as the need for greater data protection has grown along with the use of wireless networks for business-critical applications. In this section we will describe the different methods for providing encryption to a wireless transmission after association and authentication have occurred.

Wired Equivalent Privacy (WEP)

Wired Equivalent Privacy (WEP) was introduced as part of the IEEE 802.11 standard in the infancy of wireless networks as a method for simultaneously providing authentication and encryption to a wireless LAN. WEP is based on the *RC4 stream cipher*, which specifies a keystream that includes a 24-bit IV (Initialization Vector) and a specific static WEP key. This static key is specified and shared by both the wireless client and the Wireless LAN Controller or access point. There are two specific levels of WEP encryption that are most often used by wireless networks. The level of encryption is determined by the length of the WEP key and the IV.

- 64-bit encryption (40-bit WEP key + 24 bit IV)
- 128-bit encryption (104-bit WEP key + 24 bit IV)

Depending on the client type, you may see the configuration for a WEP key specified as 64- or 40-bit or 128- and 104-bit, respectively. Regardless of what is used, it is important to take into account the initialization vector's contribution to the WEP key's length.

As wireless technologies became more widely adopted, glaring vulnerabilities in WEP key technology were exposed. This was largely because of the arithmetic nature of the WEP encryption process. When WEP encryption occurs, the WEP key and initialization vector are combined to create a keystream using the RC4 stream cipher. The resulting keystream is then used to encrypt *plaintext* (unencrypted) data using an XOR cipher (a simple encryption algorithm) to produce the transmitted encrypted data, known as *ciphertext*. The problem with this process, shown in Figure 5-7, was that the IV was sent in the clear and could be used to determine the keystream with known ciphertext data. This known ciphertext data could be easily generated by purposely sending known data to a valid wireless client and sniffing for the ciphertext (*WEP key reuse*) or purposely generating a known ciphertext error message transmission by transmitting an erroneous altered packet (*bit flipping*). Today, WEP is not considered a proper security method for wireless networks and has fallen out of use in virtually every enterprise environment. Remnants of WEP encryption can still be seen in home wireless networks, but even that has been largely replaced by WPA and WPA2 pre-shared key (PSK) technology.

FIGURE 5-7 WEP key encryption process

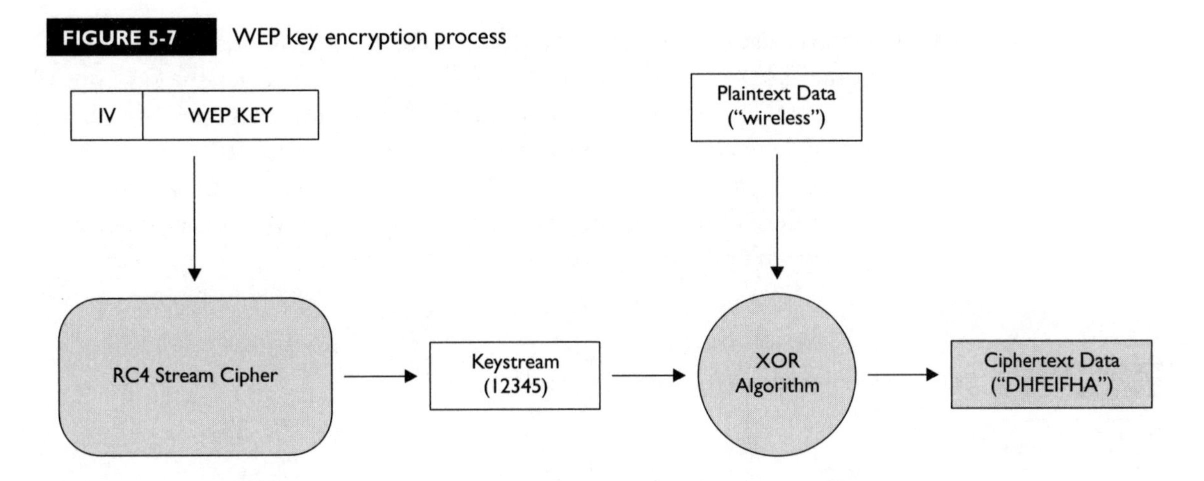

WPA with Temporal Key Integrity Protocol (TKIP)

As the IEEE 802.11i standard for security was being built and ratified, an intermediate solution was needed to address the severe weaknesses of the WEP encryption standard. Consequently, the Wi-Fi Alliance (discussed in Chapter 2) created the *Wi-Fi Protected Access (WPA)* standard. The WPA standard leverages the *Temporal Key Integrity Protocol (TKIP)* encryption method. TKIP is very similar to WEP encryption in that it also operates on the RC4 stream cipher. TKIP has the following improvements:

- **Stronger initialization vector** 48-bit IV instead of the weaker 24-bit IV found in WEP.
- **Per-packet keying (PPK)** Instead of simply concatenating the cleartext IV with the encryption key, TKIP hashes the IV and encryption key before it is sent, making the WEP key that is sent to the RC4 different with each packet. The process of PPK is shown in Figure 5-8.
- **Message integrity check (MIC)** TKIP resists bit-flipping by inserting a randomly calculated MIC value in the data packet and causing the wireless network to drop any packets that are received out of sequence. The MIC process also drops any other packets where alterations to the hashed MIC value are detected.

- **Countermeasures** TKIP *countermeasures* are designed to prevent continuous key recovery attacks. When someone tries to crack the key, two unsuccessful attempts (seen as a two invalid MICs within a minute of each other) will put clients in a timeout list for 60 seconds before they are allowed to connect again. This feature is a deterrent to attackers but can also create a denial-of-service threat as an attacker can purposely provoke client timeouts to slow down or disable a TKIP-secured network.

FIGURE 5-8 TKIP improvements over WEP with per-packet keying

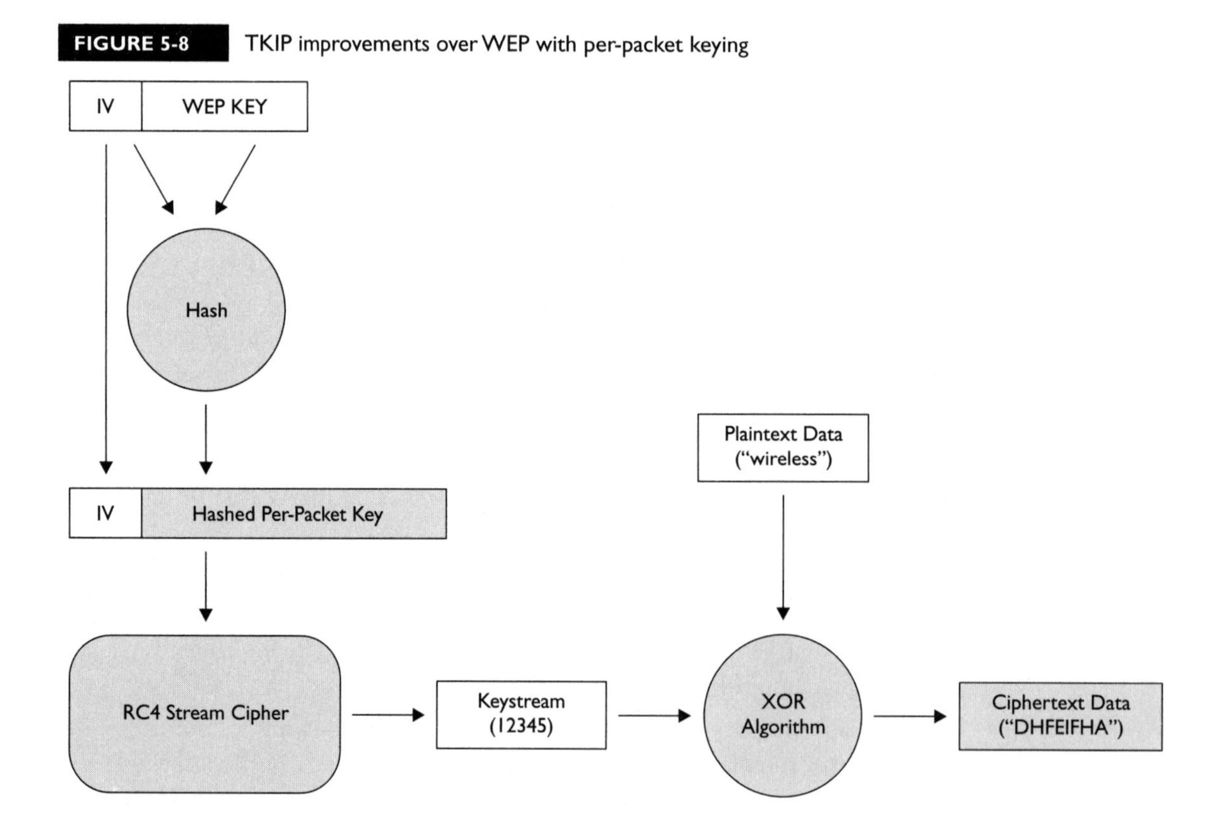

In addition to specifying TKIP encryption, the WPA standard also specifies the use of authentication types. WPA has the following categories of authentication:

- **WPA Enterprise Mode** The use of EAP-based, RADIUS 802.1X authentication paired with WPA standard encryption.
- **WPA Personal Mode** Incorporates the use of a *pre-shared key (PSK)* string which, although less secure than 802.1X authentication, is easier to deploy for home use or environments where a RADIUS server and/or PKI infrastructure is not feasible or cost effective.

To understand how encryption in WPA occurs, it is important to see how it ties in to the authentication process in both WPA modes. In the final step of the 802.1X authentication process, identical session keys are generated for both the authenticator (Wireless LAN Controller/access point) and the supplicant (client). This session key, or *pairwise master key (PMK)*, is what starts the encryption process. In WPA Personal Mode, the pre-shared key string is the key that is used as the PMK, provided that it is properly configured in the supplicant and authenticator. In either case, the PMK is put into a four-step handshake process, which is shown in Figure 5-9, to generate a *pairwise transient key (PTK)* and a *groupwise transient key (GTK)*:

1. The authenticator generates a random number on its side and sends this random number (called the *ANonce*) to the supplicant.
2. The supplicant also generates a random number (called the *SNonce*) and then combines it with the random number (ANonce) from the authenticator. The PTK is constructed by the supplicant using these numbers. The supplicant then sends its random number (SNonce) to the authenticator along with a message integrity check (MIC) that is built from the PMK.
3. When the authenticator receives the supplicant's SNonce, it constructs a PTK with that and its own ANonce. The authenticator validates the MIC using its own PMK.

FIGURE 5-9 WPA encryption four-way handshake

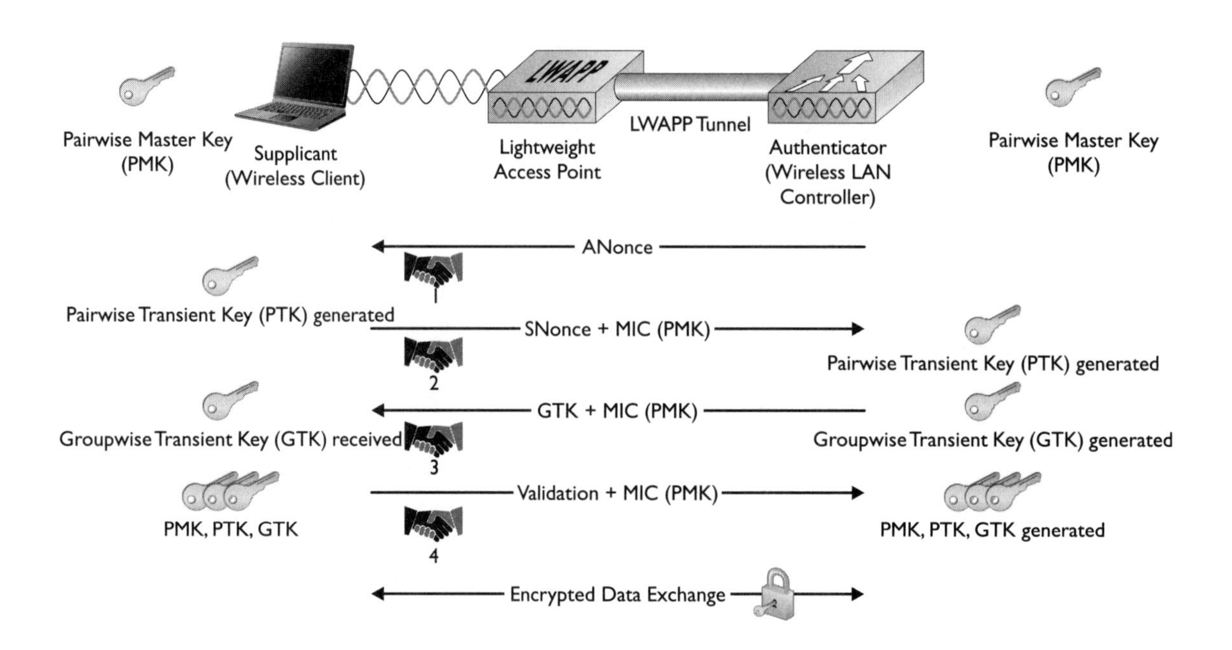

4. If validation is successful, the authenticator generates and sends a *groupwise transient key (GTK)* that is used as the encryption key for broadcast and multicast traffic. This is then sent back to the supplicant along with a MIC.
5. Once the supplicant receives and validates the GTK using the MIC, it installs the PTK and GTK and sends back a confirmation message with a MIC.
6. When the authenticator validates this confirmation message it installs the PTK.
7. The client and access point have verified that they share the same PMK, PTK and GTK, and traffic is then transmitted and encrypted using the PTK and GTK.

exam watch

The WPA four-way handshake is the foundation for how authentication and encryption are tied together into one process. It is for this reason that, with 802.1X authentication, there is no longer the need to manually enter or create encryption keys. For home and lower-security networks, WPA/WPA2 PSK is the natural progression from traditional WEP encryption.

WPA2 with Advanced Encryption Standard (AES)

WPA2, which is otherwise known in the IEEE 802.11i standard as *Robust Security Network (RSN)*, is the second version of the WPA standard. WPA2 is the Wi-Fi Alliance's designation for the 802.11i wireless security standard that was ratified in 2004 and is distinguished from WPA by the use of the *Advanced Encryption Standard (AES)* for encryption. AES, unlike TKIP and WEP, does not use the RC4 stream cipher, which only allows for encryption up to 128 bits. Instead, AES uses *block ciphers* that are based off the *Rijindael* algorithm. With these ciphers, AES can provide encryption at 128, 192 or 256 bits. The flavor of AES required by the IEEE 802.11i and WPA2 specifications is referred to as *AES/CCMP (Counter Mode with Cipher Block Chaining Message Authentication Code Protocol)*. Like WPA, WPA2 has an Enterprise and Personal authentication mode with the only major difference being the requirement for AES encryption. The four-way handshake process is also used in WPA2 encryption. Table 5-2 compares WPA to WPA2.

TABLE 5-2 Comparison of WPA with WPA2

	WPA	WPA2
TKIP Encryption	Mandatory	Unsupported
AES Encryption	Optional	Required
AES/CCMP Variant	Optional	Required
Enterprise Mode	802.1X + TKIP	802.1X + AES/CCMP
Personal Mode	PSK + TKIP	PSK + AES/CCMP

CERTIFICATION OBJECTIVE 5.03

Authentication Sources

Now that we have covered wireless authentication and encryption, it is important to understand some of the authentication sources that are available to validate users on a wireless network. Although there are many different authentication sources in the realm of general network security, the ones that we will concentrate on for wireless networking will be pre-shared key (PSK) and RADIUS with EAP.

Pre-shared Key (PSK)

We have already alluded to the concept of a *pre-shared key (PSK)* as a method of authenticating users in relation to the WPA and WPA2 standards. As mentioned, a pre-shared key is a central element of WPA/WPA2 personal mode authentication. Simply put, a pre-shared key is a shared passphrase that is configured on the Wireless LAN Controller and on the client device which is used to validate each device's identity and begin encryption. A PSK can be in HEX or ASCII format, but it is normally an ASCII string of between 8 and 63 characters long. A pre-shared key is not as secure as 802.1X EAP authentications, but it has the advantage of tremendous ease of deployment and maintenance. PSK is typically not recommended for enterprise environments as it is best suited for home wireless environments that typically do not carry as much sensitive data, or for smaller businesses where owning and maintaining a RADIUS server may not be a cost-effective solution. That said, there are enterprises that have been forced into using PSK for certain application-specific wireless segments due to technical or security limitations on the client supplicants.

Extensible Authentication Protocol (EAP)

After covering the 802.1X EAP types in great detail, it is also essential to note the back end infrastructure with which a wireless network is able to provide enterprise-class authentication. Client types that can support WPA/WPA2 with 802.1X RADIUS authentication have become prevalent due in large part to the adoption of the IEEE 802.11i standard and improvements in client hardware and software to support the ever-growing needs of wireless security. Because of the flexibility of the technology and adoption of more complex wireless security deployments, the demand and need for versatile RADIUS sources has grown. Fortunately, the technology has also evolved and become more flexible to meet this need.

Local RADIUS

The idea of a local RADIUS server is not a new one. Earlier Cisco IOS versions of Cisco Aironet access points existed that had the ability to perform the tasks of a RADIUS server without the management and cost overhead of a full server. The Cisco Wireless LAN Controller is no different. Within every model of Cisco controller there is a *local EAP* feature. The sole purpose of this feature is to provide a localized RADIUS server for specific EAP types. In the wireless controller, a local database of users can be added and edited for access to the network. It is important to note that this is not a fully featured RADIUS server and does not provide such capabilities as back-end database integration and full AAA functionality. In addition to this, the local EAP function is limited in its ability to pick specific attributes or perform features like VLAN assignment. Local EAP/RADIUS solutions are normally best deployed in remote offices or SOHO/small business environments where a full RADIUS server is not cost effective but pre-shared key is inadequate for security.

External RADIUS

An external RADIUS server is a dedicated server for performing AAA functions including the authentication of wireless users. External RADIUS servers such as the Cisco Access Control Server (ACS), Cisco Access Registrar (CAR) and third-party products like Funk RADIUS and FreeRADIUS are where the majority of the key configurations for the different 802.1X EAP types occur. External RADIUS servers like these are full-featured and highly customizable, so they are a far better fit for large enterprises and wireless environments where there are diverse wireless security requirements.

INSIDE THE EXAM

Wireless Authentication

For the exam, you should be able to understand the three major different categories of authentication. Open and shared authentication were both released with the IEEE 802.11 standard. Open authentication is basically a null authentication with no obstacle to getting access. Shared authentication, although it uses a shared key technology, actually compromises data and is not used because of its vulnerabilities. MAC authentication is also in use but is very vulnerable because MAC addresses on client devices can be easily spoofed.

802.1X EAP-based authentication is the primary form of authentication used in wireless networks today. LEAP is the simplest of these, but it has been found to have vulnerabilities in it which has made it less popular. EAP-TLS is the original EAP authentication type and provides high security because of mutual certificate authentication, but it is more complex and expensive to deploy. PEAP combines the convenience of password-based authentication with the security of server-side certificates, and it builds a TLS tunnel for client authentication. EAP-FAST is very similar to PEAP, except it provides a PAC key to build the tunnel. Be sure to be familiar with the packet flows associated with all of the 802.1X authentication types and understand the pros and cons of each.

In addition to 802.1X authentication, there is also the use of guest networking. Have a basic understanding of the steps involved in a Web authentication process. Understand that Web authentication is technically separate from 802.1X authentication.

Wireless Encryption

You need to understand the basics of what wireless encryption does. Encryption exists to protect data and it is separate from authentication, for the most part. Understand WEP and its weaknesses. Know about the advantages of TKIP over WEP and understand the WPA standard as well as the four-way handshake process for encryption. Know that AES provides higher level encryption by using a different cipher for encryption. Be sure to know the relationship between WPA, WPA2, and IEEE 802.11i standards. WPA2 and 802.11i are largely parallel. WPA was created as an intermediate standard while 802.11i was being prepared.

Authentication Sources

The only thing you need to pay attention to on wireless authentication sources is the difference between PSK and local and external RADIUS. PSK (pre-shared key) does not involve any kind of server at all. Local RADIUS and EAP provide limited support for 802.1X EAP types, and external RADIUS is the most robust authentication source.

CERTIFICATION SUMMARY

Wireless security is best understood when broken down into its key components: authentication and encryption. Authentication basically verifies that a user or device belongs on a network. Encryption is the method by which transmitted data is protected, using an encryption algorithm of some kind.

In the infancy of the IEEE 802.11 standard, open and shared authentication were introduced. Open authentication is essentially null authentication and largely depended on a pairing with WEP encryption to provide wireless security. Shared authentication attempted to improve authentication security by using keys, but this actually led to more data compromise. MAC authentication was used in an attempt to improve security as well, but the ease of MAC address spoofing eliminated that as a valid authentication method.

To improve on authentication weaknesses, the existing wired IEEE 802.1X standard was adapted to wireless networks. Out of that was born several different 802.1X Extensible Authentication Protocols (EAP). These were LEAP, EAP-TLS, PEAP, and EAP-FAST. EAP-TLS was the original of the EAP protocols and continues to provide the highest level of security due to its use of a PKI infrastructure with server and client side certificate authentication. LEAP was introduced by Cisco as a simpler password-based authentication type, but it has fallen out of favor due to a dictionary attack vulnerability that was addressed in its predecessor EAP-FAST. PEAP combines the best of both worlds by providing server-side certificate authentication to create an encrypted Transport Layer Security (TLS) tunnel that allows client authentication with either password (MS-CHAPv2) or token (GTC) based client authentication. EAP-FAST, which was built to replace LEAP, uses a similar TLS tunnel, but does not require server certificates, instead using Protected Access Credential (PAC) key provisioning. All of the EAP authentication types have their pros and cons and should be deployed based on needs and requirements.

As with authentication, encryption has evolved since the ratification of the IEEE 802.11 standard as well. WEP key encryption, which came in 40 and 104 bit encryptions, was originally designed to provide both encryption and authentication to wireless networks. By using an RC4 stream cipher, a specific key could be used to encrypt data and also verify the identity of client machines. Unfortunately, WEP technology had many vulnerabilities and could be easily hacked. These vulnerabilities eventually led to the ratification of the IEEE 802.11i security standard, but not before the introduction of Wi-Fi Protected Access (WPA) by the Wi-Fi Alliance. WPA specified a substantially improved form of WEP encryption known as Temporal Key Integrity Protocol (TKIP). TKIP enhanced WEP by having a stronger Initialization Vector and improved encryption algorithm based on the same RC4 stream cipher. TKIP was also resistant to many of the attacks that WEP was vulnerable to with technologies that strengthened encrypted transmissions (such as per-packet keying or PPK) and verified the authenticity of data (message integrity check, or MIC). As a part of the WPA standard, authentication subtypes are specified. These are WPA-Enterprise, which includes 802.1X EAP authentication, and WPA-Personal, which includes pre-shared key (PSK).

The WPA standard tied wireless authentication and encryption together with a four-way handshake process to generate encryption keys for wireless transmissions. Using a PMK (pairwise master key) that is generated during the authentication process, an MIC is used for integral communications between the supplicant (client) and authenticator (access point/Wireless LAN Controller). Using an exchange of generated random numbers (or nonces) the authenticator and supplicant each produce a PTK (pairwise transit key) and utilize a GTK (groupwise transit key) to encrypt different classes of wireless data traffic.

With the ratification of the IEEE 802.11i and parallel WPA2 standard, the Advanced Encryption Standard (AES) was introduced. AES provides substantially higher level encryption and uses a more complex and effective algorithm for protecting data by using a replacement Rijindael algorithm for encrypting data.

The backbone of wireless authentication, however, lies in its authentication sources. Pre-shared key (PSK) is an alternative to 802.1X security within the WPA/WPA2 standard. A pre-shared key is usually an 8–63 character string that can be configured on the client and Wireless LAN Controller. It is typically used in home and small business networks that may not require as much security. Local RADIUS servers can be configured on certain APs and are most prominently featured as the local EAP function on Cisco Wireless LAN Controllers. Local RADIUS sacrifices the capabilities of a dedicated RADIUS server but can provide EAP-type authentication in environments that may not have the infrastructure to properly support full RADIUS/AAA capabilities using a localized database of users. External RADIUS servers are the most robust choice for any 802.1X EAP authentication scheme. Cisco ACS is most often used, but third-party authentication servers such as Funk RADIUS and FreeRADIUS are also available. External RADIUS affords a great deal of customization and also supports authentication using back-end databases such as Microsoft Active Directory (AD) and LDAP servers.

TWO-MINUTE DRILL

Wireless Authentication

- Open and shared authentications were introduced along with the IEEE 802.11 standard. Open is the only authentication that still has a practical application today, mostly in unsecured wireless networks or guest access networks. Open authentication is also still used in the initial phases of 802.1X authentication processes.
- MAC authentication was introduced to work with open authentication, but it provides little security because MAC addresses can be easily spoofed.
- The 802.1X EAP authentication process contains three major components: the supplicant, the authenticator, and the authentication server.
- LEAP is a password-based 802.1X EAP authentication type that utilizes mutual authentication. It has become obsolete due to a dictionary attack vulnerability but is still seeing limited use in enterprises where strong password policies are in place.
- EAP-TLS was the original EAP type and is considered the most secure. EAP-TLS is distinguished by its use of client and server side certificates. It is heavily used by government and financial institutions.
- PEAP is an 802.1X EAP type that provides the security of certificates with the convenience of passwords. It does this by allowing client authentication inside a tunnel that is encrypted in the server authentication process. Clients can use a variety of methods to authenticate and servers use a server-side certificate.
- EAP-FAST is similar to PEAP except it does not require server-side certificates and instead builds a tunnel using a PAC provisioning process. EAP-FAST is the designated replacement for LEAP and is not hampered by the dictionary attack vulnerability.
- Guest access networks utilize a feature called Web authentication, which puts the user through a security process after they have fully associated to the wireless network. Web authentication restricts available resources until the user has authenticated.

Wireless Encryption

- WEP encryption is a part of the 802.11 standard and works by using a 40- or 104-bit WEP key. WEP uses the RC4 stream cipher and is no longer considered a valid encryption method because it is very easy to break.
- TKIP encryption was released as a part of the WPA standard to address the weaknesses of WEP. TKIP improves on WEP while continuing to use the RC4 stream cipher. It accomplishes this with a stronger Initialization Vector (IV), per-packet keying (PPK), Message Integrity Check (MIC), and countermeasures.
- The WPA standard also specified authentication subtypes. These are WPA-Enterprise, which includes 802.1X EAP authentication, and WPA-Personal, which includes pre-shared key (PSK).
- The WPA encryption process uses a PMK (pairwise master key) that is generated during the authentication process. This key is used as a message integrity check in a four-way handshake between the supplicant (client) and authenticator (access point/Wireless LAN Controller) to produce the PTK (pairwise transit key) and GTK (groupwise transit key) used to encrypt wireless traffic.
- The WPA2 standard includes the use of AES encryption, which uses a different Rijindael algorithm for encrypting data. WPA2 also has authentication subtypes like WPA but provides for higher level encryption. WPA2 is directly related to the IEEE 802.11i wireless security standard.

Authentication Sources

- Pre-shared key (PSK) is an alternative to 802.1X security within the WPA/WPA2 standard. A pre-shared key is usually an 8–63 character string that can be configured on the client and Wireless LAN Controller. It is not recommended for enterprise wireless networks.
- Local RADIUS servers can be configured on certain APs and are also seen in the local EAP function on Cisco Wireless LAN Controllers. Local RADIUS is far less feature rich than a dedicated RADIUS server but can provide EAP-type authentication in an environment that requires it.
- External RADIUS servers are the cornerstone of any 802.1X EAP authentication scheme. Cisco ACS is most often used, but third-party authentication servers are also available. External RADIUS is the most feature rich and scalable solution for growing secure wireless networks.

SELF TEST

The following self test questions will help you measure your understanding of the material presented in this chapter. Read all the choices carefully, as there may be more than one correct answer. Choose all correct answers for each question.

Wireless Authentication

1. Which of the following authentication types utilizes server-side only certificates?

A. EAP-FAST

B. PEAP

C. LEAP

D. EAP-TLS

2. Which of the following EAP types exhibits the dictionary-attack vulnerability?

A. EAP-FAST

B. PEAP

C. LEAP

D. EAP-TLS

3. When deploying a guest access network, which of the following is *not* an authentication type that you would be likely to use?

A. Open authentication

B. Shared authentication

C. Web authentication

D. All of the above are authentications you would use for guest access.

4. Which of the following characteristics is a valid reason for the limited deployment of EAP-TLS?

A. EAP-TLS is more expensive to deploy than PEAP or EAP-FAST due to supplicant licensing costs.

B. EAP-TLS is not sanctioned by government agencies due to its relative lack of security.

C. EAP-TLS is more challenging to deploy because it requires mutual certificate authentication.

D. The 802.1X standard recognizes EAP-TLS as a legacy technology.

5. Which of the following devices can be classified as an authenticator?
   A. A Cisco ACS RADIUS server
   B. A non ad hoc laptop with properly configured wireless software
   C. A Cisco lightweight AP
   D. A Cisco Wireless Controller

Wireless Encryption

6. Which of the following is not an improvement of WPA TKIP over WEP?
   A. Larger Initialization Vector (IV) provides strong security.
   B. TKIP bit-flips packets to avoid interception by attackers.
   C. TKIP is a part of the WPA standard.
   D. TKIP provides per-packet keying, which helps strengthen encryption.
7. According to WPA and WPA2 standards, which of the following is not a valid WPA and encryption pairing?
   A. WPA with TKIP
   B. WPA with AES/CCM
   C. WPA2 with TKIP
   D. WPA2 with AES/CCM
8. Which of the following encryption types does not operate with an RC4 stream cipher?
   A. WEP
   B. TKIP
   C. AES
   D. All of the above utilize the RC4 stream cipher in one form or another.
9. Which of the following is not an example of something that a pairwise master key (PMK) can be directly derived from?
   A. A user password entered in authentication and stored in the RADIUS server
   B. A session key created by 802.1X authentication
   C. A WPA pre-shared key entered on the client and access point
   D. All of the above can be used to derive a PMK.

Authentication Sources

10. Which of the following is a valid authentication source for a large hospital running PEAP?

A. A WPA2 AES encrypted pre-shared key

B. A Cisco ACS server with certificates installed

C. A 128-bit encrypted WEP key

D. A properly configured Local RADIUS server with usernames of valid staff

SELF TEST ANSWERS

Wireless Authentication

1. Which of the following authentication types utilizes server-side only certificates?

A. EAP-FAST

B. PEAP

C. LEAP

D. EAP-TLS

☑ **B.** PEAP is the only authentication type in the list that provides server-side only certificates.

☒ **A** and **C** are incorrect because both LEAP and EAP-FAST operate with passwords and don't use certificates. **D** is incorrect because EAP-TLS uses both server and client certificates.

2. Which of the following EAP types exhibits the dictionary-attack vulnerability?

A. EAP-FAST

B. PEAP

C. LEAP

D. EAP-TLS

☑ **C.** LEAP has become far less popular because it has the dictionary attack vulnerability and requires strong passwords.

☒ **B, C,** and **D** are incorrect because they represent EAP types that have addressed the dictionary attack vulnerability using either certificates or PAC (Protected Access Credential) technology.

3. When deploying a guest access network, which of the following is *not* an authentication type that you would be likely to use?

A. Open authentication

B. Shared authentication

C. Web authentication

D. All of the above are authentications you would use for guest access.

☑ **B.** Shared authentication is very undesirable in any wireless deployment and has fallen severely out of use.

☒ **A, C,** and **D** are incorrect because open authentication is often paired with Web authentication in guest access deployments.

4. Which of the following characteristics is a valid reason for the limited deployment of EAP-TLS?

A. EAP-TLS is more expensive to deploy than PEAP or EAP-FAST due to supplicant licensing costs.

B. EAP-TLS is not sanctioned by government agencies due to its relative lack of security.

C. EAP-TLS is more challenging to deploy because it requires mutual certificate authentication.

D. The 802.1X standard recognizes EAP-TLS as a legacy technology.

☑ C. EAP-TLS is very challenging to deploy because of the requirement to install client certificates on every machine along with server certificates.

☒ **A**, **B**, and **D** are incorrect because none of these statements about EAP-TLS are true.

5. Which of the following devices can be classified as an authenticator?

A. A Cisco ACS RADIUS server

B. A non ad hoc laptop with properly configured wireless software

C. A Cisco lightweight AP

D. A Cisco Wireless Controller

☑ D. The only valid authenticator in this list is the Cisco Wireless Controller.

☒ **A** and **B** are incorrect because they represent an authentication server and supplicant, respectively. **C** is incorrect because, although it is very close to being an authenticator, in a lightweight wireless environment the controller plays the role of authenticator.

Wireless Encryption

6. Which of the following is not an improvement of WPA TKIP over WEP?

A. Larger Initialization Vector (IV) provides strong security.

B. TKIP bit-flips packets to avoid interception by attackers.

C. TKIP is a part of the WPA standard.

D. TKIP provides per-packet keying, which helps strengthen encryption.

☑ B. Bit-flipping is a vulnerability in WEP encrypted networks, not an improvement.

☒ **A**, **C**, and **D** are incorrect because they are all ways in which TKIP is an improvement over WEP.

7. According to WPA and WPA2 standards, which of the following is not a valid WPA and encryption pairing?
   A. WPA with TKIP
   B. WPA with AES/CCM
   C. WPA2 with TKIP
   D. WPA2 with AES/CCM

☑ C. The WPA2 standard requires AES/CCM and does not support the use of TKIP.

☒ **A**, **B**, and **D** are incorrect because they are all valid according to the WPA and WPA2 standards.

8. Which of the following encryption types does not operate with an RC4 stream cipher?
   A. WEP
   B. TKIP
   C. AES
   D. All of the above utilize the RC4 stream cipher in one form or another.

☑ C. AES uses block ciphers that are based off the Rijindael algorithm to encrypt data, not RC4.

☒ **A** and **B** are incorrect because they both do run on the RC4 stream cipher. **D** is incorrect because it includes A and B.

9. Which of the following is not an example of something that a pairwise master key (PMK) can be directly derived from?
   A. A user password entered in authentication and stored in the RADIUS server
   B. A session key created by 802.1X authentication
   C. A WPA pre-shared key entered on the client and access point
   D. All of the above can be used to derive a PMK.

☑ **A.** User passwords are not involved in the creation of a PMK in any way.

☒ **B** and **C** are incorrect because they are examples of elements that can create a PMK. **D** is incorrect because it states that **A** can be involved in PMK creation, which it does not.

Authentication Sources

10. Which of the following is a valid authentication source for a large hospital running PEAP?

A. A WPA2 AES encrypted pre-shared key

B. A Cisco ACS server with certificates installed

C. A 128-bit encrypted WEP key

D. A properly configured Local RADIUS server with usernames of valid staff

☑ **B.** WPA2 AES is the only valid authentication source for a large enterprise healthcare environment like a hospital, especially with the requirement for PEAP.

☒ **A** and **C** are valid authentication sources but are inappropriate for such a large environment with security concerns. **B** is not even a valid authentication source, and WEP, although it performs an indirect authentication function, is unfit for enterprise use.

Part III

Cisco Wireless LAN Architecture

CHAPTERS

6

Understanding the Cisco Unified Wireless Network Architecture

CERTIFICATION OBJECTIVES

6.01	The Basics of the Cisco Unified Wireless Network Architecture
6.02	The Modes of Controller-based AP Deployment
6.03	Controller-based AP Discovery and Association
6.04	Roaming
✓	Two-Minute Drill
Q&A	Self Test

As discussed in previous chapters, wireless networks consist of access points as basic building blocks. When the wireless network is small, access points can be easily managed and all the intelligence and configurations usually reside on access points. Access points usually provide standalone 802.11 PHY and MAC services and control client authentication and access to the wireless LAN. Cisco refers to standalone access points as *autonomous* APs because each access point manages an autonomous wireless network on its own. As wireless LAN technology matures and becomes widely adopted, companies are starting to build large and more complex wireless LAN to help serve different business needs. Large wireless networks that comprise hundreds and sometimes thousands of access points are no longer rare in a large and complex enterprise network. Autonomous access point–based wireless architecture quickly runs into scalability issues as the network expands, and administrative overheads become a major challenge as the wireless network grows. To simplify design and reduce administrative challenges, Cisco Unified Wireless Network (CUWN) was created based on a centralized architecture to help better manage AP configurations, monitor RF environment, and control devices in the wireless network. A new type of device was introduced in CUWN called a Wireless LAN Controller (WLC), which houses intelligence centrally and works with access points based on the Cisco Unified wireless architecture. Access points in CUWN maintain minimal information in its memory and do not store configurations or process many services, such as authentication and radio frequency management. The CUWN based access points are called lightweight APs, also referred to as *thin* APs. In contrast, autonomous APs are sometimes referred to as *fat* APs or thick APs. In this chapter, we will provide an overview of the WLC, services provided by WLCs, and interactions between WLCs and lightweight APs. We will also discuss different types of service modes a thin AP can perform in CUWN, different protocols used by thin APs to communicate with a WLC, and different communications and design concerns in CUWN. In the next chapter, we will discuss the Cisco Express Unified Wireless Network solution design, which is based on CUWN but specifically targets small and medium types of wireless networks.

CERTIFICATION OBJECTIVE 6.01

The Basics of the Cisco Unified Wireless Network Architecture

The Cisco Unified Wireless Network (CUWN) is designed to centrally provide 802.11 wireless services to clients and APs for large and complex wireless networks. The CUWN simplifies deploying and managing large-scale wireless LANs and enables a unique best-in-class security architecture that meets critical business requirements in a complex environment. The CUWN consists of the following components:

- Cisco Wireless LAN Controllers (WLCs)
- Cisco Lightweight Access Points (LAP)
- Cisco Wireless Control System (WCS)

Figure 6-1 illustrates how the different components fit in a large enterprise network from a high level. We will discuss in more detail how each component interacts with one another and with other network devices. It should be noted that the figure does not include all network components such as the firewall, servers, and other network services devices.

FIGURE 6-1 A high level view of CUWN and enterprise network

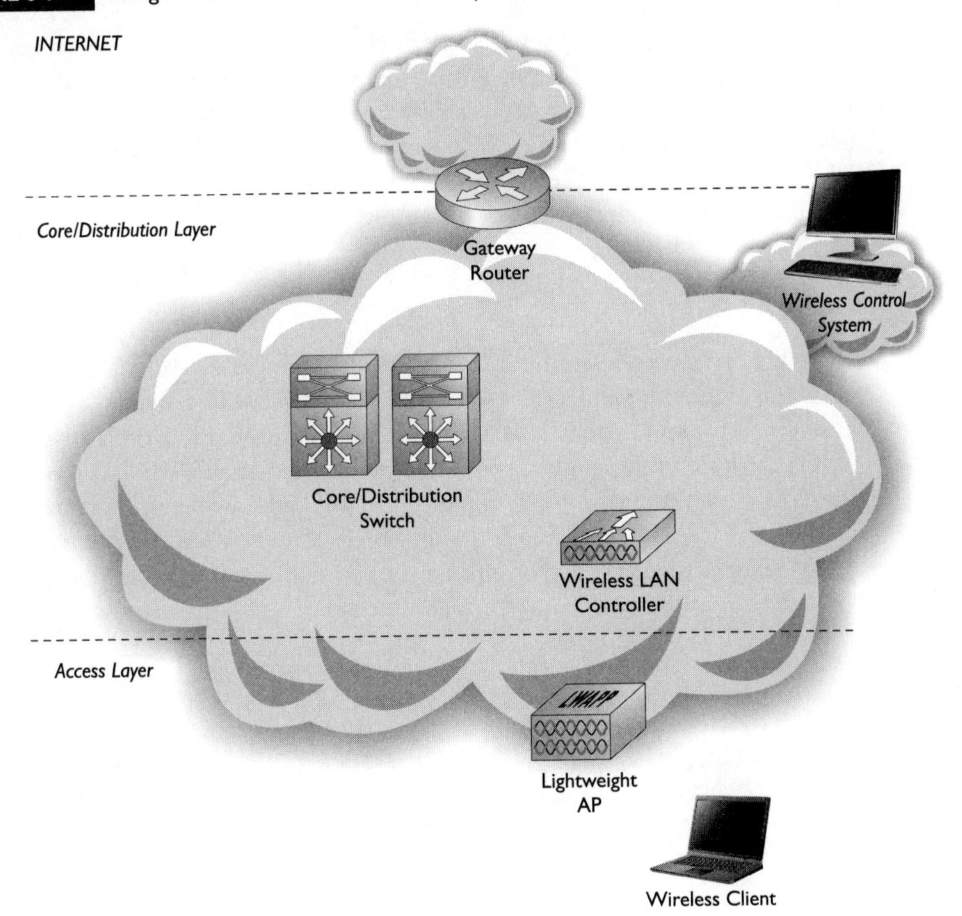

Cisco Wireless LAN Controllers

Cisco Wireless LAN Controllers (WLCs) are responsible for system-wide wireless LAN functions, such as security policies, intrusion detection and prevention, RF management, quality of service (QoS), and mobility services. WLCs work in conjunction with Cisco Lightweight Access Points and the Cisco Wireless Control System (WCS) to support business-critical wireless applications, such as voice, data, and location-based asset tracking. The CUWN provides the control, monitoring, and security operations that network managers need to build a secure enterprise class wireless networks—from branch offices to main campuses. In an essence, the WLC is the brain of CUWN.

Cisco offers several different flavors of WLC that can manage between 6 and 300 APs in different form factors, including dedicated WLC appliances, integrated WLC network modules for 2800/3800 Integrated Services Routers (ISR), integrated Wireless Service Modules (WiSM) for Catalyst 6500 series switch, and integrated WLC in a Catalyst 3750 switch. WLCs can be grouped together to form a larger wireless network while being managed by a central management system, WCS, to form a unified wireless solution. WLCs can also share information among a group of WLCs to derive optimized channel and power settings for APs. Table 6-1 highlights the different WLC hardware platforms.

Cisco Lightweight APs

Cisco lightweight APs are also called the *thin* APs. Lightweight APs cannot operate and provide full services to clients without a WLC. Cisco APs can operate in both lightweight AP and autonomous AP modes, depending on the version of IOS software running on the AP. All wireless client traffic will be relayed to the WLC and, through a WLC, client traffic will be forwarded to the intended network destination. Cisco offers a wide variety of APs; some are specifically designed for outdoor use and others may be appropriate for both indoor and

TABLE 6-1 Overview of Cisco Wireless LAN Controller Platforms

Wireless LAN Controller	Description	AP Support
4402 WLAN Controller	4400 Series WLC with 2 SFP Gigabit Ethernet uplinks	Cisco 4402 series WLC has 3 variants and supports up to 12, 25, and 50 APs
4404 WLAN Controller	4400 Series WLC with 4 SFP Gigabit Ethernet uplinks	Cisco 4404 series WLC supports up to 100 APs
2106 WLAN Controller	2100 Series WLC with 8 10/100 Ethernet ports	Cisco 2106 series WLC supports up to 6 APs
2112 WLAN Controller	2100 Series WLC with 8 10/100 Ethernet ports	Cisco 2112 series WLC supports up to 12 APs
2125 WLAN Controller	2100 Series WLC with 8 10/100 Ethernet ports	Cisco 2125 series WLC supports up to 25 APs
3750 G Integrated WLAN Controller	3750 Series WLC with 24 10/100/1000 Ethernet ports with IEEE 802.3af and Cisco prestandard PoE	Cisco 3750G series WLC has 2 variants and supports up to 25 and 50 APs
3750 G Integrated WLAN Controller	3750 Series WLC with 24 10/100/1000 Ethernet ports with IEEE 802.3af and Cisco prestandard PoE	Cisco 3750G series WLC has 2 variants and supports up to 25 and 50 APs
Wireless Services Module (WiSM)	Cisco Wireless Services Module with 2 WLCs for Cisco Catalyst 6500/7600 series	WiSM contains 2 WLCs with support up to 300 APs
Wireless LAN Controller Module (WLCM)	Cisco Wireless LAN Controller Module for the 2800 and 3800 Integrated Services Router (ISR)	WLCM contains 1 WLC with support for 6, 8, 12, and 25 APs
5500 WLAN Controller	5500 Series WLC with 8 SFP gigabit Ethernet uplinks and is first supported in WLC release 6	Cisco 5500 series WLC supports 12, 25, 50, 100, and 250 APs based on AP license

outdoor use. More specific lightweight AP operations will be discussed later in the chapter. Table 6-2 highlights the different AP hardware platforms supported in thin AP mode.

TABLE 6-2 Overview of Cisco Aironet Series AP Platforms

Access Point	Description	Mode of Operations
Aironet 1230 Series Access Point	■ 1230 Series AP with dual band (2.4 1GHz and 5 GHz) radios with external antenna support and 10/100 Ethernet port ■ 1230 Series AP supports 802.11 a/b/g and Cisco prestandard PoE	Cisco 1230 Series AP supports both lightweight and autonomous modes.
Aironet 1240 Series Access Point	■ 1240 Series AP with dual band (2.4 GHz and 5 GHz) radios with external antenna support and 10/100 Ethernet port ■ 1240 Series AP supports 802.11 a/b/g and IEEE 802.3af PoE	Cisco 1240 Series AP supports both lightweight and autonomous modes.
Aironet 1250 Series Access Point	■ 1250 Series AP with dual band (2.4 GHz and 5 GHz) radios with external antenna support and 10/100/1000 Ethernet port ■ 1250 Series AP supports 802.11 a/b/g/n DRAFT2 and IEEE 802.3af PoE and Cisco prestandard ePoE	Cisco 1250 Series AP supports both lightweight and autonomous modes.
Aironet 1130 Series Access Point	■ 1130 Series AP with dual band (2.4 GHz and 5 GHz) radios with integrated antenna support and 10/100 Ethernet port ■ 1130 Series AP supports 802.11 a/b/g and IEEE 802.3af PoE	Cisco 1130 Series AP supports both lightweight and autonomous modes.
Aironet 1140 Series Access Point	■ 1140 Series AP with dual band (2.4 GHz and 5 GHz) radios with integrated antenna support and 10/100 Ethernet port ■ 1140 Series AP supports 802.11 a/b/g/n Draft 2 and IEEE 802.3af PoE	Cisco 1140 Series AP supports only lightweight mode at the time of this writing. Autonomous mode support for Cisco 1140 series AP is planned for future releases.
Aironet 1300 Series Outdoor Access Point/ Bridge	■ 1300 Series AP/Bridge with 2.4 GHz radio with integrated antenna support and 10/100 Ethernet port ■ 1300 Series AP/Bridge supports 802.11 b/g only	Cisco 1300 Series AP/Bridge supports both lightweight and autonomous modes. Cisco 1300 series AP/Bridge supports AP and Bridge modes; hence, it can simultaneously support wireless bridge link(s) and client connectivity.

TABLE 6-2 Overview of Cisco Aironet Series AP Platforms *(continued)*

Access Point	Description	Mode of Operations
Aironet 1400 Series Outdoor Bridge	■ 1400 Series Bridge with 5 GHz radio with external antenna support and 10/100 Ethernet port	Cisco 1400 Series Bridge supports only autonomous mode and cannot be used as an AP.
Aironet 1520 Series Outdoor Mesh Access Point	■ 1520 Series Outdoor Mesh Access Point with dual band (2.4 GHz and 5 GHz) radios with external antenna support and 10/100 Ethernet port	Cisco 1520 Series Outdoor Mesh Access Point supports only lightweight mode.

You should be very familiar with the different models of AP and WLC platforms from Tables 6-1 and 6-2.

Cisco markets lightweight AP as "zero touch" as there are build-in mechanisms in the CUWN that will allow WLCs to manage APs as soon as APs are powered up and establish network connectivity. Details of these mechanisms will be discussed later in this chapter.

Wireless Control System (WCS)

The Wireless Control System (WCS) is a wireless LAN planning, configuration, and management solution for CUWN. The WCS allows multiple WLCs and APs to be controlled, configured, and monitored from a centralized location using a web-based graphic user interface (GUI). The WCS simplifies wireless network operations, monitors the RF environment, manages policy provisioning, and provides a single management platform for normal network management tasks, including troubleshooting, user tracking, security monitoring, and WLC systems management. The WCS also includes tools for advanced wireless applications, such as wireless LAN planning and design, RF management, location tracking, and wireless IDS/IPS.

The WCS runs on both Windows 2003 and Red Hat Linux servers and is capable of scaling up to support up to 3000 lightweight APs, 1250 autonomous APs, and 750 WLCs. Cisco has provided server specifications for the WCS supporting different number of APs in a wireless LAN.

To scale a wireless LAN beyond 3000 lightweight APs, the WCS Navigator is needed. The WCS Navigator can be considered the "manager of managers" in CUWN, and it can support up to 20 WCS platforms and manage up to 30,000 lightweight APs from a single WCS Navigator console. The WCS Navigator greatly increases the flexibility and scalability of CUWN and can be located on the same LAN as other WCSes, on separate routed subnets, or across a WAN connection.

Lightweight AP Protocol (LWAPP)

WLCs integrate with existing enterprise networks and communicate with Cisco lightweight APs over any Layer 2 (Ethernet) or Layer 3 (IP) infrastructure using Lightweight Access Point Protocol (LWAPP) and support automation of numerous WLAN configuration and management functions across all enterprise locations. LWAPP facilitates configuration and management of a wireless network; it is the facility to tunnel client traffic to and from WLCs and APs.

on the job

LWAPP is a Cisco proprietary communications protocol between a WLC and an AP. LWAPP was created and used by Cisco in the absence of a standards-based solution to answer increasing demands from customers to design and manage large wireless networks under a centralized network architecture. Since LWAPP was created, the IETF Control and Provisioning of Wireless Access Points (CAPWAP) working group has adopted many of the LWAPP features and published CAPWAP as a standards-based wireless protocol. In WLC release 5.2, Cisco officially supports CAPWAP. Prior to release 5.2, Cisco WLC supports only LWAPP.

exam watch

The CCNA Wireless exam will focus on the features and operations of LWAPP, not CAPWAP. CAPWAP includes additional features in the standard protocol and is not covered by LWAPP. The CCNA Wireless exam will also focus primarily on releases 5.1 and prior.

The LWAPP was designed to encapsulate information using LWAPP headers, and information is exchanged between a WLC and an AP for control plane functions and data plane functions. For control plane functions, LWAPP facilitates exchange of control messages between the WLC and AP for the controller discovery/join process, AP firmware updates, AP configuration management, RF information gathering, wireless IDS/IPS information exchange, and security services. LWAPP control messages are marked with a "control bit" in the LWAPP header and encrypted using AES. In data plane functions, client traffic is encapsulated by LWAPP and tunneled between a WLC and an AP. This is referred to as *data encapsulation* by LWAPP.

Split MAC

LWAPP not only simplifies the AP management, it also decouples 802.11 MAC services between a WLC and an AP to support issues commonly seen in a large wireless network. The WLC replaces a conventional ESS, where multiple APs are connected together through network. LWAPP distributes all MAC functions between WLCs and APs that are linked by LWAPP across a network and provides the same ESS services in a manner that is easier to deploy and simpler to manage.

The AP interacts with clients through 802.11 MAC with real-time (time-sensitive) services, such as transmission of beacon frames, buffer frames for clients in power save mode, response to probe request frames, and encryption and decryption of 802.11 packets. Additional 802.11 MAC services provided by APs include monitoring channel and radio utilization, noises, and interferences. The client information, data traffic, and RF information will then be forwarded back to the WLC via LWAPP where client traffic will enter the network through the WLC.

WLCs, on the other hand, process MAC services that are less time sensitive (non-real-time), such as 802.11 authentication, association, and reassociation. The split MAC architecture reduces CPU and memory utilization on the AP while allowing the wireless network to grow. This architecture also allows system-wide intelligence to be processed in a single WLC or multiple WLCs in a coordinated fashion supporting tens and hundreds of APs in a wireless network. Furthermore, because RF information from each AP is forwarded back to the WLC, the centralized architecture is more optimized to manage radio resources and configure

FIGURE 6-2

LWAPP split MAC

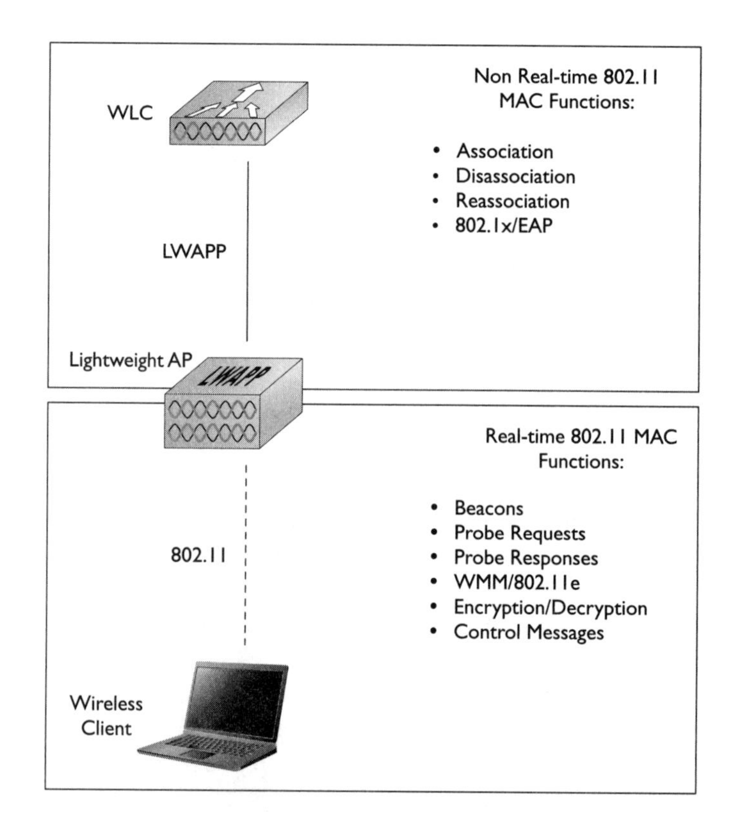

advanced QoS services where the WLC centrally makes resources scheduling decisions. Figure 6-2 illustrates the split MAC and services provided by a WLC and an AP. Details of radio resource management services will be discussed later in this chapter.

CERTIFICATION OBJECTIVE 6.02

The Modes of Controller-based AP Deployment

In CUWN, APs can be configured into different modes of operations for special purposes. The following modes of operations are supported:

- Local
- Hybrid REAP (H-REAP)
- Monitor
- Rogue detector
- Sniffer
- Bridge

Local Mode

AP in local mode is the default operation for a lightweight AP. Local mode AP will serve clients using predefined radio and WLAN (SSID) policies and scan unused channels and RF for monitoring and advanced applications, such as location-based services (LBS) and security services, such as wireless IDS and WIPS.

Hybrid REAP (H-REAP) Mode

Remote edge AP (REAP) is a special AP mode that is designed for a lightweight AP to be deployed at a remote locations. The REAP allows WLANs (SSIDs) to be terminated locally to one VLAN while managed by a WLC in a central location. REAP is only supported by Cisco 1030 Series AP, which has been discontinued by Cisco. Instead, hybrid REAP (H-REAP) succeeds REAP by Cisco 1240, 1250, 1130, and 1140 Series AP. H-REAP supports all REAP functionality plus more. The WLC can provide centralized authentication services by taking advantage of the central authentication server and corporate directory and network services. The WAN link connecting H-REAP from remote site to central site should have speeds higher than 128 Kbps, with roundtrip delay less than 100 ms.

H-REAP Modes of Operation

H-REAP is typically deployed at remote sites and has two modes of operations: connected mode and standalone mode. In connected mode, H-REAP maintains connectivity with the WLC. In standalone mode, the WLC is unreachable by H-REAP. This usually happens when the WAN link is down.

Switching Modes

One main advantage of REAP/H-REAP is the flexibility of deploying CUWN without deploying a WLC at remote locations. REAP/H-REAP provide local switching functionality so that local traffic, such as a print job or a local data store, does not unnecessarily consume valuable WAN bandwidth by forwarding traffic to a WLC at a central site and back to the remote site, also known as the *tromboning effect*. H-REAP supports two switching modes: local switched and central switched. In local switched mode, H-REAP is capable of mapping users to local VLANs on a per WLAN (SSID) basis. An 802.1Q trunk will be established between H-REAP and the local switch, and traffic destined for the remote network will be forwarded by standard IP routing by the remote site router. Support of 802.1Q trunk and local VLANs is available only on H-REAP in WLC release 4 and later, and not on REAP. In central switched mode, wireless traffic will be tunneled back to the central WLC via LWAPP, where user traffic will be mapped to VLANs at central site at the WLC.

H-REAP Operation

With the combination of operation mode and switching mode, H-REAP is usually in one of five states: authentication-central/switch-central, authentication-central/switch-local, authentication-local/switch-local, authentication-down/switch-local, or authentication-down/switch-down.

- **Authentication-central/switch-central** A centralized authentication service is used, and all user traffic is forwarded to the WLC. Figure 6-3 illustrates this H-REAP state.
- **Authentication-central/switch-local** A centralized authentication service is used, but user traffic is switched locally by local switch(es) and router(s). Figure 6-4 illustrates this H-REAP state.

FIGURE 6-3 Authentication-central and switch-central H-REAP state

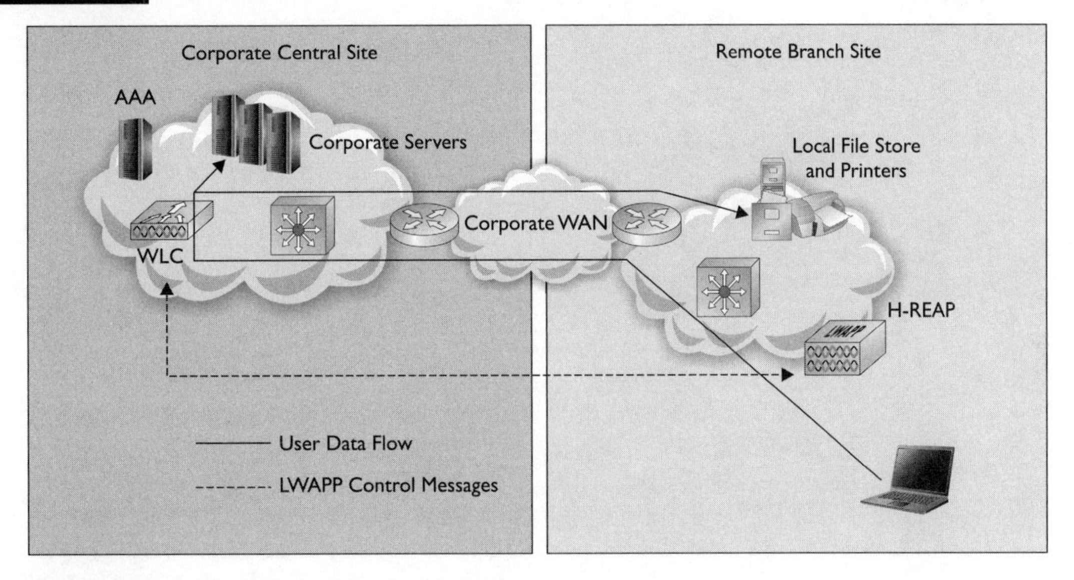

FIGURE 6-4 Authentication-central and switch-local H-REAP state

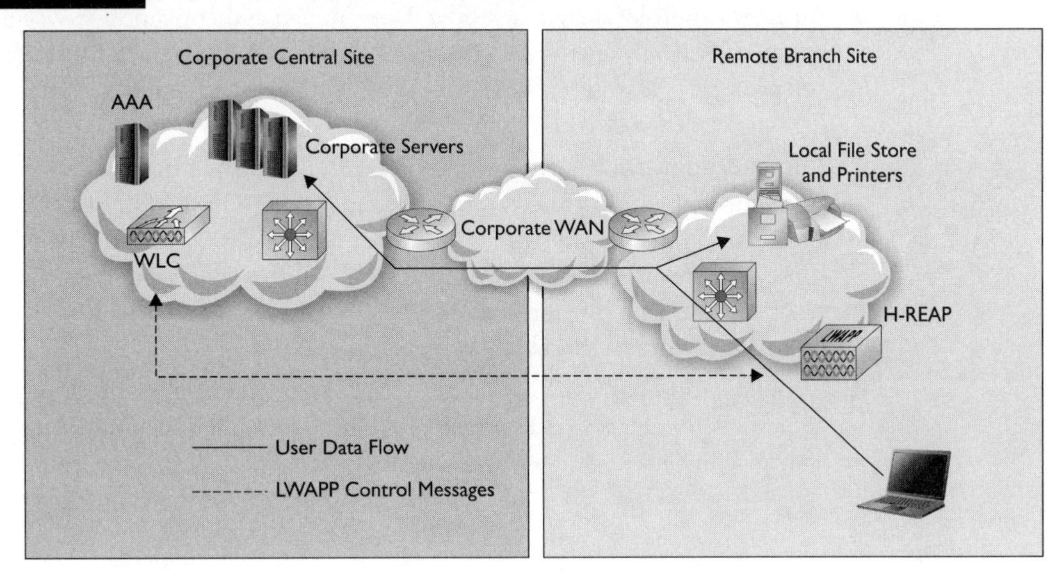

FIGURE 6-5 Authentication-local and switch-local H-REAP state

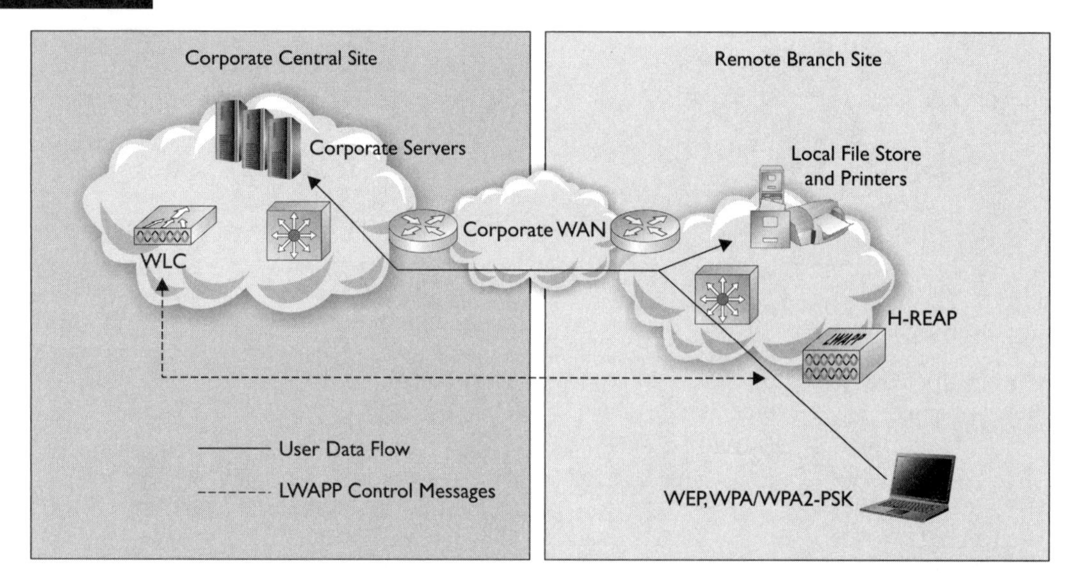

- **Authentication-local/switch-local** A centralized authentication service is not used. Typically clients use static WEP or WPA/WPA2-PSK for security, and user traffic is switched locally by local switch(es) and router(s). Figure 6-5 illustrates this H-REAP state.
- **Authentication-down/switch-local** A centralized authentication service fails while existing clients continue to send traffic to local switch(es) and router(s). No new clients can connect with H-REAP because the authentication service is unavailable. Typically, this state represents H-REAP in standalone mode and the WAN connection may be lost. Figure 6-6 illustrates this H-REAP state.
- **Authentication-down/switch-down** H-REAP is in standalone mode, and no new clients can connect with H-REAP because authentication service is unavailable. Figure 6-7 illustrates this H-REAP state.

FIGURE 6-6 Authentication-down and switch-local H-REAP state

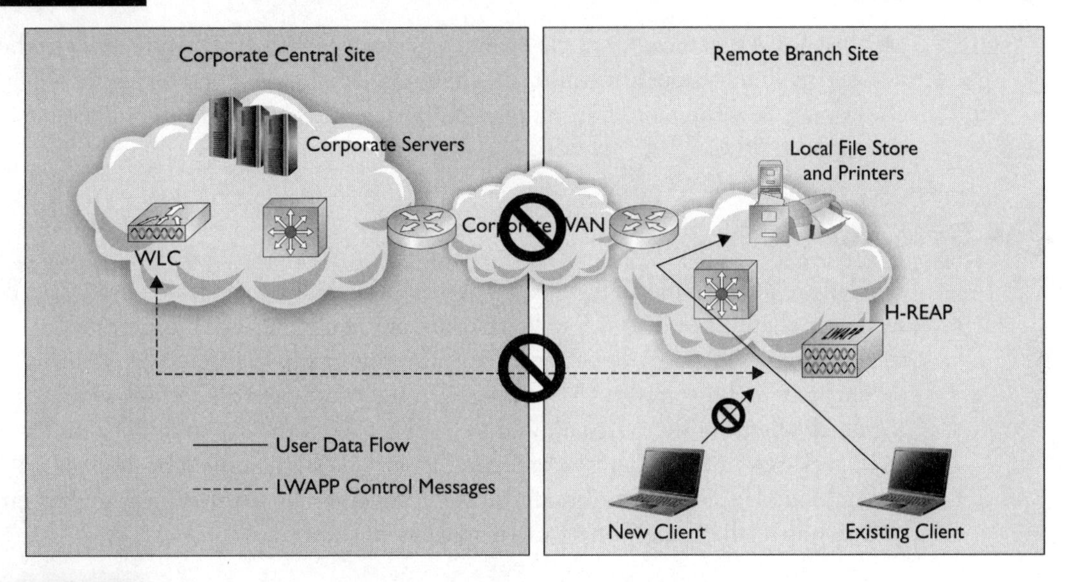

FIGURE 6-7 Authentication-down and switch-down H-REAP state

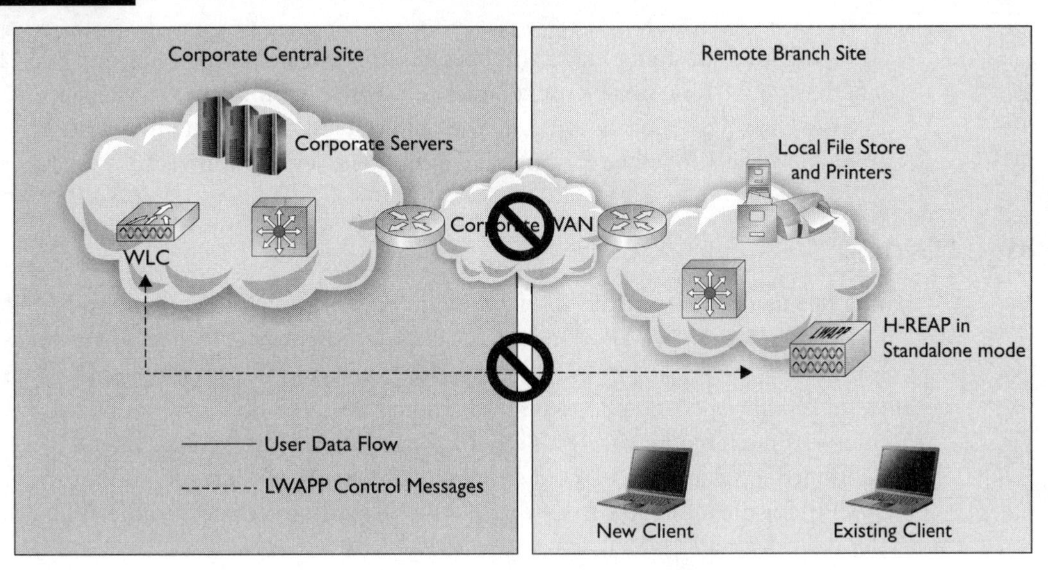

Monitor Mode

Lightweight AP in monitor mode operates in passive mode by listening only to the wireless medium through scanning all channels. AP does not actively serve any clients, nor does the radio transmit any traffic. Channel scanning is configurable through the WLC.

Rogue Detector Mode

Similarly to monitor mode, lightweight AP in rogue detector mode does not actively serve any clients, nor does the radio transmit any traffic. Unlike monitor mode, rogue detector AP does not perform channel scanning. Rogue detector AP turns off its radios and listens for ARP messages on the wired network for any suspicious rogue AP/client traffic. AP is allowed to be placed on a trunk port and listens for ARP packets for MAC addresses of rogue clients and APs identified by the WLC. If a matching MAC address is found, the WLC generates an alarm by sending out an SNMP trap to the WCS or network management station.

Sniffer Mode

Lightweight AP in sniffer mode also operates in passive mode by capturing 802.11 data that can be reviewed by packet analyzer, such as Omnipeek, AirMagnet, or WireShark. This mode is typically used in troubleshooting, and information gathered by the AP will be forwarded to a specified device for review.

Bridge Mode

In bridge mode, lightweight APs act as point-to-point or point-to-multipoint bridges. Outdoor mesh AP or indoor AP in Enterprise mesh mode (also known as indoor mesh mode) support the bridge mode operation. The 1500 Series Outdoor Mesh AP will support outdoor mesh mode, and the 1240 Series AP will support Enterprise mesh mode, which is primarily deployed in indoor environments. As mentioned in Chapter 2, Cisco developed the Adaptive Wireless Path Protocol (AWPP) for the mesh APs to determine the best path to connect to the WLC.

CERTIFICATION OBJECTIVE 6.03

Controller-based AP Discovery and Association

Lightweight AP does not store any configurations in its NVRAM; thus, when the AP is powered up, it needs to discover and associate to a WLC before it can operate normally. AP will download its configuration and firmware, if available, from the WLC before it becomes operational. This architecture allows the WLC to support many APs in a large wireless network, while still managing AP configurations and firmware. Lightweight AP will go through a predefined process to discover the WLC and associate (join) the WLC to download AP specific configurations before it can provide wireless services to clients.

Lightweight AP Discovery Process

After the lightweight AP is powered up, it enters the LWAPP Discovery state. Two LWAPP control messages are exchanged between an AP and the WLC: LWAPP Discovery Request and LWAPP Discovery Response. The AP sends LWAPP Discovery Request message to one or more WLCs, and the WLCs respond with LWAPP Discovery Response messages back to the AP when they receive Discovery Request messages. The lightweight AP will select a WLC to join from the WLCs that respond to the AP's Discovery Request messages. The AP will run through the following steps to discover WLCs to send LWAPP Discovery Request messages:

1. When the AP boots up, it will attempt to acquire an IP address through the DHCP process unless a static IP address has been configured for the AP.
2. AP broadcasts an LWAPP Discovery Request on the local subnet. If a WLC resides on the same subnet, it responds with an LWAPP Discovery Response to the AP in unicast.
3. WLCs support a feature called Over-the-Air Provisioning (OTAP). When OTAP is enabled on the WLC, APs that have joined to the WLC will advertise their known WLCs and send over the wireless medium. The AP

in LWAPP discovery mode will listen for these messages and unicast an LWAPP Discovery Request to each WLC learned through OTAP. When the LWAPP Discovery Request messages are received, WLCs respond with LWAPP Discovery Responses to the AP in unicast.

4. If the AP has previously joined or discovered WLCs, up to 10 WLC IP addresses are stored in the AP NVRAM. The AP unicasts LWAPP Discovery Request messages to WLCs whose IP addresses are stored in the AP NVRAM. When the LWAPP Discovery Request messages are received, the WLCs respond with LWAPP Discovery Responses to the AP in unicast.
5. DHCP Option 43 can be used to provide the WLC IP addresses in the DHCP offer to lightweight AP. When AP receives an IP address via DHCP, it also examines for the WLC IP addresses in Option 43 in the DHCP offer. The AP then sends a unicast LWAPP Discovery Request to each WLC in DHCP option 43. When the LWAPP Discovery Request messages are received, the WLCs respond with LWAPP Discovery Responses to the AP in unicast.
6. Lightweight AP will attempt to resolve a default DNS name "CISCO-LWAPP-CONTROLLER.localdomain." A DNS server can resolve this name and point to one or more IP addresses. The AP sends a unicast LWAPP Discovery Request to the resolved IP addresses, and when LWAPP Discovery Request messages are received, the WLCs respond with LWAPP Discovery Responses to the AP in unicast.
7. If the AP does not receive LWAPP Discovery Response messages after steps 2–6, the AP restarts and repeats the discovery process until it successfully joins a WLC.

DHCP and DNS are usually used to provide seed WLC IP addresses for AP, and the seed WLC will then provide a full list of WLCs available to APs. Lightweight AP always completes all the LWAPP discovery steps before it determines which WLC to join. Figure 6-8 illustrates the process of LWAPP discovery.

The LWAPP discovery process is important when a brand new AP is powered up and connects to the network. Once the AP has joined a WLC, it caches the WLC IP address information of all the WLCs within the mobility group of the WLC that it has joined. OTAP is rarely used as it may inadvertently create undesired effects for a brand new AP that has never joined a WLC.

FIGURE 6-8

The LWAPP discovery process

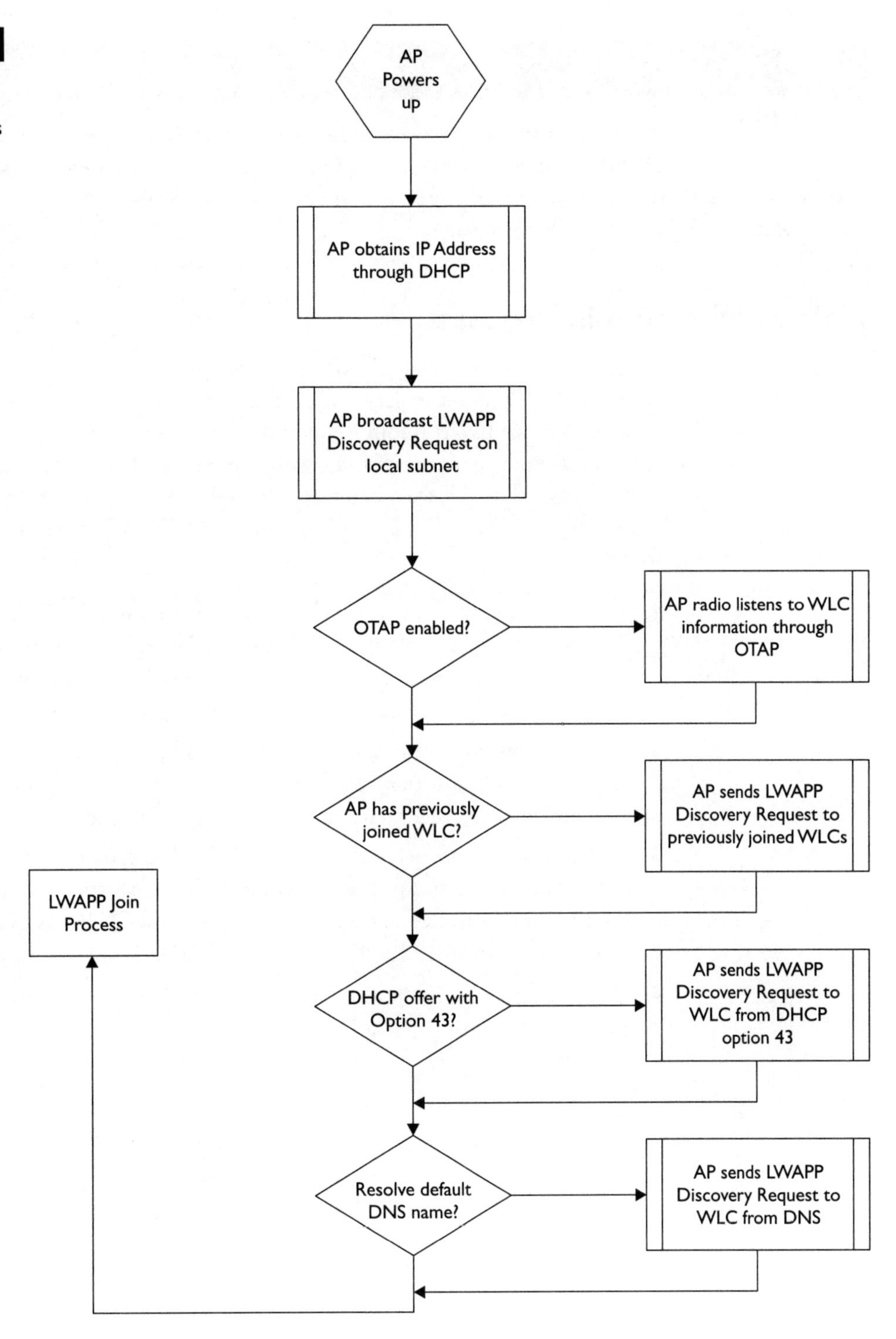

Make sure that you understand how LWAPP discovery works. Remember, a lightweight AP will always complete the whole LWAPP discovery process before it attempts to join a WLC. The LWAPP discovery process will create a candidate WLC list.

Lightweight AP Join Process

After lightweight AP receives LWAPP Discovery Response messages, it builds a "candidate WLC" list. It selects a WLC from the list and attempts to join the WLC by sending an LWAPP Join Request message. The WLC responds with an LWAPP Join Response message after it validates and authenticates the LWAPP Join Request message. The AP then completes the LWAPP join process by validating and authenticating the LWAPP Join Response message. The following steps outline the AP join process:

1. The AP acquires information about the WLC in an LWAPP Discovery Response message, including the WLC sysName, WLC type, WLC AP capacity, WLC current loads (the number of APs currently associated with the WLC), and something called Master Controller status.
2. If the AP has previously joined the WLC and been configured with a primary, secondary, and/or tertiary WLC, the AP will attempt to join these WLCs based on primary, secondary, and/or tertiary precedence.
3. If no primary, secondary, and/or tertiary WLC has been configured or none of the WLCs respond to the LWAPP Join Request, the AP will check the Master Controller status of the candidate WLCs. If a WLC is configured as Master Controller, the AP will send an LWAPP Join Request to this WLC.
4. If steps 2 and 3 both fail, the AP will join the WLC with the greatest excess capacity (or least loaded WLC).

During the join process, the WLC and AP will mutually validate and authenticate each other and derive a key to encrypt LWAPP Control messages using AES-CCMP encryption. The AP and WLC both carry X.509 certificates and public keys. When the AP sends an LWAPP Join Request message, it also sends its X.509 certificate to the WLC. The WLC will validate the certificate and derive an AES encryption key and send a Join Response message along with its own X.509 certificate. The AP will follow the same steps to validate the WLC's certificate and derive the same key for control message encryption to complete the join process. Cisco Aironet 1130, 1230, 1240, and 1300 Series Access Points manufactured after July 18, 2005 carry a manufacturing installed certificate. Sometimes, manufacturing installed certificate is also referred to as MIC for short. It should not be confused with message integrity check discussed in Chapter 4 and Chapter 5. Any AP manufactured prior to July 15, 2005 will require you to generate a self-signed certificate (SSC) in order to operate in lightweight mode and join a WLC.

exam watch

Make sure that you understand how the LWAPP join process works. Remember, a lightweight AP creates a candidate WLC list at the end of the LWAPP discovery process. During LWAPP join process, the AP will then attempt to join a WLC based on information saved in the AP and information obtained from discovery.

Wireless LAN Redundancy

The CUWN architecture provides flexibility and resiliency in its design by default. The LWAPP was designed to address WLC redundancy from a system perspective. To validate connectivity, AP sends an LWAPP heartbeat message to the WLC in a per-defined interval, with a default at 30 seconds, and the WLC responds with an LWAPP heartbeat response. If the LWAPP heartbeat response is not received, the AP will resend the heartbeat message five times at one-second intervals. The AP determines the WLC unreachable after five retries and will restart the LWAPP join process using the candidate WLC list from the LWAPP discovery. The AP can dynamically join a WLC based on the WLC's AP loads and capacity, or it can carry static WLC assignments with primary, secondary, and/or tertiary WLCs for the AP, which is also referred to as priming. With static WLC assignments, once the AP determines its primary WLC is unreachable, it attempts to join the secondary WLC. If the secondary WLC is also unreachable, the AP attempts to join the tertiary WLC, if configured. When none of the preconfigured WLCs are reachable, the AP will resort back to the dynamic join process by attempting to join the least loaded WLC. There are several common designs for WLC redundancy:

- **1+1** In this design, there is always a designated backup WLC configured to provide redundancy to a given WLC. This design requires little planning as there will be a standby WLC designated for an active WLC.
- **N+1** In this design, there is one single spare WLC designated to provide redundancy for multiple WLCs. The spare WLC can assume the role of an active WLC when any WLC fails. However, in the event of multiple WLC failures simultaneously, the spare WLC may become overwhelmed by APs attempting to join from the failed WLCs. This design requires little planning, as there will be one spare WLC designated to back up multiple WLCs in the system. Figure 6-9 illustrates the N+1 redundancy design for the WLC.

FIGURE 6-9 N+1 redundancy

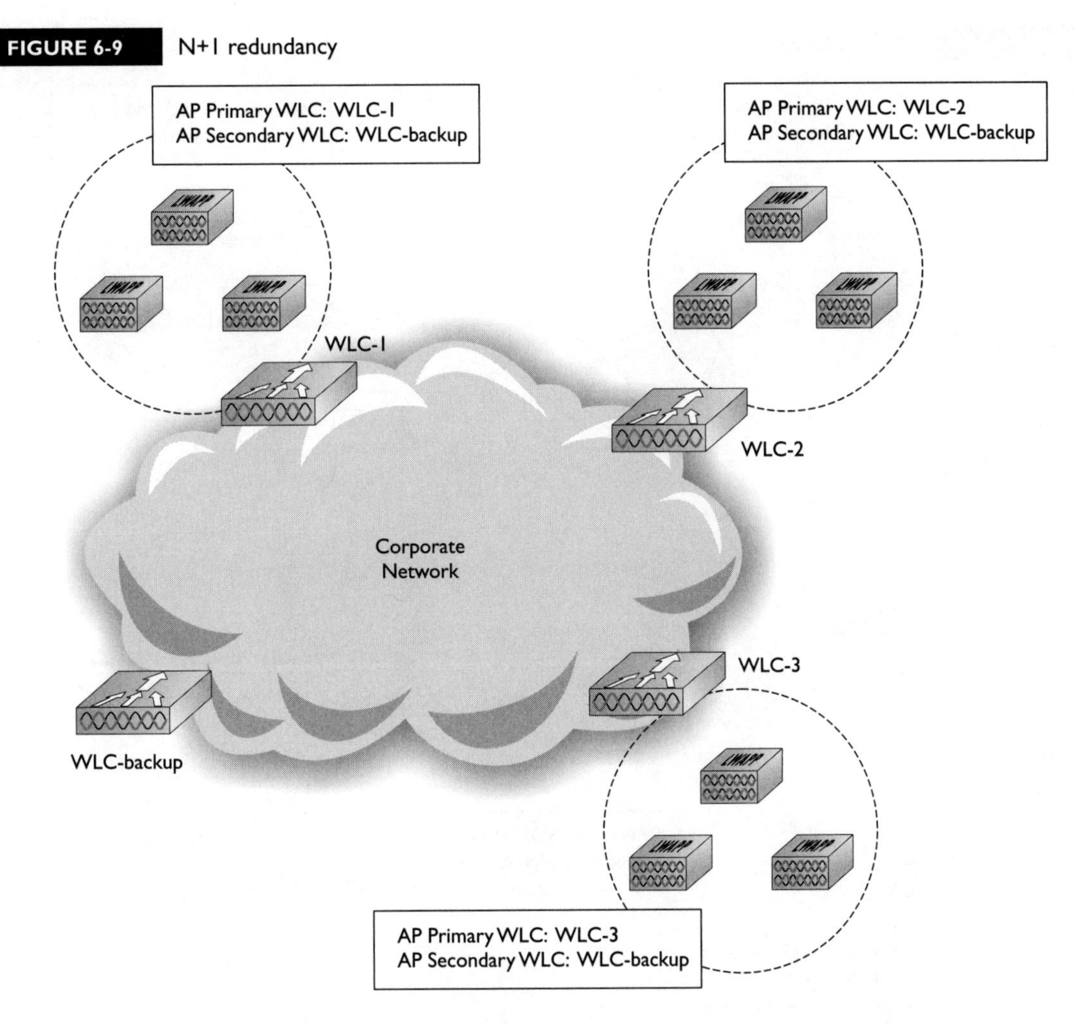

- **N+N** In this design, each WLC reserves some capacity to back up another WLC. For example, APs on WLC-1 may be configured to join WLC-1 as primary WLC and WLC-2 as secondary WLC while APs on WLC-2 are

FIGURE 6-10 N+N redundancy

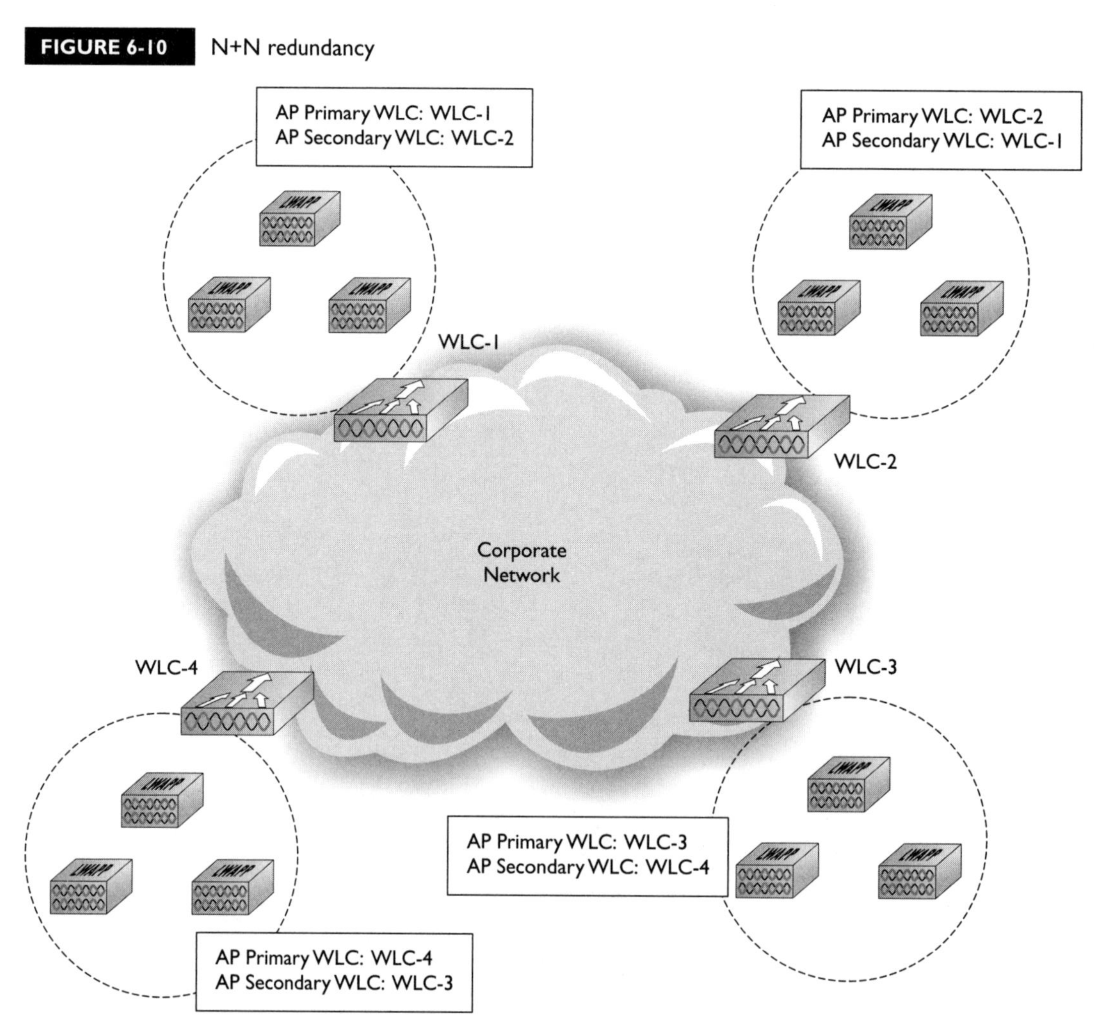

configured to join WLC-2 as primary WLC and WLC-1 as secondary WLC. This design requires good planning and coordination so that no one WLC will be overwhelmed. Figure 6-10 illustrates the N+N redundancy design for the WLC.

- **N+N+1** Similar to N+N, each WLC reserves some capacity and provides redundant support for other WLCs; there is also one spare WLC designated as a backup WLC for all WLCs in the system. All WLCs can use this spare WLC as a tertiary WLC. Figure 6-11 illustrates the N+N+1 redundancy design for WLC.

FIGURE 6-11 N+N+1 redundancy

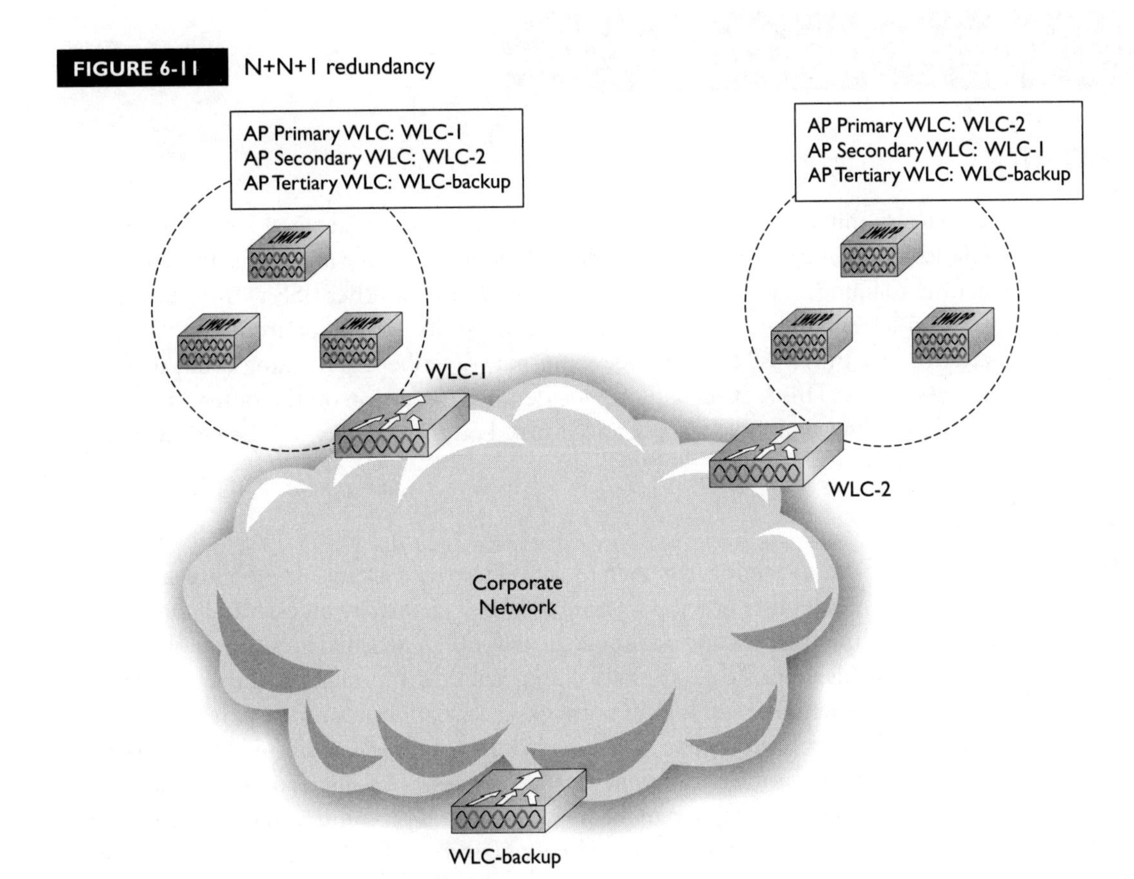

WLC redundancy is an important design consideration. Because different design options require different numbers of WLCs, it may impact the overall budget for the project. Review the application requirements and tolerance for failure for the network before deciding on the redundancy option.

CERTIFICATION OBJECTIVE 6.04

Roaming

One major difference between wireless LAN and traditional wired LAN is that clients are not tied to a network cable and are free to move around as desired. 802.11 defines roaming as a client moving from one BSS to another BSS within the same ESS, which simply means that the client changes its association from one AP to another AP. In CUWN, several scenarios may describe this roaming event in a large wireless LAN. This section will provide detailed discussion on the different types of roaming events and design implications based on the AP and WLC relationship.

Mobility Group

Before we discuss roaming, we need to introduce the concept of "mobility group" in CUWN. A mobility group is a group of WLCs that share an essential client, an AP, and RF information and act as a logical WLC in a virtual group. A mobility group allows member WLCs to share client roaming information, authentication and client credentials, and logical network and location information. It provides a well rounded approach to manage a large wireless network supported by multiple WLCs that share a same set of policies and services in a complex network. Each WLC within a mobility group makes decisions based on information it receives from the mobility group and acts in a coordinated fashion. Members of a mobility group create fully meshed connections with every other WLC within the same mobility group using Ethernet over IP (EoIP) tunneling technology, which is described in IETF RFC 3378. LWAPP is not used by mobility groups for sending group messages to mobility group members. Figure 6-12 illustrate the concept of meshed mobility group connections among its member WLCs.

FIGURE 6-12

Fully meshed mobility group connectivity

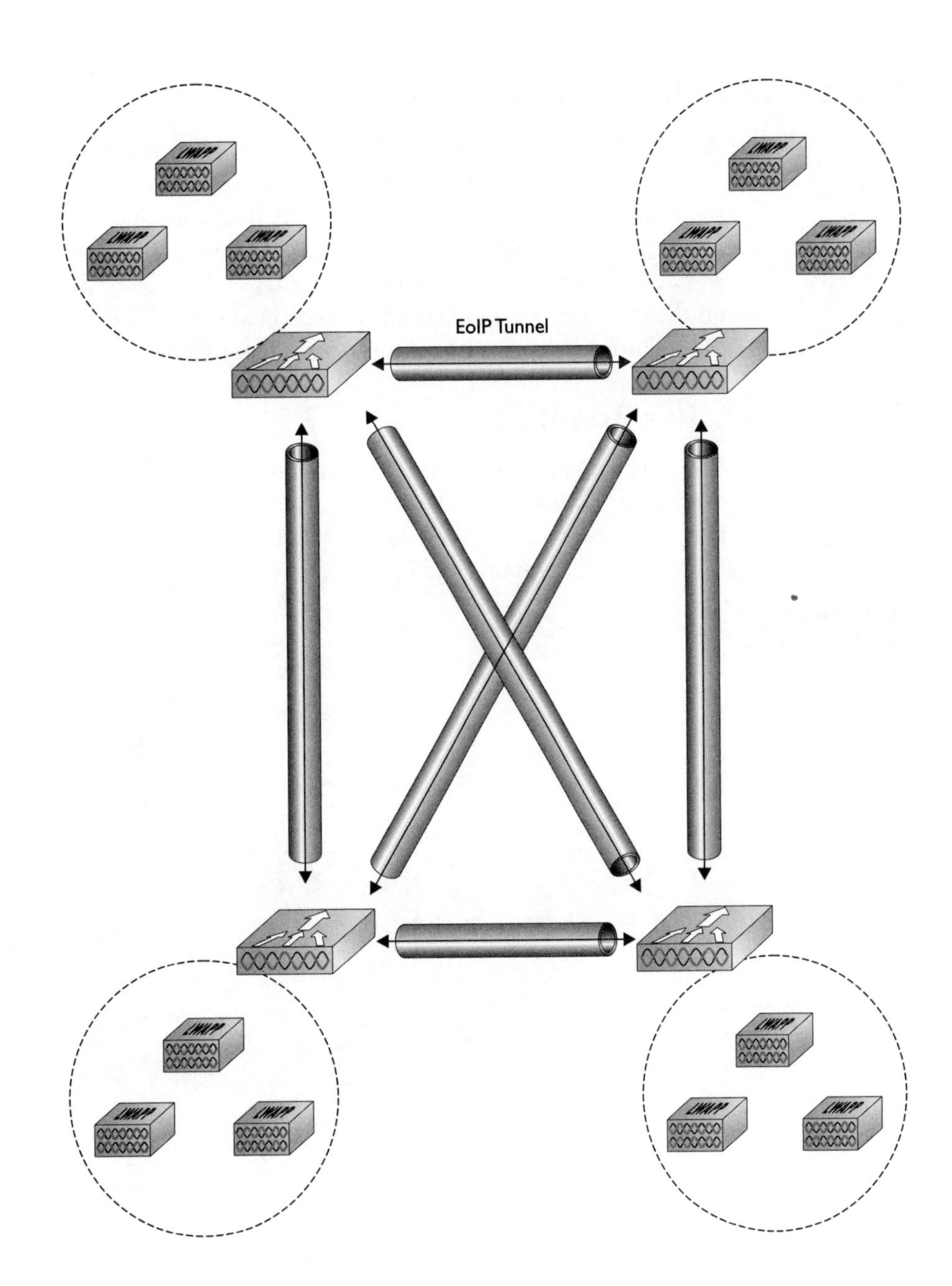

Intra-Controller Roaming

When a client roams from one AP to another AP, which has joined the same WLC as the first AP, it is referred to as an intra-controller roaming. In an intra-controller roaming event, the client will maintain its IP address and the WLC will simply update its client database by associating the client with a new AP. Typically, an intra-controller roaming is described as a Layer 2 roaming where the client remains on the same Layer 2 and Layer 3 network before and after the roaming event. Figure 6-13 illustrates the concept of intra-controller roaming.

Inter-Controller Roaming

When a client roams from one AP to another AP, which has joined a different WLC as the first AP, it is referred to as an inter-controller roaming. In a large wireless network, inter-controller roaming may fall under two different scenarios: Layer 2 inter-controller roaming and Layer 3 inter-controller roaming.

FIGURE 6-13 Intra-controller roaming

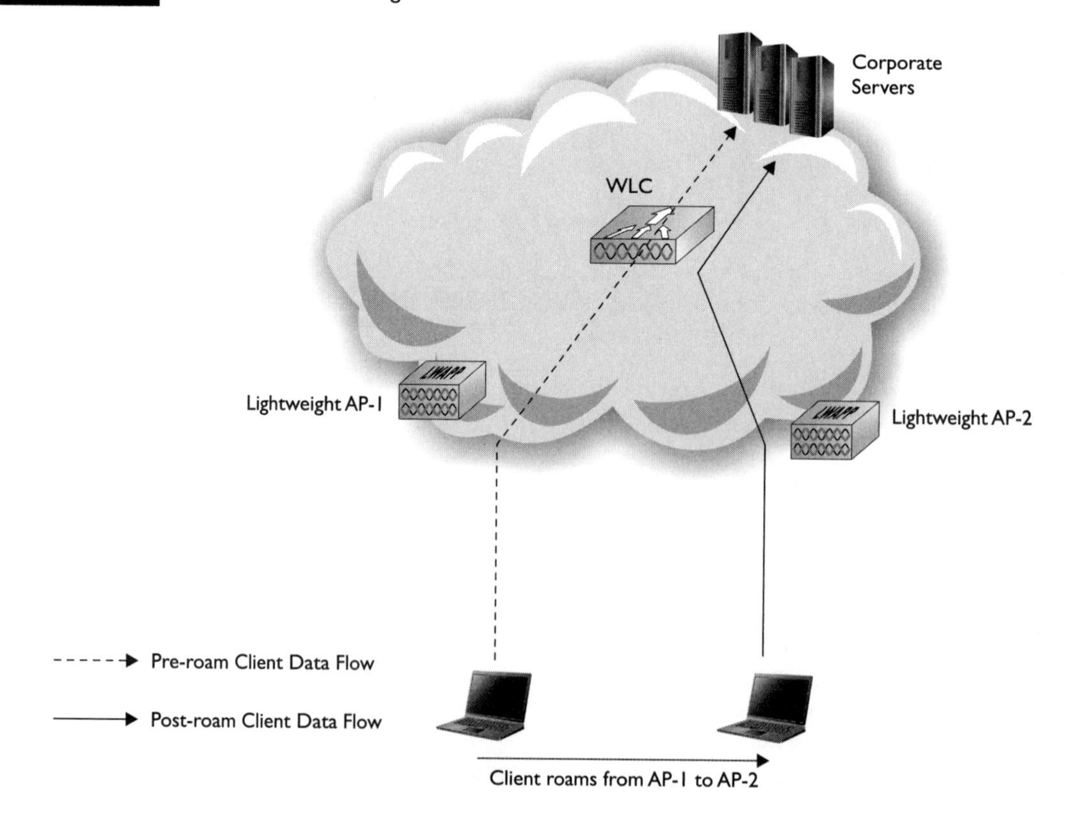

Layer 2 Inter-Controller Roaming

In a Layer 2 inter-controller roaming event, a client roams from one AP (AP-1), that has joined a WLC (WLC-1), to another AP (AP-2), that has joined from a different WLC (WLC-2). The two WLCs bridge client wireless traffic to the same VLAN and IP subnet on the wired network. The two WLCs should be configured with the same mobility group so that when the client reassociates to AP-2, WLC-2 exchanges client information with WLC-1, and the client database entry is moved from WLC-1 to WLC-2. The client will reauthenticate to the network and the client entry will be updated in WLC-2 for AP-2. If the two controllers are not in the same mobility group, a normal 802.11 roaming will occur, but no client information will be exchanged between the two controllers; hence, client IP connectivity may break if reassociation takes longer than expected. Figure 6-14 illustrates the concept of Layer 2 Inter-controller roaming.

FIGURE 6-14 Inter-controller roaming

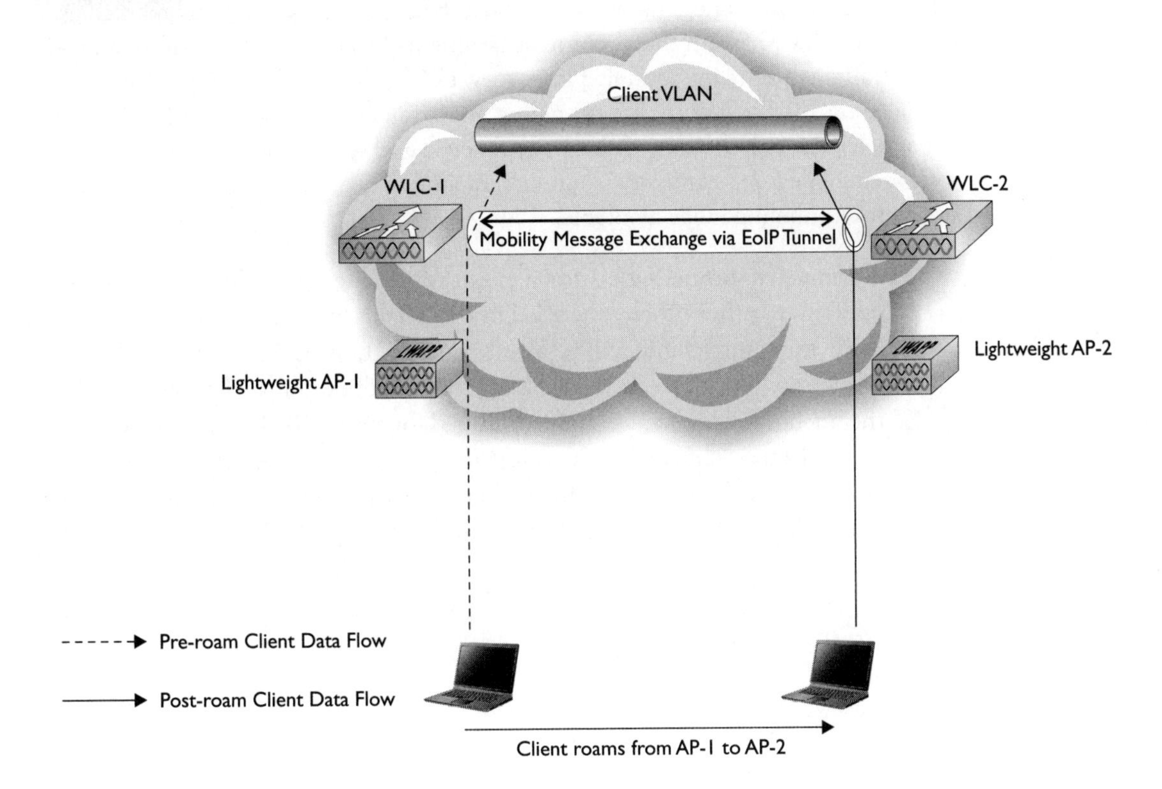

Layer 3 Inter-Controller Roaming

In a Layer 3 inter-controller roaming event, a client roams from one AP (AP-1), which has joined a WLC (WLC-1), to another AP (AP-2), which has joined from a different WLC (WLC-2). The two WLCs bridge client wireless traffic to a different VLAN and IP subnet on the wired network. While the two WLCs map the same WLAN (SSID) traffic to a different VLAN and IP subnet, the client will maintain its IP identity from the original AP (AP-1) and WLC (WLC-1). The two WLCs should be configured with the same mobility group. When the client reassociates to AP-2, WLC-2 exchanges client information with WLC-1, and the client database entry will be copied to WLC-2. WLC-1 will maintain the original client database entry and mark it as "anchor," which signifies that the client has originated initial connection from WLC-1. The client database entry is copied to WLC-2 and will be marked as "foreign" by WLC-2, which simply means that the client roams to an AP on a new WLC while the client maintains its IP identity at the original WLC (WLC-1), which is described as the client anchors its connectivity at WLC-1. If a subsequent Layer 3 inter-controller roam occurs and the client roams to a new AP (AP-3) on a new WLC (WLC-3), the client database entry on WLC-2 will be removed, and a new client database entry will be copied to WLC-3 and marked as foreign on WLC-3. With the client maintaining its IP identity anchored to WLC-1, client traffic is also being tunneled back to WLC-1. There are two tunneling modes available on WLCs: asymmetric tunneling and symmetric tunneling. WLC software releases 4.1 through 5.1 support both asymmetric and symmetric tunneling; in WLC software release 5.2, asymmetric tunneling mode has been removed and leaving only symmetric tunneling mode available.

- **Asymmetric tunneling** After a client roams to a foreign WLC, the client's outbound traffic will be forwarded to the default gateway of the foreign WLC while the return traffic will be sent to anchor the WLC because the client's IP identity resides on the VLAN and the subnet on the anchor WLC. When the anchor WLC receives traffic destined for a roamed client, it looks up the client database and marks the traffic for foreign WLCs. An EoIP tunnel will be created and the client traffic will be forwarded from the anchor WLC to a foreign WLC through the EoIP tunnel. The foreign WLC will forward traffic to AP via LWAPP and then to the client. The asymmetric tunneling may work well for some but not all scenarios. For example, when there is a firewall in the path of client traffic to a server, the firewall may block the asymmetric traffic flow as it may fail firewall's stateful traffic inspection and be mistaken as an attack. Figure 6-15 illustrates the concept of Layer 3 inter-controller roaming using asymmetric tunneling.

FIGURE 6-15 Layer 3 inter-controller roaming using asymmetric tunneling

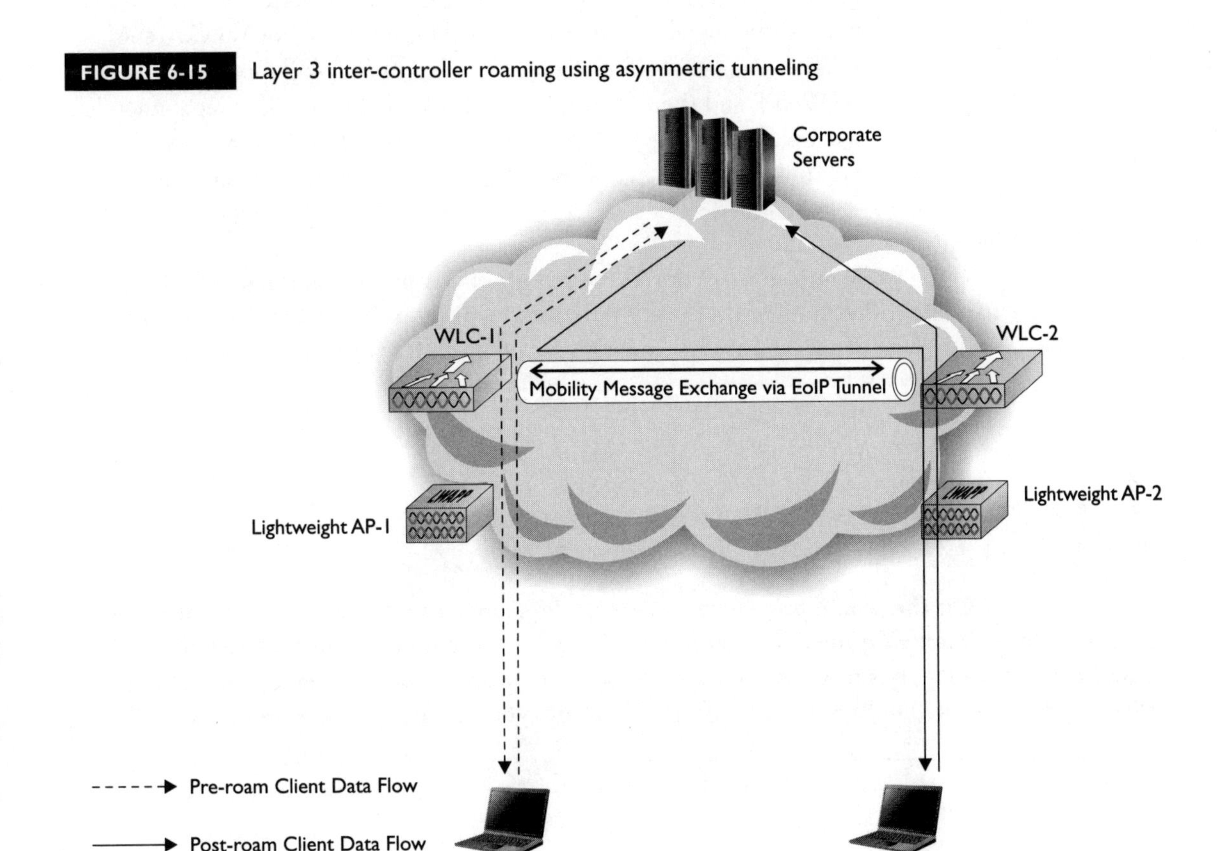

- **Symmetric tunneling** After a client roams to a foreign WLC, the client's outbound traffic will be forwarded back to the anchor WLC using EoIP tunneling and then forwarded to the default gateway of the anchor WLC. The return traffic will be sent to the anchor WLC as the client's IP identity still resides on the VLAN and subnet on the anchor WLC. When the anchor WLC receives traffic destined for a roamed client, it looks up the client database and marks the traffic for the foreign WLC. An EoIP tunnel will be used and the client traffic will be forwarded from the anchor WLC to the foreign WLC through EoIP tunnel. The foreign WLC will forward traffic to AP via LWAPP and then to the client. This tunneling mode is recommended by Cisco and has become the only tunneling mode after WLC release 5.2. Figure 6-16 illustrates the concept of Layer 3 inter-controller roaming using asymmetric tunneling.

If the two controllers are not in the same mobility group, a normal 802.11 roaming will occur, but no client information will be exchanged between the two controllers; hence, client IP connectivity may break if reassociation takes longer than expected.

exam watch

On the exam, pay close attention to questions on the client traffic flow from different types of roaming. The client traffic flow will help you understand the difference in different types of roaming. The wired network design (VLAN and subnet) usually holds the key to questions about types of roaming in a wireless LAN.

FIGURE 6-16 Layer 3 inter-controller roaming using symmetric tunneling

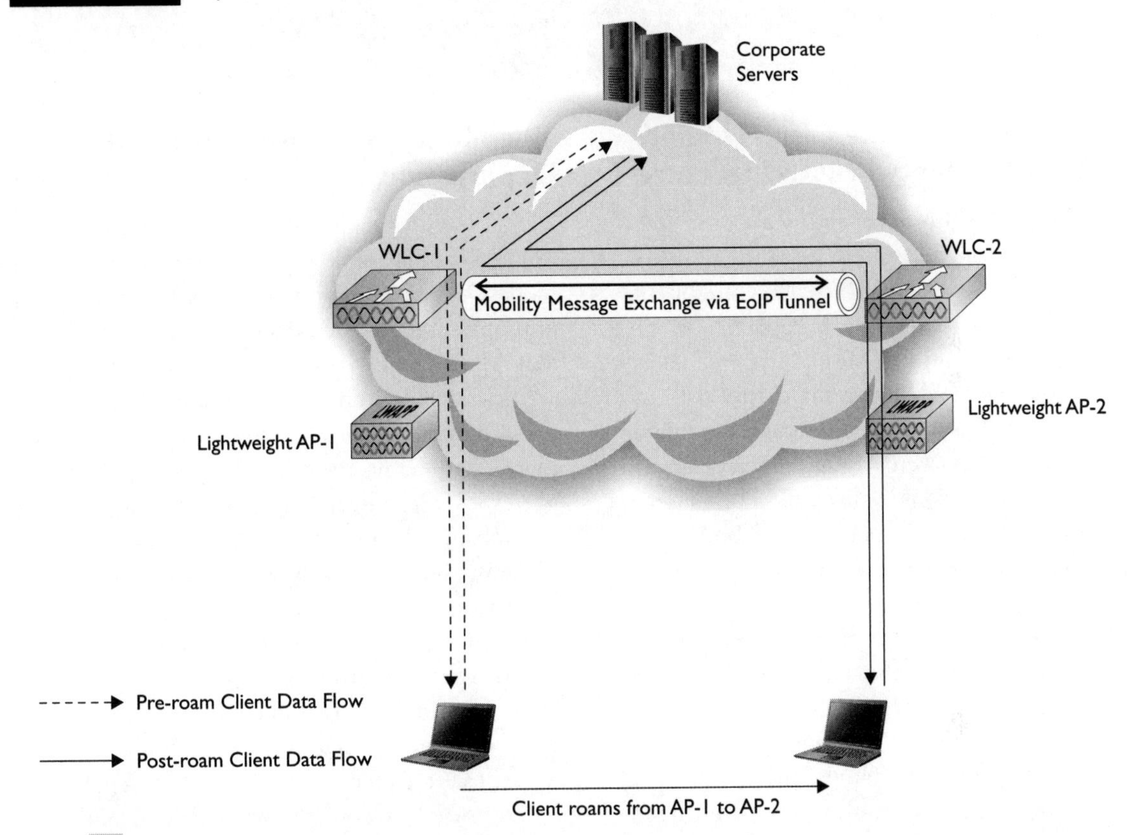

on the job

The LAN and network design usually dictate what type of inter-controller roaming will occur in CUWN. In a typical office environment, Layer 2 inter-controller roaming is normally used, where clients may roam from one AP to another AP within the same floor or building. Layer 3 inter-controller roaming is normally seen where clients may roam from one AP to another AP across floor or building boundaries.

INSIDE THE EXAM

The Basics of the Cisco Unified Wireless Network Architecture

The goal of this exam is to make sure you have a basic understanding of the CUWN architecture and the role each component plays in the solution. It is important to have a basic understanding of the different hardware platform and what they are capable of.

LWAPP is designed to facilitate communications between APs and WLCs. LWAPP breaks away from traditional WLAN deployment model by decoupling 802.11 MAC functionality into real-time sensitive functions and non-real-time sensitive functions and allowing the intelligence to be centralized at the WLC. This is referred to as split MAC architecture. AP will serve real-time MAC services, including transmitting beacons, buffering frames, encrypting and decrypting 802.11 traffic, and gathering real-time RF information. The WLC is mainly responsible for non-real-time MAC services, including authentication, association, and reassociation. The LWAPP is currently being developed in IETF and has been renamed CAPWAP protocol. The exam will be focusing on LWAPP operations, not CAPWAP. It is important to note that WLC release 5.2 and later will be CAPWAP based. This exam will focus on WLC releases 5.0 and 5.1 of operation.

The Modes of Controller-based AP Deployment

CUWN also provides added flexibility to how AP can be deployed. There are several modes APs can be configured to serve special purposes. Local mode APs will provide normal AP services. Hybrid-REAP mode APs provide local switching functionality that allows remote sites to provide local traffic switching to conserve WAN bandwidth while aggregating intelligence on the WLC at the central site. Monitor mode turns APs into RF sensors that gather RF information and provide more granular monitoring of the RF environment. Rogue detector mode turns an AP into a wired sensor to help the system track down rogue APs and clients on the wired network. Sniffer mode APs allow APs to assist with packet capturing and are a great tool for 802.11 troubleshooting. Bridge mode supports mesh AP operations in CUWN.

Controller-based AP Discovery and Association

A lightweight AP cannot operate without a WLC as it does not carry any configurations. LWAPP has defined a discovery and join process for lightweight AP to join a WLC on the network. The AP will follow through a series of steps, including local broadcasting, reviewing

INSIDE THE EXAM

cached WLC information, over-the-air-provisioning, DHCP, and DNS to discover WLCs on the network and build a candidate WLC list. After the discovery process is complete, the AP will go through a series of steps to join a WLC based predefined criteria, including static primary, secondary, and tertiary WLC assignment, the availability of Master Controller status, and the WLC's AP capacity and load factor. The WLC and AP will authenticate each other before completing a join process and encrypt all control messages using the AES algorithm. There are also several redundancy options available for WLCs: 1+1, N+1, N+N, and N+N+1.

Roaming

Mobility group is a setting on WLCs that allows WLCs to share information among a logical group of WLCs. WLCs will create full connections with every other WLC in the mobility group. When the client roams from one AP to another, the WLC will facilitate sharing client information with other WLCs if the new AP has joined a different WLC. WLCs use Ethernet over IP (EoIP) as tunneling technology to exchange mobility information as well as client information. LWAPP is not used by mobility groups. Depending on the WLC design, client inter-controller roaming may be either Layer 2 or Layer 3. In Layer 2 inter-controller roaming, WLCs share the same client VLAN. Client information is copied from the original WLC to the new WLC and removed from the original WLC client database. In Layer 3 inter-controller roaming, the original WLC will mark itself as the anchor WLC and copy the client information to a new WLC, which will then be marked as the foreign WLC. All mobility information exchange is carried by EoIP tunnel. There are two ways of deploying EoIP tunneling: asymmetric tunneling and symmetric tunneling. Cisco recommends symmetric tunneling, and support of asymmetric tunneling is removed from WLC release 5.2.

CERTIFICATION SUMMARY

The foundation of CUWN is based on LWAPP architecture, which breaks up the 802.11 MAC services to allow a more flexible way of designing and deploying wireless networks. The more real-time based 802.11 MAC functions remain on the AP, and the less real-time based 802.11 MAC services are centralized on the WLC.

This allows the wireless network to scale better to support tens and hundreds of AP. Intelligent decisions, such as radio frequency resources management, authentication, and client management, can be processed and optimized by the WLC. Cisco offers a wide variety of WLC and AP platforms that are designed to serve different sized and various types of wireless networks.

As wireless networks continue to grow, scalability becomes a management nightmare to administer tens and hundreds of AP configurations. CUWN utilizes lightweight APs and WLCs to address configuration and scalability concerns in a large wireless network. A lightweight AP by default does not carry any configurations, and it cannot operate without a WLC. Once powered up, a lightweight AP will use discovery and join processes defined in LWAPP to locate a WLC and download a copy of configuration and supported firmware from the WLC. Lightweight AP sends LWAPP Discovery Requests to local subnet, to cached WLCs, to WLCs learned from over-the-air-provisioning, DHCP and DNS to discover WLCs. After LWAPP discovery, AP will attempt to join a WLC based on static primary, secondary, and tertiary WLC assignment, availability of Master Controller, and the WLC's AP capacity.

LWAPP not only simplifies AP configuration and firmware management, it also facilitates 802.11 events, such as client roaming. A mobility group is a group of WLCs, sharing the clients and RF information with one another. Within a mobility group, client information will be copied when roaming occurs across WLC boundaries. In a Layer 2 inter-controller roaming, the original WLC will copy client information to a new WLC and remove the client entry from its client data base. In a Layer 3 inter-controller roaming, the original WLC will copy client information to a new WLC and the new WLC will be referred to as a "foreign" WLC. The original WLC is marked as the "anchor" WLC, and it keeps the client entry in its client data base.

TWO-MINUTE DRILL

The Basics of the Cisco Unified Wireless Network Architecture

- CUWN consists of three major components: a lightweight AP, a Wireless LAN Controller, and a wireless control system.
- Lightweight AP is usually referred to as thin AP, and standalone AP is usually referred to as fat AP.
- Cisco offers a wide variety of WLC platforms, including a standalone WLC appliance and integrated WLCs with a Cisco switch. Cisco also offers a wide variety of AP platforms, including APs with integrated antennas and external antenna connectors. Some APs support 802.3af POE and some are purpose built mesh AP and bridge.
- LWAPP is designed to provide services and messaging that ties CUWN components together. LWAPP is a Cisco proprietary protocol while CAPWAP is the standard based solution created by IETF.
- LWAPP splits 802.11 MAC functions between the AP and WLC. AP handles real-time 802.11 MAC services while the WLC offers non-real-time based MAC services, including authentication, association, and reassociation.

The Modes of Controller-based AP Deployment

- Lightweight AP can be configured in the following modes: local, Hybrid-REAP, monitor, rogue detector, sniffer, and bridge.
- Local mode AP offers normal 802.11 services to clients and scans unused channels and RF.
- Hybrid REAP allows wireless clients to access local resources while allowing the WLC to process authentication and RF information from a central location.
- Hybrid REAP can be deployed in five different states: authentication-central and switch-central; authentication-central and switch-local; authentication-local and switch-local; authentication-down and switch-local; authentication-down and switch-down;
- Monitor mode AP listens to RF traffic passively and does not serve any wireless clients.

- Rogue detector mode AP allows AP to listen to the wired network for MAC addresses of rogue AP and clients. When a rogue AP or client is detected, the WLC will generate an alarm through SNMP trap.
- Sniffer mode AP can act as an RF sniffer and capture 802.11 data to be reviewed by packet analyzer.
- Bridge mode AP acts as a point to point or point to multipoint bridge.

Controller-based AP Discovery and Association

- Lightweight AP does not store configuration in its NVRAM. When AP is powered up it goes through LWAPP discovery and join processes to download configuration and firmware from the WLC.
- The LWAPP discovery process consists of the following steps: broadcast Discovery Request to local subnet; send Discovery Request to the WLC learned over the air; send Discovery Request to the cached WLC; send Discovery Request to the WLC learned from DHCP Option 43; send Discovery Request to the WLC learned through DNS resolution. The AP will build a WLC candidate list and start the LWAPP join process after it completes the LWAPP discovery process.
- Lightweight AP selects a WLC to join based on the following: static assignment of primary, secondary, and/or tertiary WLCs; Master Controller; the WLC with the least AP loads.
- APs and WLCs authenticate each other using an X.509 certificate, and they encrypt LWAPP control messages using AES encryption.
- WLCs can be deployed with the following redundancy options: 1+1, N+1, N+N, N+N+1.

Roaming

- Mobility groups allow WLCs to share information among a logical group of WLCs. The WLC will create fully meshed Ethernet over IP tunnels to connect to all other WLCs within a mobility group, exchange mobility messages, and allow for roaming without the loss of the IP configuration.
- With intra-controller roaming, a client roams from one AP to another AP, and both APs join the same WLC.

- With inter-controller roaming, client roams from one AP to another AP, and the APs join different WLCs.
- There are two types of inter-controller roaming: Layer 2 inter-controller roaming and Layer 3 inter-controller roaming. WLCs exchange mobility messages using EoIP tunneling to facilitate sharing of client information for the roaming event.
- When a client roams from one AP to another AP on different WLCs but bridges client traffic to the same VLAN, it is referred to as a Layer 2 inter-controller roaming.
- When a client roams from one AP to another AP on different WLCs but bridges client traffic to a different VLAN, it is referred to as a Layer 3 inter-controller roaming. The client will maintain its IP identity and EoIP tunneling will be used to exchange client mobility information and forward client traffic.
- Layer 3 inter-controller roaming can use asymmetric EoIP tunneling or symmetric EoIP tunneling. Cisco recommends symmetric EoIP tunneling.

SELF TEST

The following self test questions will help you measure your understanding of the material presented in this chapter. Read all the choices carefully, as there may be more than one correct answer. Choose all correct answers for each question.

The Basics of the Cisco Unified Wireless Network Architecture

1. What type of AP is part of CUWN and managed by the WLC?
   - A. Fat AP
   - B. Autonomous AP
   - C. Standalone AP
   - D. Lightweight AP
2. Which protocol facilitates communications between the AP and WLC and transport client traffic between the AP and WLC?
   - A. TCP
   - B. LWAPP
   - C. HTTP
   - D. STP
3. What kind of 802.11 services do WLCs provide? (Choose all that apply.)
   - A. Beaconing
   - B. Association
   - C. Reassociation
   - D. Encryption
   - E. Authentication
4. What kind of 802.11 services do AP provide? (Choose all that apply.)
   - A. Probe response
   - B. Authentication
   - C. Reassociation
   - D. Decryption
   - E. Association

5. Which LWAPP feature improves wireless network scalability and allows the network to manage more APs?
   A. Split MAC
   B. 802.11i
   C. EoIP Tunneling
   D. LWAPP tunneling

The Modes of Controller-based AP Deployment

6. Which AP mode puts AP in passive mode and scanning only?
   A. Hybrid-REAP
   B. Local
   C. Bridge
   D. Monitor
7. Which AP mode will serve client and collect RF information?
   A. Normal mode
   B. Monitor mode
   C. Operation mode
   D. Local mode

Controller-based AP Discovery and Association

8. Which DHCP option is used to push WLC information to a lightweight AP?
   A. Option 150
   B. Option 43
   C. Option 82
   D. Option 60
9. Which is not a step of the LWAPP discovery process for an AP to learn the IP address of a WLC?
   A. Learn the WLC IP address through DHCP Option 43.
   B. Learn the WLC IP address through Master Controller.
   C. Learn the WLC IP address through the DNS default WLC name of CISCO-LWAPP-CONTROLLER.localdomain.
   D. Learn the WLC IP address through OTAP.

10. What does AP do with the WLC information it receives through the LWAPP discovery process and prior to entering LWAPP join process?
    A. AP tries to match the WLC with its primary, secondary, and/or tertiary WLC assignments.
    B. AP builds a candidate list of WLCs.
    C. AP tries to determine which WLC is the Master Controller.
    D. AP tries to resolve IP addresses and hostnames of WLCs.
11. A lightweight AP connects to the network and obtains an IP address through DHCP. What step does the AP go through next in the LWAPP discovery process after it sends out the LWAPP Discovery Request in local subnet?
    A. Send Discovery Request to the WLC learned from OTAP
    B. DHCP Option 43
    C. Layer 2 LWAPP broadcast
    D. DNS resolution for CISCO-LWAPP-CONTROLLER.localdomain
12. A lightweight AP connects to the network and obtains an IP address through DHCP. What step does the AP go through next in the LWAPP discovery process after it sends out the LWAPP Discovery Request to the WLC learned from OTAP?
    A. Send Discovery Request to the WLC AP previously learned
    B. Layer 3 LWAPP broadcast
    C. Layer 2 LWAPP broadcast
    D. DNS resolution for CISCO-LWAPP-CONTROLLER.localdomain
13. What server platform(s) will support WCS for managing CUWN and its components? (Choose all that apply.)
    A. Red Hat Linux
    B. MAC OS X
    C. Windows Server
    D. SUSE Linux Enterprise
14. What completes an LWAPP join process for an AP to join a WLC?
    A. The AP and WLC authenticate each other and encrypt traffic using AES.
    B. The AP and WLC complete 802.1x authentication and communicate using WPA2.
    C. The AP and WLC communicate per 802.11i and encrypt the communications using AES.
    D. The AP and WLC create an EoIP tunnel and encrypt traffic using AES.

15. How do an AP and a WLC authenticate each other in the LWAPP join process?
    A. They use an x.509 certificate.
    B. They use IPSec preshare secret.
    C. They use RSA nounce exchange through a RADIUS server.
    D. They verify a hash using MD5.

Roaming

16. What type of tunneling protocol is used to facilitate exchange of mobility messages?
    A. Ethernet over IP tunneling
    B. LWAPP tunneling
    C. IPSec tunneling
    D. GRE tunneling
17. What type of inter-controller roaming is where client traffic will be tunneled to the anchor WLC and forwarded to the network?
    A. Layer 2 roaming
    B. Asymmetric tunneling
    C. Layer 3 roaming
    D. LWAPP tunneling
18. What type of inter-controller roaming is where client traffic to the network will be forwarded by the foreign WLC and returned traffic will be sent to the anchor WLC and forwarded to the foreign WLC?
    A. Layer 2 roaming
    B. Asymmetric tunneling
    C. Symmetric tunneling
    D. LWAPP tunneling

SELF TEST ANSWERS

The Basics of the Cisco Unified Wireless Network Architecture

1. What type of AP is part of CUWN and managed by the WLC?

A. Fat AP

B. Autonomous AP

C. Standalone AP

D. Lightweight AP

☑ **D.** In CUWN, lightweight AP is managed by the WLC.

☒ **A, B,** and **C** are incorrect. Fat AP, autonomous AP, and standalone AP all refer to the same type of AP.

2. Which protocol facilitates communications between the AP and WLC and transport client traffic between the AP and WLC?

A. TCP

B. LWAPP

C. HTTP

D. STP

☑ **B.** LWAPP facilitates communications between the AP and WLC and transports client traffic between the AP and WLC.

☒ **A, C,** and **D** are incorrect. TCP, HTTP, and STP are not used for AP and WLC communications.

3. What kind of 802.11 services do WLCs provide? (Choose all that apply.)

A. Beaconing

B. Association

C. Reassociation

D. Encryption

E. Authentication

☑ **B, C,** and **E.** The non-real-time 802.11 services are provided by the WLC.

☒ **A** and **D** are incorrect because beaconing and encryption are both real-time services provided through the AP.

4. What kind of 802.11 services do AP provide? (Choose all that apply.)

A. Probe response

B. Authentication

C. Reassociation

D. Decryption

E. Association

☑ **A** and **D**. Beaconing and decryption are both real-time services provided through AP.

☒ **B**, **C**, and **E** are incorrect. Authentication, reassociation, and association are non-real-time 802.11 services provided by the WLC.

5. Which LWAPP feature improves wireless network scalability and allows the network to manage more APs?

A. Split MAC

B. 802.11i

C. EoIP Tunneling

D. LWAPP tunneling

☑ **A**. Split MAC is the key feature that allows the network to be more scalable and support more APs.

☒ **B**, **C**, and **D** are incorrect. 802.11i, EoIP tunneling, and LWAPP tunneling are generic features used by CUWN.

The Modes of Controller-based AP Deployment

6. Which AP mode puts AP in passive mode and scanning only?

A. Hybrid-REAP

B. Local

C. Bridge

D. Monitor

☑ **D**. Monitor mode puts AP in passive mode and scans only.

☒ **A**, **B**, and **C** are incorrect. Hybrid REAP supports remote site WLAN access; bridge mode primarily supports outdoor and indoor mesh AP; and local mode is the normal operation mode for lightweight AP.

7. Which AP mode will serve client and collect RF information?
   A. Normal mode
   B. Monitor mode
   C. Operation mode
   D. Local mode

☑ **D.** Local mode is the AP's normal mode of operation. The AP will serve a client as well as scanning unused channels when configured in local mode.

☒ **A, B,** and **C** are incorrect because normal mode and operation mode are not modes of AP operations. AP in monitor mode does not serve clients.

Controller-based AP Discovery and Association

8. Which DHCP option is used to push WLC information to a lightweight AP?
   A. Option 150
   B. Option 43
   C. Option 82
   D. Option 60

☑ **B.** DHCP Option 43 provides WLC information to the AP.

☒ **A, C,** and **D** are incorrect. Options 150, 82, and 60 are not used.

9. Which is not a step of the LWAPP discovery process for an AP to learn the IP address of a WLC?
   A. Learn the WLC IP address through DHCP Option 43.
   B. Learn the WLC IP address through Master Controller.
   C. Learn the WLC IP address through the DNS default WLC name of CISCO-LWAPP-CONTROLLER.localdomain.
   D. Learn the WLC IP address through OTAP.

☑ **B.** Learning the WLC IP address through the Master Controller is not part of the LWAPP discovery process.

☒ **A, C,** and **D** are incorrect because learning the WLC IP address through DHCP Option 43, learning the WLC IP address through the DNS default WLC name of CISCO-LWAPP-CONTROLLER.localdomain, and learning the WLC IP address through OTAP are all part of the LWAPP discovery process.

10. What does AP do with the WLC information it receives through the LWAPP discovery process and prior to entering LWAPP join process?
    A. AP tries to match the WLC with its primary, secondary, and/or tertiary WLC assignments.
    B. AP builds a candidate list of WLCs.

C. AP tries to determine which WLC is the Master Controller.

D. AP tries to resolve IP addresses and hostnames of WLCs.

☑ **B.** The AP builds a candidate list of WLCs after it completes the LWAPP discovery process and prior to entering LWAPP join process.

☒ **A, C,** and **D** are incorrect. The AP tries to match the WLC with its primary, secondary, and/or tertiary WLC assignments as part of the LWAPP join process. The AP tries to determine which WLC is the Master Controller as part of the LWAPP join process. The AP tries to resolve the IP addresses of WLCs as part of the LWAPP discovery process.

11. A lightweight AP connects to the network and obtains an IP address through DHCP. What step does the AP go through next in the LWAPP discovery process after it sends out the LWAPP Discovery Request in local subnet?

A. Send Discovery Request to the WLC learned from OTAP

B. DHCP Option 43

C. Layer 2 LWAPP broadcast

D. DNS resolution for CISCO-LWAPP-CONTROLLER.localdomain

☑ **A.** The AP sends a Discover Request to the WLC learned from OTAP after it sends out a local broadcast LWAPP Discovery Request.

☒ **B, C,** and **D** are incorrect. The lightweight AP tries to discover a WLC using OTAP before it tries DHCP Option 43and DNS resolution for CISCO-LWAPP-CONTROLLER.localdomain. When a lightweight AP obtains an IP through DHCP, it does not use Layer 2 LWAPP broadcast to discover a WLC.

12. A lightweight AP connects to the network and obtains an IP address through DHCP. What step does the AP go through next in the LWAPP discovery process after it sends out the LWAPP Discovery Request to the WLC learned from OTAP?

A. Send Discovery Request to the WLC AP previously learned

B. Layer 3 LWAPP broadcast

C. Layer 2 LWAPP broadcast

D. DNS resolution for CISCO-LWAPP-CONTROLLER.localdomain

☑ **A.** The AP sends a Discovery Request to the WLC the AP previously learned was the correct answer after it sent the LWAPP Discovery Request to the WLC learned from OTAP.

☒ **B, C,** and **D** are incorrect. AP broadcasts an LWAPP Discovery Request on the local subnet before OTAP. When a lightweight AP obtains an IP through DHCP, it does not use Layer 2 LWAPP broadcast to discover a WLC. DNS resolution for CISCO-LWAPP-CONTROLLER.localdomain is used after the lightweight AP unicasts LWAPP Discovery Request messages to WLCs whose IP addresses are stored in the AP NVRAM.

13. What server platform(s) will support WCS for managing CUWN and its components? (Choose all that apply.)

A. Red Hat Linux

B. MAC OS X

C. Windows Server

D. SUSE Linux Enterprise

☑ **A** and **C**. WCS is supported on Red Hat Linux and Windows Server.

☒ **B** and **D** are incorrect because MAC OS X and SUSE Linux are not supported server platforms for WCS.

14. What completes an LWAPP join process for an AP to join a WLC?

A. The AP and WLC authenticate each other and encrypt traffic using AES.

B. The AP and WLC complete 802.1x authentication and communicate using WPA2.

C. The AP and WLC communicate per 802.11i and encrypt the communications using AES.

D. The AP and WLC create an EoIP tunnel and encrypt traffic using AES.

☑ **A**. The WLC and AP will authenticate each other and encrypt control traffic using AES to complete the LWAPP join process.

☒ **B** **C**, and **D** are incorrect because 802.1x, 802.11i, and EoIP are used by the system during various other operations, but not the LWAPP join process.

15. How do an AP and a WLC authenticate each other in the LWAPP join process?

A. They use an x.509 certificate.

B. They use IPSec preshare secret.

C. They use RSA nounce exchange through a RADIUS server.

D. They verify a hash using MD5.

☑ **A**. A x509 certificate is used by the WLC and AP to authenticate each other.

☒ **B** **C**, and **D** are incorrect because the IPSec preshared secret, RSA nounce, and MD5 hash are not used for WLC and AP authentication.

Roaming

16. What type of tunneling protocol is used to facilitate exchange of mobility messages?

A. Ethernet over IP tunneling

B. LWAPP tunneling

C. IPSec tunneling

D. GRE tunneling

☑ **A.** Ethernet over IP (EoIP) tunneling is used for exchange of mobility messages.

☒ **B**, **C**, and **D** are incorrect. LWAPP tunneling, IPSec tunneling, and GRE tunneling are not used for exchange of mobility messages.

17. What type of inter-controller roaming is where client traffic will be tunneled to the anchor WLC and forwarded to the network?

A. Layer 2 roaming

B. Asymmetric tunneling

C. Layer 3 roaming

D. LWAPP tunneling

☑ **C.** In Layer 3 roaming, client traffic will be tunneled to the anchor WLC and forwarded to the network.

☒ **A**, **B**, and **D** are incorrect. In Layer 2 roaming, anchor WLC is not used. In asymmetric tunneling, client traffic hits the network from the foreign WLC, not the anchor WLC. LWAPP tunneling is not utilized in client roaming events.

18. What type of inter-controller roaming is where client traffic to the network will be forwarded by the foreign WLC and returned traffic will be sent to the anchor WLC and forwarded to the foreign WLC?

A. Layer 2 roaming

B. Asymmetric tunneling

C. Symmetric tunneling

D. LWAPP tunneling

☑ **B.** Client traffic to the network will be forwarded by the foreign WLC, and returned traffic will be sent to the anchor WLC and forwarded to the foreign WLC when asymmetric tunneling is used.

☒ **A**, **C**, and **D** are incorrect. In Layer 2 roaming, anchor or foreign WLCs are not used. In symmetric tunneling, the anchor WLC will used for both directions of traffic to and from the network. LWAPP tunneling is not utilized in client roaming events.

7

Understanding Cisco Mobility Express Solution

CERTIFICATION OBJECTIVES

To help navigate through complex business environments and to serve the needs of small- and medium-sized businesses (SMB), Cisco offers a line of communications solution specifically designed for SMBs: the Cisco Smart Business Communications (CSBC) System. CSBC includes a suite of communications solutions designed to provide integrated voice, video, security, and wireless access to SMB. CSBC includes the following components:

- Cisco ESW 500 Series switch
- Cisco 520 Series secure router
- Cisco Unified Communications 500 Series
- Cisco 7900 IP phone
- Cisco Mobility Express Solution
- Cisco Configuration Assistant

In this chapter, we will focus our discussion on the Cisco Mobility Express Solution and its components. We will also briefly discuss configurations of Cisco Mobility Express Solution. In Chapter 8, we will discuss design, configurations, and implementing large enterprise wireless networks using CUWN with full-featured Cisco wireless LAN controllers and lightweight APs.

CERTIFICATION OBJECTIVE 7.01

Cisco Mobility Express Wireless Architecture

The Cisco Mobility Express Solution is specifically designed to offer wireless connectivity in the SMB environment using the Cisco Unified architecture. However, it does not support the full Cisco Enterprise class CUWN functionality. The Cisco Mobility Express Solution integrates with the wired infrastructure and like the full CUWN solution, the Radio Resource Management (RRM) feature delivers self-configuration, self-optimization, and self-healing, minimizing manual RF management. The concept of RRM will be discussed in more details in Chapter 8. The Cisco Mobility Express Solution also integrates with CSBC management tools for friendly user interface and simplified configurations. There are three components in the Cisco Mobility Express Solution:

- Cisco Wireless Express 521 access point
- Cisco Wireless Express Mobility 526 controller
- Cisco Configuration Assistant

exam watch

CSBC is Cisco's solution for SMB customers. It is intended to be a complete solution supporting voice, video, security, and wireless. It is important to know the different components in the CSBC and to understand the role of the Cisco Configuration Assistant and how it interoperates with all the different components in CSBC, but it is not necessary to memorize all the different products that go into the voice, video, and security solution. It is also important to keep in mind that CSBC does not interoperate with CUWN product suites. The CSBC Wireless Controller and AP cannot work with the Cisco 4400 Series and WiSM controllers. The Cisco 1240/1140/1250 Series Lightweight AP cannot join the Cisco Wireless Express Mobility 526 controller.

Cisco Wireless Express 521 Access Point

The Cisco Wireless Express 521 access point is a key part of the Cisco Mobility Express Solution and is a single-band 802.11b/g AP designed for SMB environments with carpeted offices or similar indoor environments. The 521 AP offers enterprise-class security and performance, and it can be easily deployed and managed with the Cisco Configuration Assistant. Furthermore, to accommodate a wide range of deployment and business requirements, the 521 AP can be deployed in two modes: standalone or controller mode.

- **Standalone mode** When APs are in standalone mode, up to three APs can be deployed without a controller to provide wireless connectivity between wireless devices and the wired network. In this configuration, APs have to be managed individually through the Cisco Configuration Assistant. The AP can be upgraded from standalone mode to controller mode.

- **Controller mode** When APs are in controller mode, up to 12 APs can be deployed across two Cisco Wireless Express Mobility 526 controllers; each can host up to 6 APs. Similar to CUWN lightweight APs, the Cisco Wireless Express 521 APs can monitor all wireless activities and report them to the Cisco Wireless Express Mobility 526 controller(s). APs in controller mode can also be managed through the Cisco Configuration Assistant. Because the 521 AP was purposely built for SMB, it is not capable of joining other full-function Cisco controllers.

Figure 7-1 shows a photo of the Cisco 521 AP, which shares a similar low profile design to the Cisco Lightweight AP 1130. The 521 AP is equipped with integrated diversity antennas and can be powered by IEEE 802.3af PoE and Cisco prestandard PoE. It supports both WPA and WPA2 with a hardware-based encryption engine.

At the time of this writing, there is no 802.11n-based AP in the Cisco Mobility Express Solution, and Cisco 521 AP does not interoperate with any Cisco Wireless LAN Controllers or lightweight APs other than 526 controllers.

FIGURE 7-1

Cisco Wireless Express 521 Access Point (Courtesy of Cisco Systems, Inc. Unauthorized use not permitted.)

exam watch

It is crucial to remember that the Cisco Mobility Express Solution is a complete solution by itself. While sharing the same foundation technology and architecture, the product family does not interoperate with CUWN controllers or lightweight APs. The 521 AP does not support 802.11n or 802.11a. The only way to support 802.11n and advanced AP features is to upgrade the whole solution to CUWN-based WLC and lightweight AP.

Cisco Wireless Express Mobility 526 Controller

The Cisco 526 Wireless Express Mobility controller is the heart of the Cisco Mobility Express Solution. The 526 controller is a network appliance, and it supports both basic wireless connectivity and advanced mobility services. It is specifically designed to manage 521 AP and it fits most needs of SMB environments. The 526 controller supports WEP, WPA, WPA2, and 802.1X with EAP. Because the 526 controller only supports a subset of CUWN, it cannot communicate or interoperate with full-featured WLC or lightweight AP. Figure 7-2 shows a photo of Cisco 526 controller.

FIGURE 7-2 Cisco Wireless Express Mobility 526 controller (Courtesy of Cisco Systems, Inc. Unauthorized use not permitted.)

The 526 controller supports a subset of CUWN features and services supported by full functioned Cisco WLC, including wireless guess access, voice over WLAN, Radio Resource Management (RRM), and a wide range of security services. Here is a partial list of features that are *not* supported on the Cisco 526 controller:

- RADIUS accounting
- TACACS+
- Link Aggregation (LAG)
- Management Frame Protection (MFP)
- DHCP Server
- SNMP v3
- Wireless IDS

Cisco Configuration Assistant

The Cisco Configuration Assistant (CCA) is an integral component of the CSBC that includes voice, security, and wireless solutions. SMB organizations can provide employees and customers with highly secure and easy access to information for communications needs. The CCA provides configuration, deployment, and network management support for the Wireless Mobility Express Solution, including the Wireless Express Mobility 526 controller and the Cisco Wireless Express 521 AP, in both standalone and controller modes. CCA is a shared component by CSBC, but it is not interoperable with CUWN components.

Implementing Cisco Mobility Express Solution

Cisco Mobility Express Solution implements a subset of the LWAPP protocol; thus, the solution is similar to the CUWN solution, but it does not have full features and functionalities of the full enterprise class version of the CUWN solution. The configuration and implementation of Cisco Mobility Express Solution is slightly different from CUWN because CCA plays an important role interacting with all components of the solution.

Configuring Cisco 526 Controller and 521 Access Point

There are three different ways you can configure the Cisco Express Mobility Solution: with the controller console command line interface (CLI), the controller graphic user interface (GUI), or the Cisco Configuration Assistant (CCA). The 521 AP has no CLI interface, so it can only be managed through 526 controller or Cisco Configuration Assistant. The 521 AP is a plug-and play device and will obtain an IP address through DHCP. In controller mode, 521 AP should use DHCP option 43 for controller discovery. The AP will use DHCP option 60, Vendor Class Identifier (VCI), to inform the DHCP server that it is a Cisco 521 AP, so that the DHCP server can correctly return controller information using option 43. The VCI is a text string that uniquely identifies a type of vendor device, and AP 521 uses "Cisco AP c520" as its VCI.

Configure the Cisco 526 Controller Using CLI The CLI interface of the Cisco 526 controller provides similar functionality to CLI for a full-featured WLC. The default IP address for the management interface is 192.168.1.1. After connecting to the 526 controller console port, you can use the same **show sysinfo** and **show interface summary** commands to review basic information about the 526 controller.

```
Cisco Controller) >show sysinfo
Manufacturer's Name.............................. Cisco Systems Inc.
Product Name..................................... Cisco Controller
Product Version.................................. 4.2.61.8.7
RTOS Version..................................... 4.1.154.22
Bootloader Version............................... 4.0.191.0
Build Type....................................... DATA + WPS
System Name...................................... WLC-520-1
System Location..................................
System Contact...................................
System ObjectID.................................. 1.3.6.1.4.1.9.1.828
IP Address....................................... 192.168.18.151
System Up Time................................... 0 days 22 hrs 32 mins 39 secs
Configured Country............................... US  - United States
Operating Environment............................ Commercial (0 to 40 C)
```

```
Internal Temp Alarm Limits....................... 0 to 65 C
Internal Temperature............................. +51 C
State of 802.11b Network......................... Enabled
State of 802.11a Network......................... Enabled
--More-- or (q)uit
Number of WLANs.................................. 1
3rd Party Access Point Support................... Disabled
Number of Active Clients......................... 0
Burned-in MAC Address............................ 00:1B:D5:00:4C:20

(Cisco Controller) >show interface summary

Interface Name          Port  Vlan Id   IP Address       Type     Ap Mgr
--------------------    ----  --------  ---------------  -------  ------
ap-manager               1    untagged  192.168.18.152   Static   Yes
management               1    untagged  192.168.18.151   Static   No
virtual                 N/A   N/A       1.1.1.1          Static   No
```

The 526 controller CLI can also be accessed using SSH and Telnet. SSH is enabled by default, and Telnet is disabled by default. Although the 526 controller shares the same CLI as full-featured CUWN controllers, some functionalities are not available on the 526 controller CLI, and it is not recommended for configurations. Instead, the GUI and Cisco Configuration Assistant should be the primary methods for managing configurations on a 526 controller.

While it is command driven and similar to the Cisco IOS CLI, the Cisco controller uses a different style CLI. The command menu differs significantly from an IOS-style CLI, and there are no User EXEC and Privileged EXEC modes in the controller CLI. The use `tab` ***function and question mark (***`?`***) are still available, but the controller CLI does not edit users to change part of the CLI input in each command entry.***

Configure the Cisco 526 Controller Using GUI The GUI interface of the Cisco 526 controller provides the same look and feel as the GUI for a full-featured WLC. The GUI can be accessed by using HTTPS and a web browser. Enter the IP address of the management interface of the 526 controller into a web browser. The 526 controller supports both HTTP and HTTPS, but HTTP is disabled by default. Figure 7-3 illustrates the GUI interface of a 526 controller.

Figure 7-4 illustrates the GUI interface of a WLAN (SSID) configuration and some of the options available for the SSID.

FIGURE 7-3 Cisco Wireless Express Mobility 526 controller GUI summary page

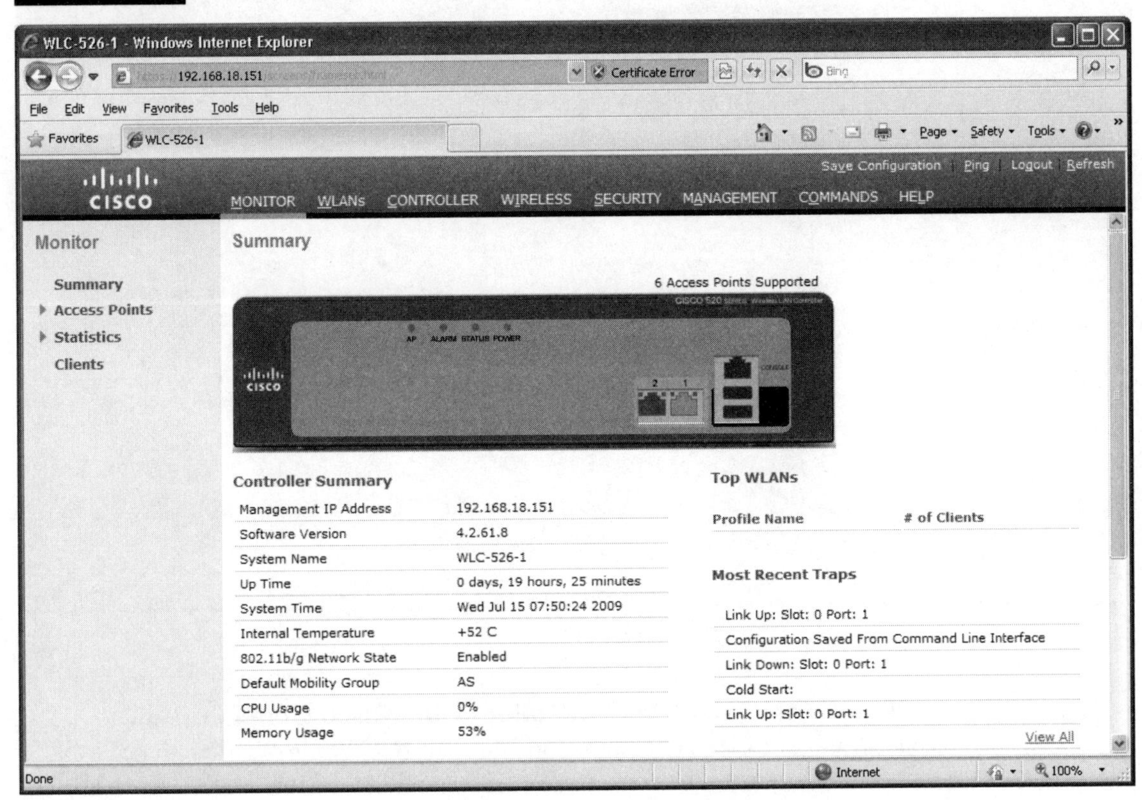

FIGURE 7-4 Cisco Wireless Express Mobility 526 controller GUI WLAN page

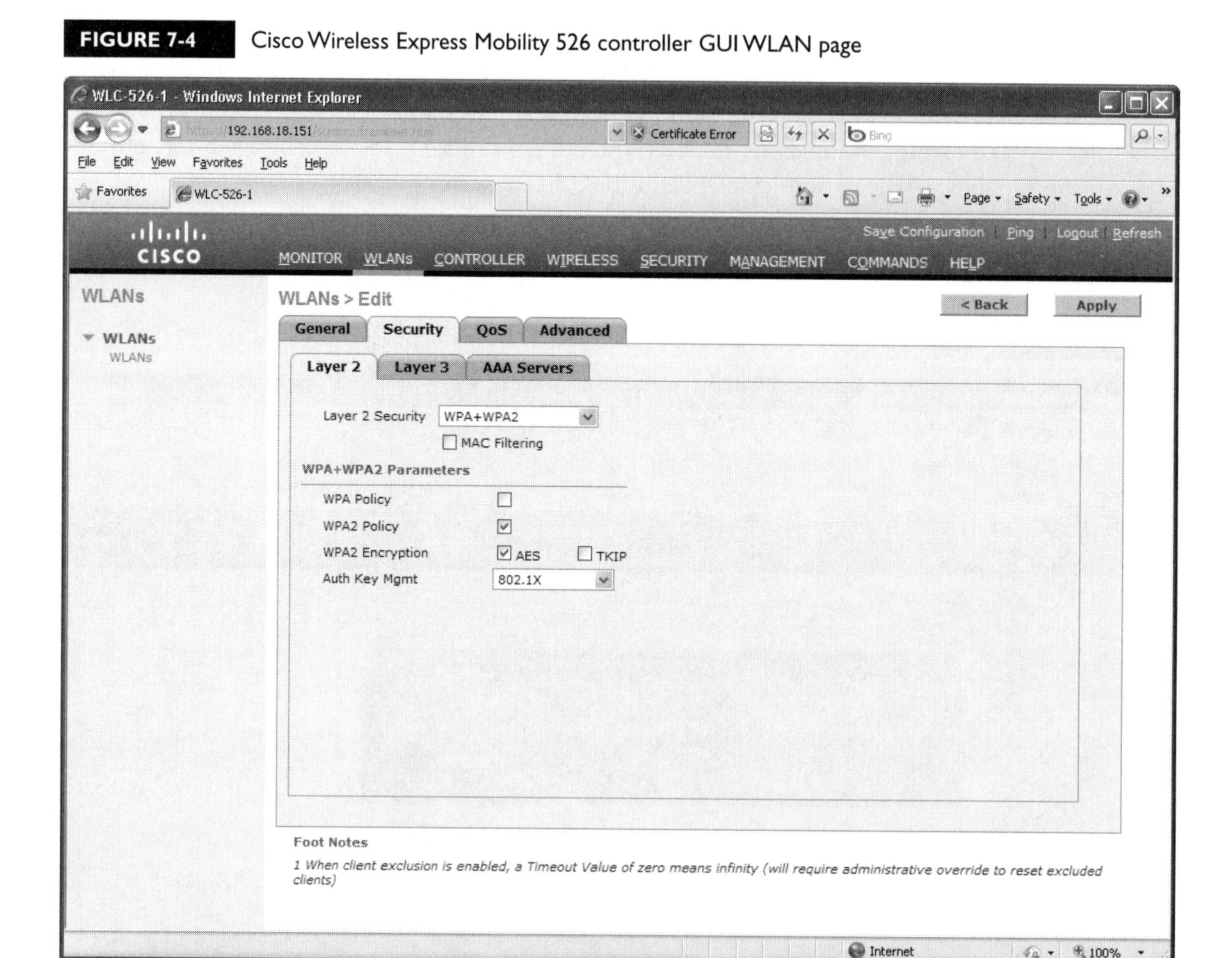

Configure the Cisco 526 Controller Using Cisco Configuration Assistant The Cisco Configuration Assistant (CCA) is a Windows-based application that can be used to manage CSBC and Cisco Mobility Express Solution. The CCA is the only tool that can manage 521 AP for both standalone and controller modes. In addition, the CCA can also migrate standalone 521 AP to controller mode. The CCA provides a graphics interface and wizard tools to lead the user through different configuration tasks for both the 526 controller and 521 AP. Users can drag and drop devices to build a network topology or right-click on the device to perform configuration tasks. Figure 7-5 illustrates the device wizard that steps through the setup process for connecting an AP 521 and a 526 controller to the CSBC.

FIGURE 7-5 Cisco Configuration Assistant Device Setup Wizard

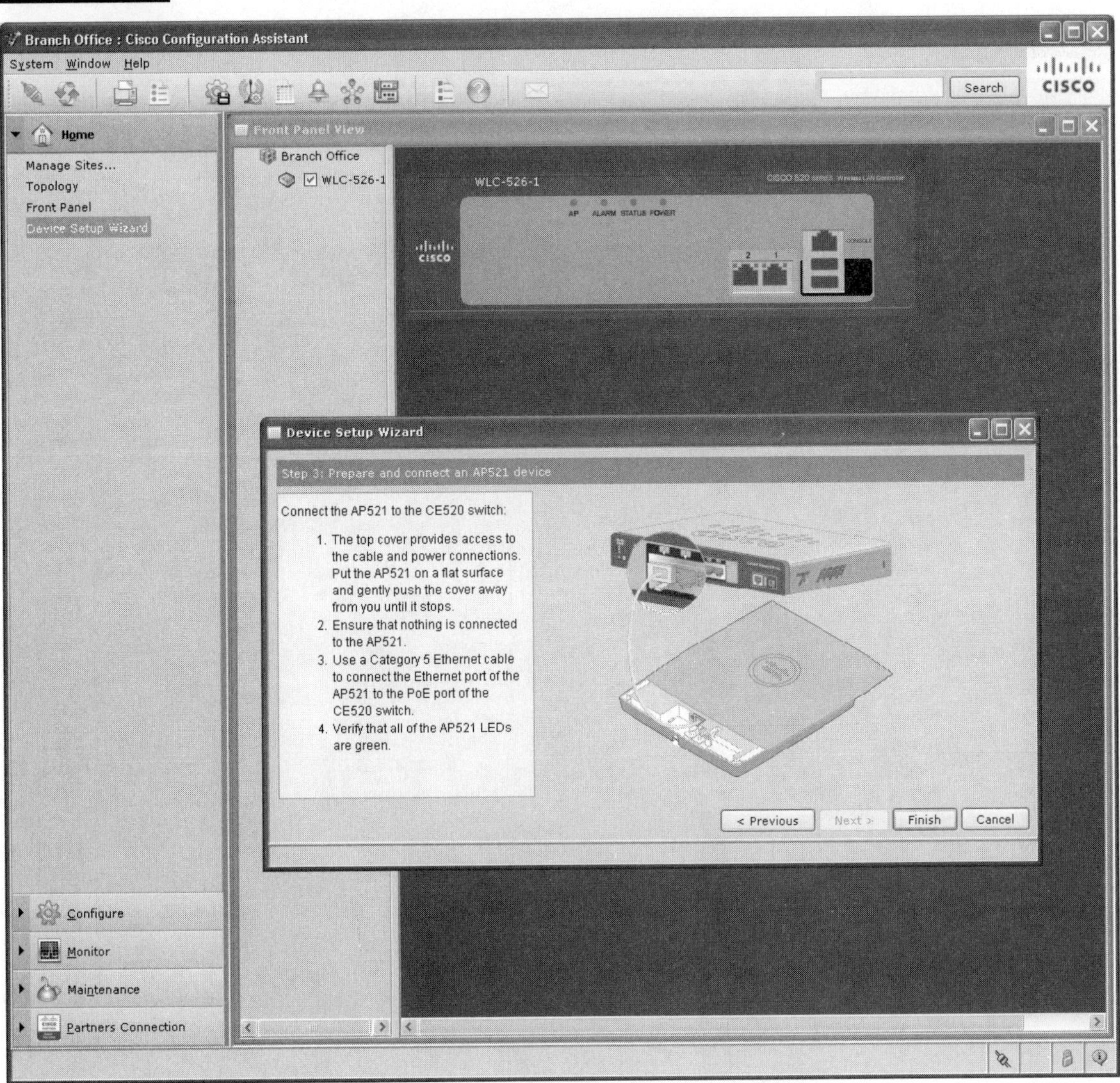

Similar configuration tools will also manage 526 controller configurations. Figure 7-6 illustrates the WLAN (SSID) setup for the 526 controller through CCA. The WLAN setup configurations illustrated are comparable to the configurations illustrated in Figure 7-4. The same configuration options are presented in CCA for configuring WLAN in the 526 controller.

FIGURE 7-6 Cisco Configuration Assistant WLAN configuration

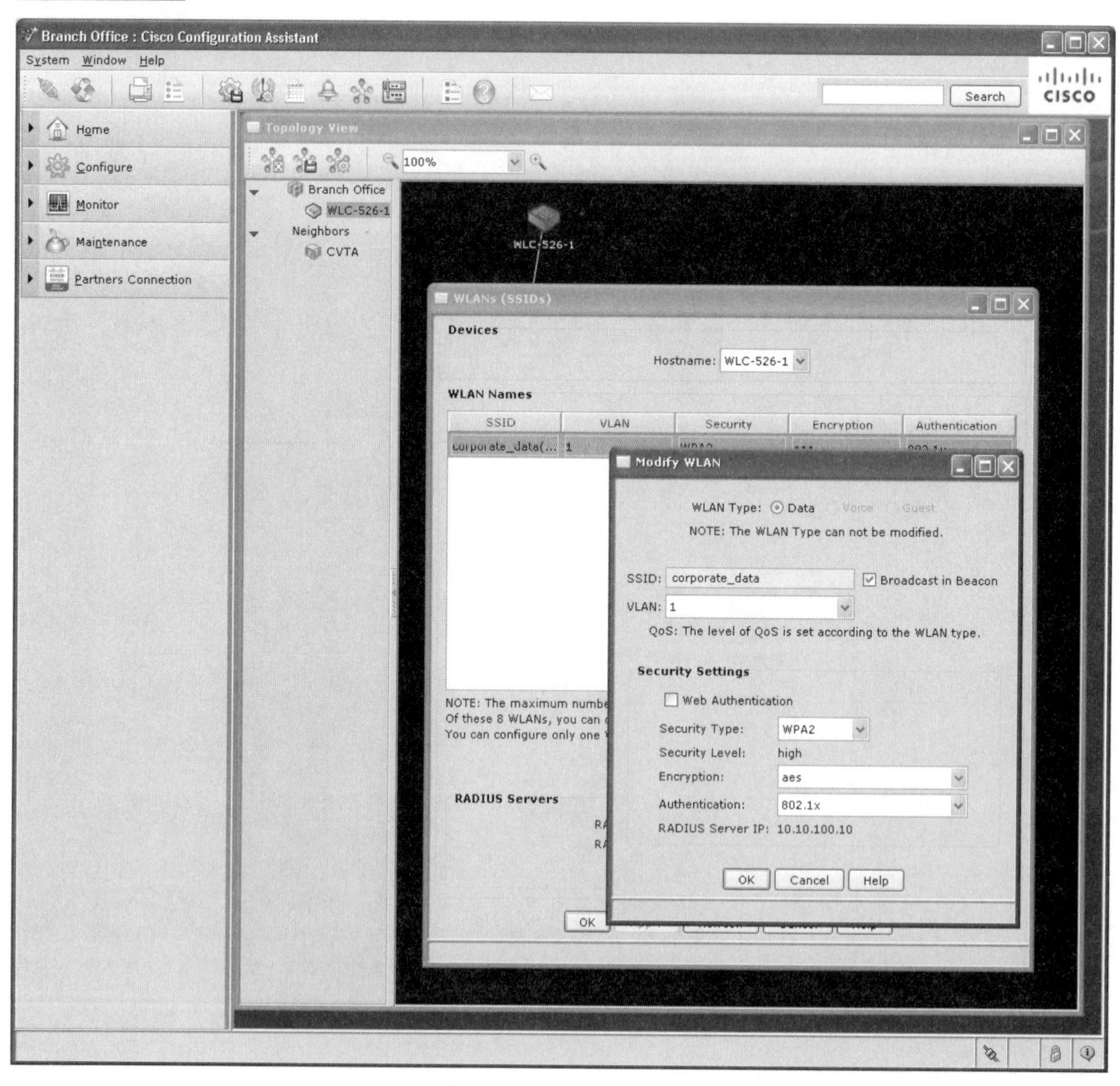

In addition to configuration and device management, the CCA is also the network management tool for CSBC components. Figure 7-7 illustrates the event reporting and system monitoring view of Cisco Mobility Express components on the CCA.

FIGURE 7-7 Cisco Configuration Assistant system management and event monitoring

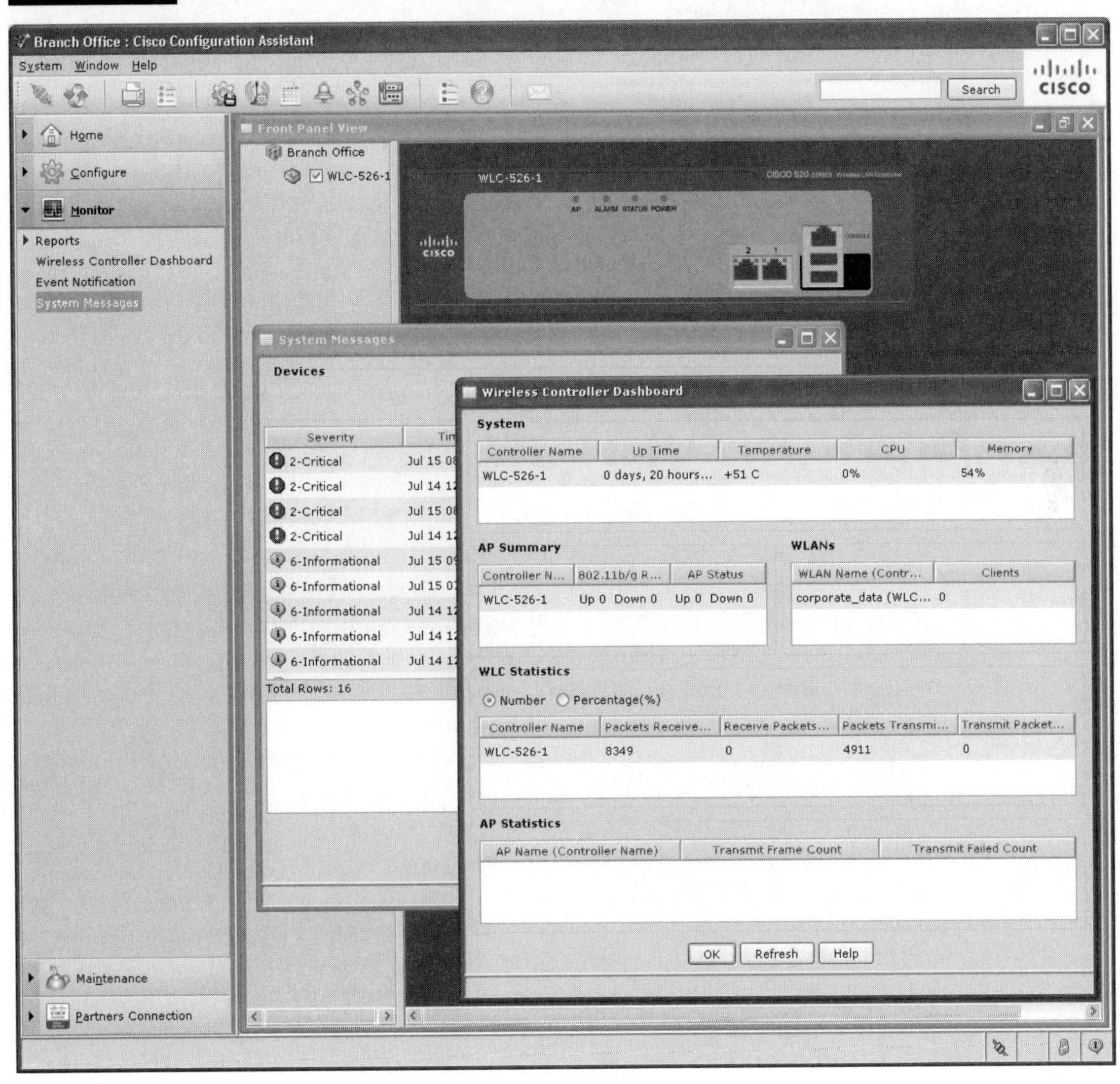

INSIDE THE EXAM

Cisco Mobility Express Wireless Architecture

For the exam, you should have a basic understanding of the Cisco Smart Business Communications System (CSBC) solution and how Cisco positions this solution to serve the small- and medium-sized business (SMB) market. You should also know the three different components in Cisco Mobility Express Solution and how they compare to the full-featured CUWN wireless products.

It is important to know that the 521 AP does not interoperate with any controllers other than the 526 controller. The 526 controller can support up to six 521 APs and cannot interoperate with other CUWN controllers or lightweight AP. The Cisco Configuration Assistant is the only tool for 521 AP configurations, and it is also the management tool for 526 controllers and other CSBC components.

CERTIFICATION SUMMARY

Cisco offers a full line of communications solutions specifically designed for SMB: Cisco Smart Business Communications System (CSBC). The CSBC helps customers in this particular market segment navigate through complex business environments by offering a complete suite of communications solutions covering services in voice, video, security, and wireless. Cisco Mobility Express Solution is the wireless solution of CSBC and includes three components: the Cisco 526 controller, Cisco 521 AP, and Cisco Configuration Assistant (CCA).

The Cisco 521 AP is an 802.11 b/g AP and can operate in two modes: standalone and controller. The Cisco 521 can be configured either through a 526 controller or CCA; it does not have a GUI or CLI interface of its own. The Cisco 521 shares the same form factor as the Cisco 1130 lightweight AP and can be powered by IEEE 802.2af standard PoE or Cisco prestandard PoE.

A Cisco 526 controller shares the same CLI and GUI as a full-featured CUWN controller, but it only offers a subset of the services available on the CUWN controller. The 526 controller can be managed through CLI, GUI, and CCA, but it is recommended you use GUI and CCA, as the CLI does not allow access to all configuration tasks.

The CCA is a Windows-based application and the primary configuration and network management tool for CSBC. The CCA provides device configuration, system monitoring, and reporting functions for Cisco Mobility Express Solution.

TWO-MINUTE DRILL

Cisco Mobility Express Wireless Architecture

- CSBC is specifically designed for the SMB market and offers a full range of communications solutions, including voice, video, security, and wireless.
- Cisco Mobility Express Solution is the wireless solution in CSBC and includes three primary components: the Cisco 526 controller, Cisco 521 AP, and CCA. Cisco Mobility Express supports other CSBC solutions while being managed by a common management tool, CCA.
- The Cisco 521 AP is 802.11b/g only and does not support 802.11a or 802.11n.
- The Cisco 521 AP can operate in standalone or controller mode; it can only be managed through CCA in standalone mode. The 521 AP cannot join any other CUWN controllers.
- The Cisco 526 controller is 802.11b/g only and does not support 802.11a or 802.11n.
- The Cisco 526 controller shares the same CLI and GUI as CUWN controllers but does not offer all features offered by CUWN controller.
- The Cisco 526 controller can be managed using CLI, GUI, and CCA.
- The CCA is a Windows-based configuration and network management tool for CSBC.
- The CCA can configure and manage both 526 controllers and 521 APs.

SELF TEST

The following self-test questions will help you measure your understanding of the material presented in this chapter. Read all the choices carefully, as there may be more than one correct answer. Choose all correct answers for each question.

Cisco Mobility Express Wireless Architecture

1. Cisco Mobility Express Solution does not support which protocols? (Choose all that apply.)

A. 802.3af

B. 802.11n

C. 802.11b

D. 802.11g

2. Which management tool can help manage Cisco Mobility Express Solution?

A. WCS

B. ACS

C. CCA

D. LMS

3. Which management tool can manage and configure a Cisco 521 AP?

A. WCS

B. ACS

C. CCA

D. LMS

4. Which platform supports the CCA to manage CSBC Solutions?

A. Windows

B. Linux

C. Apple OS X

D. Symbian

5. Which devices are part of Cisco Mobility Express?

A. Cisco ACS RADIUS server

B. Cisco 526 wireless controller

C. Cisco 2106 wireless controller

D. Cisco 521 lightweight access point

E. Cisco 1130 lightweight access point

6. Which device can the Cisco 521 AP work with to provide 802.11a and 802.11n support?
   A. Cisco 1250 802.11n radio
   B. Cisco 4404 controller
   C. WCS
   D. Cisco 521 AP does not support 802.11a and 802.11n
7. What tools can be used to manage and configure Cisco 521 AP in standalone mode?
   A. Cisco Configuration Assistant
   B. Cisco 2106 controller
   C. Cisco 526 controller
   D. WCS
8. What tools can be used to manage and configure Cisco 526 controller? (Choose all that apply.)
   A. Cisco Configuration Assistant
   B. WCS
   C. Console CLI
   D. HTTPS
9. Which devices will Cisco Configuration Assistant manage? (Choose all that apply.)
   A. Cisco ESW 500 Series switch
   B. Cisco 520 Series Secure router
   C. Cisco 3800 Series Secure router
   D. Cisco 526 Series controller
10. Company ACME is migrating their Cisco Mobility Express Solution with six 521 APs to the CUWN with a 4402 WLC and 1130 AP. When the Cisco 526 controller is powered down, the 521 APs are not joining the 4402 WLC. What is stopping 521 AP from joining 4402 WLC?
    A. The x.509 certificate is not valid on the 521 AP.
    B. 521 AP does not support 802.11a and 802.11n.
    C. The PoE does not supply enough power for the 521 AP.
    D. The 521 AP is not capable of joining 4402 WLC.

SELF TEST ANSWERS

Cisco Mobility Express Wireless Architecture

1. Cisco Mobility Express Solution does not support which protocols? (Choose all that apply.)

A. 802.3af

B. 802.11n

C. 802.11b

D. 802.11g

☑ **B.** 802.11a and 802.11n are not supported by AP 521.

☒ **A, C,** and **D** are incorrect. The 521 AP supports 802.11a and 802.11n, and 521 AP can be powered by 802.3af.

2. Which management tool can help manage Cisco Mobility Express Solution?

A. WCS

B. ACS

C. CCA

D. LMS

☑ **C.** Cisco Mobility Express can be managed by the Cisco Configuration Assistant (CCA).

☒ **A, B,** and **D** are incorrect. Cisco Mobility Express cannot be managed by WCS, ACS, or LMS. WCS manages only full-featured Cisco WLC. ACS is an AAA server, not a management tool. LMS is the flagship network management tool for Cisco campus switches.

3. Which management tool can manage and configure a Cisco 521 AP?

A. WCS

B. ACS

C. CCA

D. LMS

☑ **C.** The Cisco 521 AP can be managed by the Cisco Configuration Assistant (CCA).

☒ **A, B** and **D** are incorrect. Cisco Mobility Express cannot be managed by WCS, ACS, or LMS. WCS manages only full-featured Cisco WLC. ACS is an AAA server, not a management tool. LMS is the flagship network management tool for Cisco campus switches.

4. Which platform supports the CCA to manage CSBC Solutions?
   - A. Windows
   - B. Linux
   - C. Apple OS X
   - D. Symbian

☑ **A.** The Cisco Configuration Assistant is a Windows-based application.

☒ **B, C,** and **D** are incorrect. Linux, Apple OS, and Symbian do not support CCA.

5. Which devices are part of Cisco Mobility Express?
   - A. Cisco ACS RADIUS server
   - B. Cisco 526 wireless controller
   - C. Cisco 2106 wireless controller
   - D. Cisco 521 lightweight access point
   - E. Cisco 1130 lightweight access point

☑ **B** and **D.** The Cisco 526 wireless controller and the 521 AP are both part of Cisco Mobility Express.

☒ **A, C,** and **E** are incorrect. The Cisco ACS RADIUS server, 2106 WLC, and 1130 lightweight AP are not part of the Cisco Mobility Express Solution.

6. Which device can the Cisco 521 AP work with to provide 802.11a and 802.11n support?
   - A. Cisco 1250 802.11n radio
   - B. Cisco 4404 controller
   - C. WCS
   - D. Cisco 521 AP does not support 802.11a and 802.11n

☑ **D.** The AP521 does not support 802.11a and 802.11n.

☒ **A, B,** and **C** are incorrect. AP 1250 802.11n radio cannot be used by AP521, and the 4404 controller does not support AP 521. WCS is the network management platform for CUWN components, not Cisco Mobility Express components.

7. What tools can be used to manage and configure Cisco 521 AP in standalone mode?
   - A. Cisco Configuration Assistant
   - B. Cisco 2106 controller
   - C. Cisco 526 controller
   - D. WCS

☑ **A.** AP521 in standalone mode can be managed only by CCA.

☒ **B, C** and **D** are incorrect. Controller 2106 does not manage AP 521. The Cisco 526 controller manages only AP 521 in controller mode, not standalone mode. WCS does not manage Cisco Mobility Express components.

8. What tools can be used to manage and configure Cisco 526 controller? (Choose all that apply.)

A. Cisco Configuration Assistant

B. WCS

C. Console CLI

D. HTTPS

☑ **A, C,** and **D**. Cisco 526 controller can be managed by Console CLI, Cisco Configuration Assistant, and Web GUI through HTTPS.

☒ **C** is incorrect. WCS is designed to manage full-featured CUWN controller but is not capable of managing the Cisco 526 controller.

9. Which devices will Cisco Configuration Assistant manage? (Choose all that apply.)

A. Cisco ESW 500 Series switch

B. Cisco 520 Series Secure router

C. Cisco 3800 Series Secure router

D. Cisco 526 Series controller

☑ **A, B,** and **D**. The Cisco Configuration Assistant can manage CSBC components that include Cisco 500 Series switches, 520 Series secure routers, 526 Series controllers, and 521 Series APs.

☒ **C** is incorrect. The Cisco Configuration Assistant cannot manage a Cisco 3800 Series router.

10. Company ACME is migrating their Cisco Mobility Express Solution with six 521 APs to the CUWN with a 4402 WLC and 1130 AP. When the Cisco 526 controller is powered down, the 521 APs are not joining the 4402 WLC. What is stopping 521 AP from joining 4402 WLC?

A. The x.509 certificate is not valid on the 521 AP.

B. 521 AP does not support 802.11a and 802.11n.

C. The PoE does not supply enough power for the 521 AP.

D. The 521 AP is not capable of joining 4402 WLC.

☑ **D.** AP 521 is not capable of joining WLC 4402.

☒ **A, B,** and **C** are incorrect. AP 521 can join only the 526 controller.

Part IV

Implementation of the Cisco Wireless LAN

CHAPTERS

8

Deploying Cisco Wireless LAN Components

CERTIFICATION OBJECTIVES

8.01	Standalone Access Point Configuration	✓	Two-Minute Drill
8.02	Configuration of Cisco Wireless LAN Components	Q&A	Self Test
8.03	Radio Resource Management (RRM) Concepts		

In Chapter 6, we covered the concepts of autonomous and lightweight access points as well as the role of Wireless LAN Controllers in a Cisco Unified Wireless Network (CUWN) architecture. An important part of the CCNA Wireless exam is the hands-on portion, which involves basic configuration of the elements needed to create a wireless network. There are a variety of terms and concepts involved in controller and access point configuration that will be an important part of the exam, and it is necessary to fully comprehend the steps of deploying a wireless network using Cisco Wireless LAN controller. In this chapter, we will cover the configuration and monitoring of the wireless network using Cisco Wireless LAN Controllers and access points.

CERTIFICATION OBJECTIVE 8.01

Standalone Access Point Configuration

Before configuring a controller-based wireless network, it is important to understand the mechanism by which Cisco standalone (or autonomous) IOS APs are configured. In the infancy of wireless networks, the majority of wireless access points had independent configurations and ran an independent operating system without the need for a controller. Although this model had inherent advantages, the rapid increase in breadth and size of enterprise wireless LANs necessitated a more centralized approach, which took us toward the modern unified controller-based model. There are still many quality applications for standalone access points, and you will often see standalone/autonomous configurations in home use, SOHO environments, and even some large enterprises. To meet these needs, Cisco access points continue to be sold in both autonomous IOS and unified LWAPP and CAPWAPP (Control and Provisioning of Wireless Access Points Protocol) flavors. In varied customer deployments, you will likely encounter standalone access point configurations, which is part of the reason it is included in the CCNA Wireless exam. In this section we will describe the configuration of a Cisco standalone IOS access point using the graphical user interface (GUI).

Basic Configuration

All modern Cisco access points run the Cisco Internetwork Operating System (IOS). In the case of autonomous APs, the IOS plays a far more important role in actual configuration, whereas in the unified model, the IOS is simply the shell operating system that lightweight access points perform basic operations on. As with all IOS devices, a standalone Cisco IOS access point has a command line interface (CLI) that can be accessed via the console port, telnet, or Secure Shell (SSH) connection. We will not cover configuration through CLI in this book, but we will look at configuration of an IOS AP using its GUI.

Accessing the AP

A challenge faced by novice administrators is discovering the IP address of an AP out of the box when it plugs into the network. Because most Cisco APs (the early 1100 series being an exception due to their lack of a console port) do not come with a default IP address, there are specific methods by which the AP's IP address can be ascertained:

DHCP Most Cisco standalone APs are configured by factory default to accept a DHCP IP address for administration. In this scenario, it is useful to have access to either the DHCP server that the AP is pulling an address from, or the Cisco switch to which the AP is connected. If you know the MAC address of the AP—typically shown on the outside of the access point on the same barcode sticker as the serial number—you can very quickly find its appropriate address assignment/binding on your DHCP server. If the DHCP server is in a Cisco IOS switch or router, a simple `show ip dhcp bind` command will display the assigned DHCP addresses for that device.

If you have access to the switch and know which switchport the AP is connected to, you can run a `show cdp neighbors detail` command to see the IP address of the AP device, provided that the Cisco Discovery Protocol (CDP) feature is enabled. You can also run a `show arp | include <mac-address of the AP>` command to look up the AP's IP address in the switch's Content Addressable Memory (CAM) table.

Console into the AP (advanced users) Whether or not you have a DHCP server available, the most foolproof method of determining and/or setting an AP's IP address is using the command line interface via the AP's console port. It is important to understand what is required to console into an AP, and this information is covered in Chapter 12. When successfully consoled into an access point, the `show interface BVI1` command will quickly tell you the IP address that is being used by the access point. *BVI* refers to a *bridged virtual interface*, which is a sort of virtual or loopback IP address on the access point for AP management. You can also disable DHCP address assignment and manually set a static IP address by going into the AP's privileged EXEC (enable) mode and then global configuration mode (config terminal) and making changes to the interface BVI1 settings.

Cisco IP Setup Utility (IPSU) When run on a network that can talk to the AP, this utility can discover the IP address of that access point when a MAC address is entered. The IPSU is shown in Figure 8-1.

Once you have discovered the access point's IP address, you should be able to log into the access point using a standard Web browser by directing it to the IP address of the AP. In the example we will use, the AP's IP address is 192.168.1.5 and can be accessed by typing **http://192.168.1.5**. The default username and password to access a Cisco IOS AP is typically admin with a default password of Cisco. The password and username are both case sensitive. A window similar to the one in Figure 8-2 should appear.

FIGURE 8-1

Cisco IP Setup Utility (IPSU)

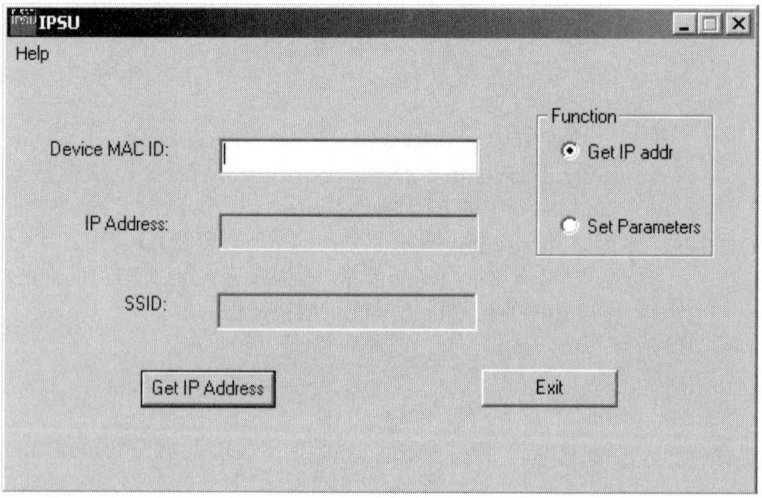

FIGURE 8-2 Initial Cisco standalone IOS AP screen

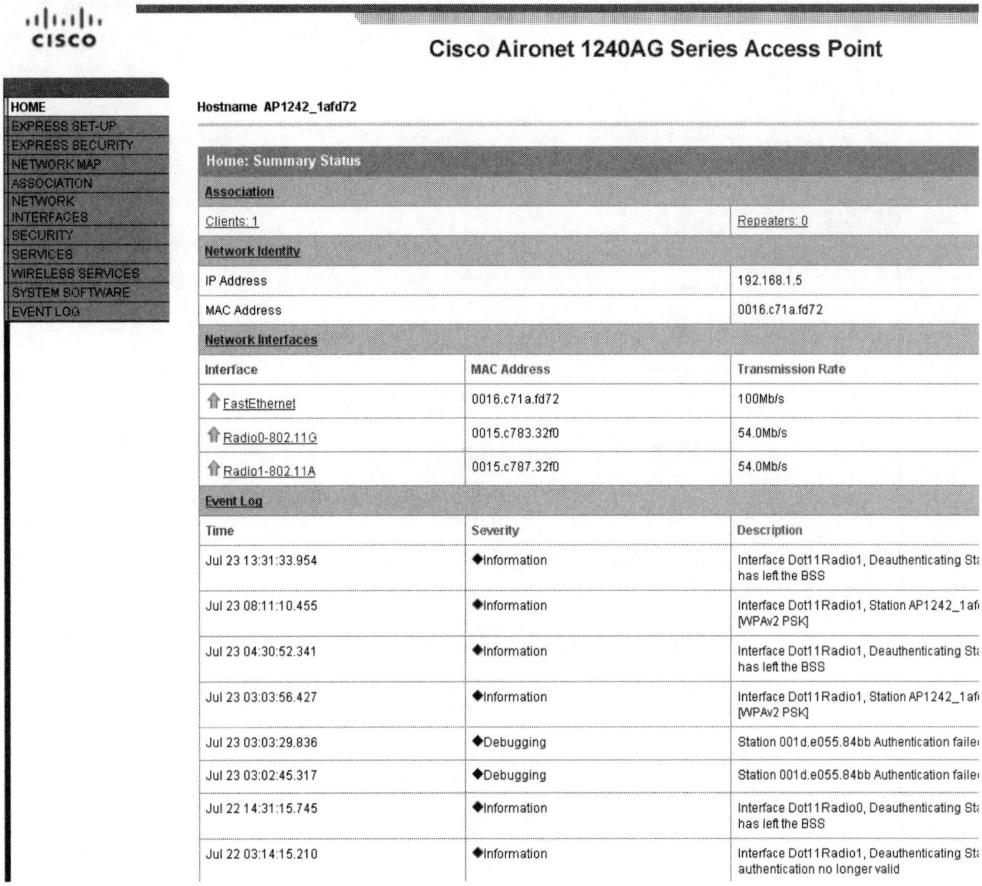

CISCO

Cisco Aironet 1240AG Series Access Point

HOME
EXPRESS SET-UP
EXPRESS SECURITY
NETWORK MAP
ASSOCIATION
NETWORK INTERFACES
SECURITY
SERVICES
WIRELESS SERVICES
SYSTEM SOFTWARE
EVENT LOG

Hostname AP1242_1afd72

Home: Summary Status

Association

Clients: 1	Repeaters: 0

Network Identity

IP Address	192.168.1.5
MAC Address	0016.c71a.fd72

Network Interfaces

Interface	MAC Address	Transmission Rate
FastEthernet	0016.c71a.fd72	100Mb/s
Radio0-802.11G	0015.c783.32f0	54.0Mb/s
Radio1-802.11A	0015.c787.32f0	54.0Mb/s

Event Log

Time	Severity	Description
Jul 23 13:31:33.954	Information	Interface Dot11Radio1, Deauthenticating St has left the BSS
Jul 23 08:11:10.455	Information	Interface Dot11Radio1, Station AP1242_1af [WPAv2 PSK]
Jul 23 04:30:52.341	Information	Interface Dot11Radio1, Deauthenticating St has left the BSS
Jul 23 03:03:56.427	Information	Interface Dot11Radio1, Station AP1242_1af [WPAv2 PSK]
Jul 23 03:03:29.836	Debugging	Station 001d.e055.84bb Authentication faile
Jul 23 03:02:45.317	Debugging	Station 001d.e055.84bb Authentication faile
Jul 22 14:31:15.745	Information	Interface Dot11Radio0, Deauthenticating St has left the BSS
Jul 22 03:14:15.210	Information	Interface Dot11Radio1, Deauthenticating St authentication no longer valid

Express Setup

The Express Setup window (shown in Figure 8-3) allows you to configure the basic settings for the standalone access point with the following menu items:

- **Hostname** You can set the hostname of the access point. The hostname is usually subject to a corporate naming convention and helps administrators identify AP characteristics such as location and role.
- **MAC Address** This is the same MAC address that you would use in the process of discovering the AP's initial IP address when using the IPSU.

FIGURE 8-3 Express Setup window

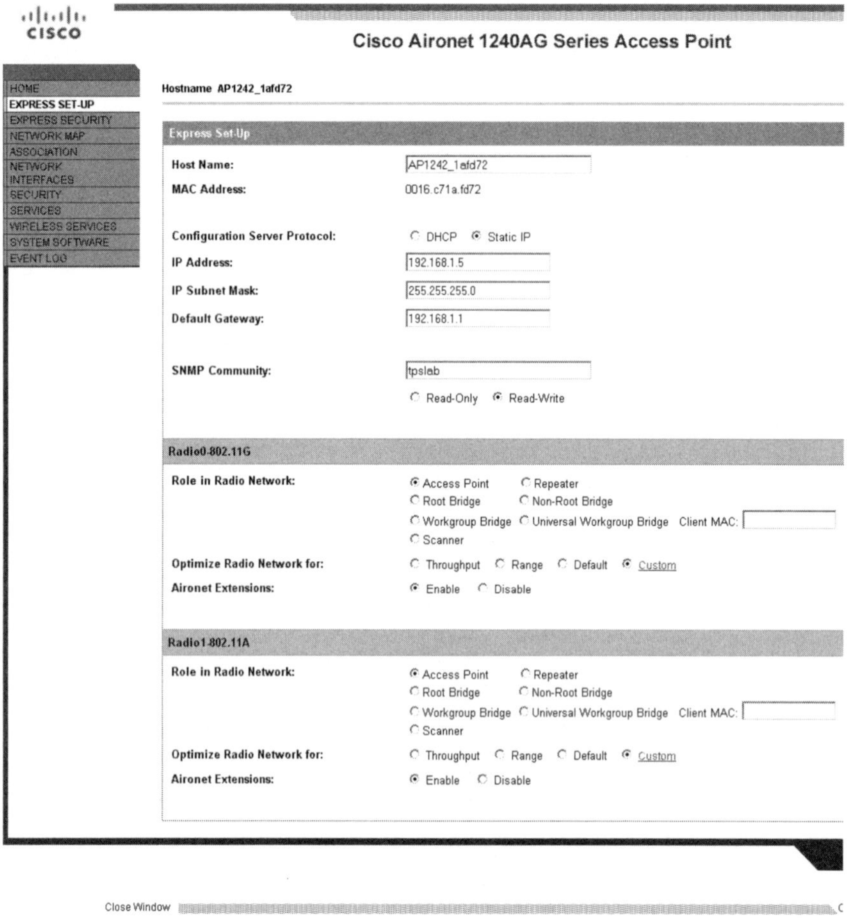

- **IP Address, Subnet Mask, and Default Gateway** This information can be statically configured or set for DHCP.
- **SNMP Community** This optional value allows communication with management servers such as the Cisco Wireless LAN Solution Engine (WLSE), which is not covered on the exam.

- **802.11A and 802.11G Radio settings** In these sections, you can set the respective radio's role in the network or configure Aironet extensions. An important submenu of these settings is the Custom hyperlink, which takes you to a more detailed screen that allows you to set channel settings, data rates, and transmit power settings, as well as enable or disable the radio.

Click Apply to save any changes you want to make to the AP.

Express Security

The Express Security window (shown in Figure 8-4) provides you with the basic options you need to set up a wireless network SSID with various security types. As with all the other windows on the Cisco IOS GUI, the Apply button saves any changes you have made to the selections. The options in the Express Security window are as follows:

- **SSID** One or more SSIDs can be configured for use on the AP. When an access point is in factory defaults, there is no SSID configured.
- **VLAN** In most standalone AP deployments, a single AP has a trunked link carrying multiple VLANs that need to be associated to each SSID. This value allows you to map a specific VLAN to the SSID. This is very similar in concept to the linking of dynamic interfaces to WLANs, which we will discuss later in this chapter in the controller configuration section.
- **Security** This section allows you specify a WEP key and/or RADIUS servers for 802.1X authentication with the access point. In some deployments, such as a guest access wireless network, you can select no security.

It is important to note that the Express Security window often does not provide the options that are needed to configure a robust network using the Cisco IOS standalone access points. The Security menu, which contains the Encryption Manager and SSID Manager submenus, allows far more specific configuration of wireless security options.

FIGURE 8-4 Express Security window

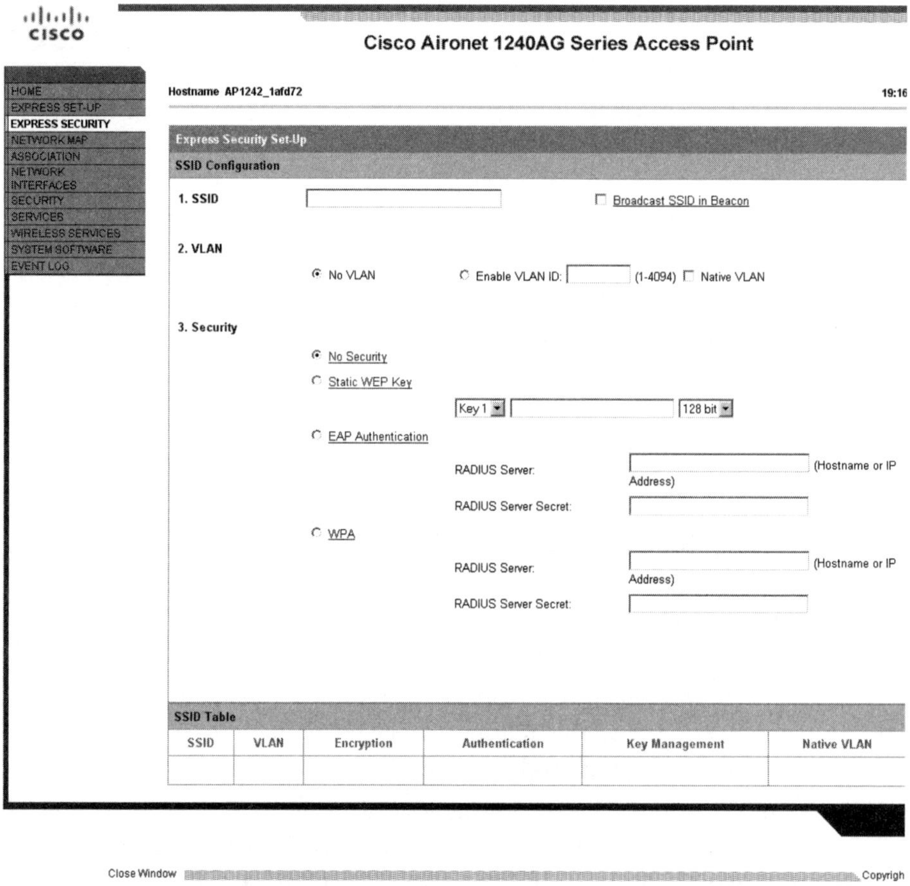

Standalone AP Software Considerations

The Cisco Standalone IOS access point, like any other network device, requires periodic software upgrades for bug fixes and general feature patches. In addition to general software upgrades, the Cisco Standalone APs can also be converted to work with a controller-based LWAPP infrastructure. In this section, we will discuss how to manage software/firmware with a Cisco standalone AP.

Software Upgrades

The process for performing software upgrades on a Cisco IOS standalone AP is relatively straightforward. Click the System Software menu item on the GUI to see a window like the one in Figure 8-5.

FIGURE 8-5 System Software window

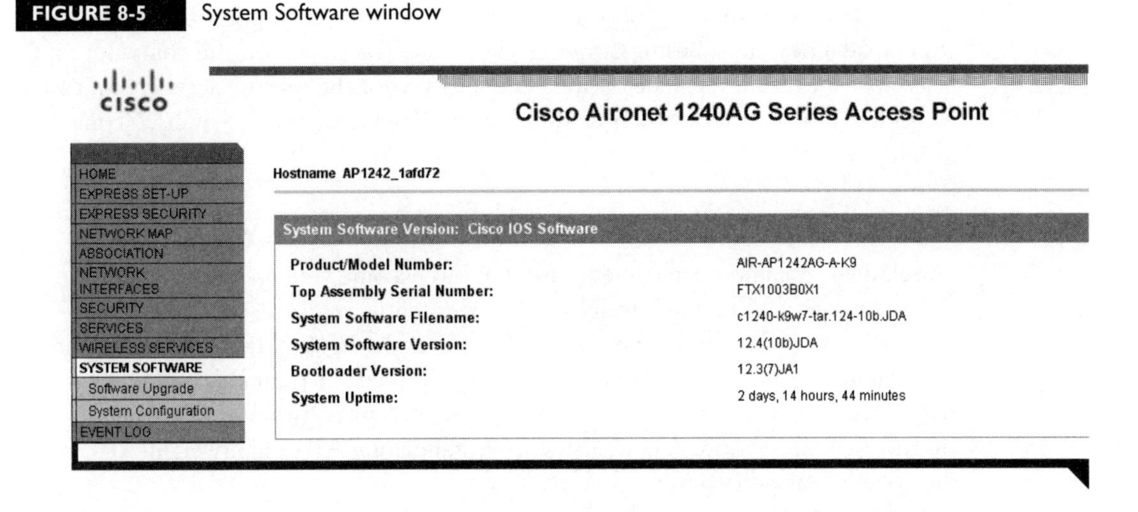

From this window, you can immediately see important status information about the access point such as the product model number, serial number, software information, and system uptime. From this menu, select the Software Upgrade submenu to see the Software Upgrade window shown in Figure 8-6.

FIGURE 8-6 Software Upgrade menu

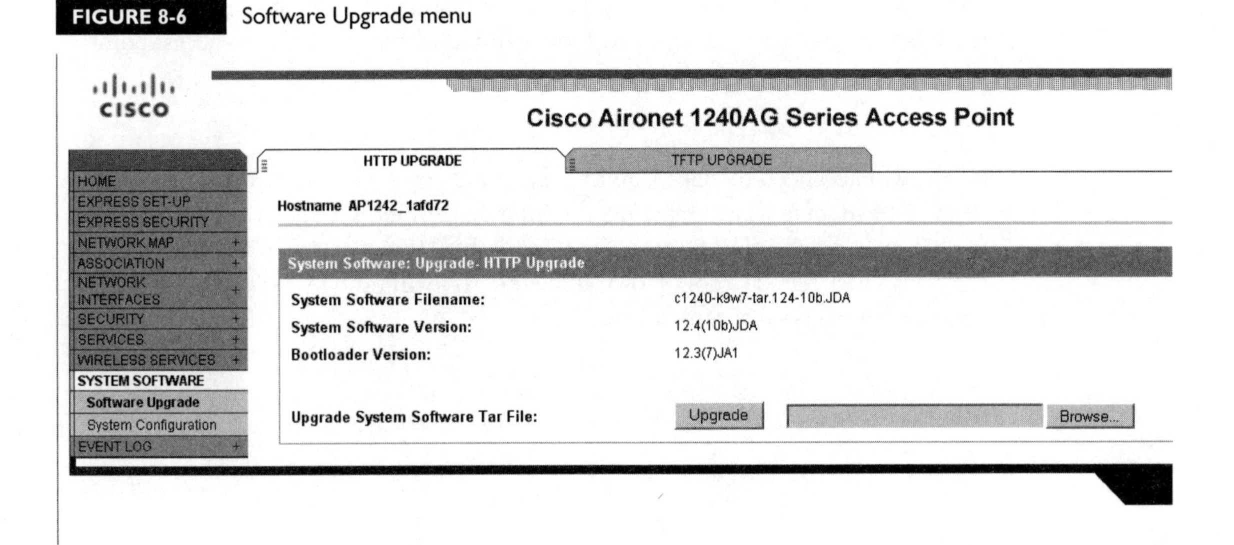

This window shows the software information on the access point and allows you to upload a new image using either HTTP with a file that is on the computer or server, or the TFTP Upgrade option. The software for the specific access point can be obtained with a Cisco Connection Online (CCO) login at www.cisco.com/public/sw-center/.

Converting an Autonomous/Standalone AP to LWAPP Mode

A relatively common deployment task that has become very prevalent with the adoption of unified lightweight AP wireless networks is the conversion of existing Cisco IOS standalone APs to LWAPP mode. In all of these scenarios, the standalone/autonomous AP must be running IOS version 12.3(7) JA software or later. The best way to verify this is to open the System Software window described in the previous section. A conversion from standalone AP to lightweight AP can be done in three different ways.

Wireless Control System (WCS) The WCS pushes the LWAPP upgrade image to autonomous APs and is technically the simplest method for converting multiple standalone APs to LWAPP mode. The LWAPP image in this method is stored on a TFTP server.

The autonomous to lightweight mode Upgrade Tool The autonomous to lightweight mode Upgrade Tool is a downloadable software package that allows you to perform upgrades on multiple Cisco IOS standalone APs at the same time. To use this tool, you must create a text format IP file with a line for each access point you want to convert in the following format:

```
IP address, Username, Password, Enable Password
```

You will also need to make sure that the standalone APs have the proper software version, have some way of reaching the intended Wireless LAN Controller after the upgrade, can reach a TFTP server containing the upgrade image, and have Telnet enabled. The Upgrade Tool (shown in Figure 8-7) will notify you whether the AP upgrade process is successful or not.

FIGURE 8-7 Autonomous to lightweight mode Upgrade Tool

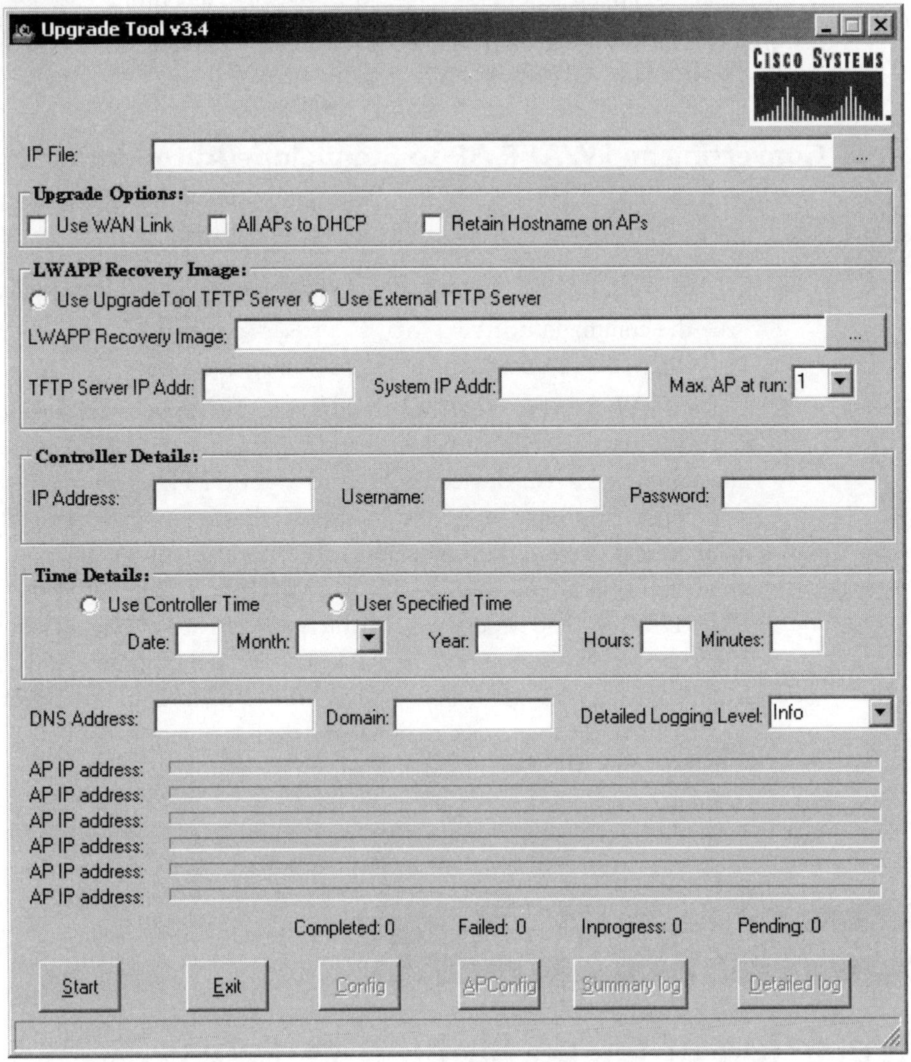

Upgrade Tool v3.4

Cisco Systems

IP File: ...

Upgrade Options:

Use WAN Link
All APs to DHCP
Retain Hostname on APs

LWAPP Recovery Image:

Use UpgradeTool TFTP Server
Use External TFTP Server

LWAPP Recovery Image: ...

TFTP Server IP Addr:
System IP Addr:
Max. AP at run: 1

Controller Details:

IP Address:
Username:
Password:

Time Details:

Use Controller Time
User Specified Time

Date:
Month:
Year:
Hours:
Minutes:

DNS Address:
Domain:
Detailed Logging Level: Info

AP IP address:
AP IP address:
AP IP address:
AP IP address:
AP IP address:
AP IP address:

Completed: 0
Failed: 0
Inprogress: 0
Pending: 0

Start
Exit
Config
APConfig
Summary log
Detailed log

Manual Upgrade via CLI or GUI You can also manually convert Cisco standalone IOS APs one at a time by simply using the LWAPP upgrade image and performing the software upgrade process shown in the previous section. This is the simplest method for AP conversion.

Converting an LWAPP AP to Standalone/Autonomous Mode

Converting an LWAPP AP back to standalone/autonomous mode is a relatively straightforward process and can be completed in two ways:

- **Use the Wireless LAN Controller** Simply log into the controller's CLI and issue the command `config ap tftp-downgrade tftp-server-ip-address filename access-point-name`.
- **Use a TFTP Server** Physically hold down the Mode button on the access point and have a TFTP server with an IP address in the 10.0.0.2–10.0.0.30 range attached to the device via Ethernet. The TFTP server must contain a proper IOS software image file for the AP that is renamed to match the name format c*xxxx*-k9w7-tar.default, where *xxxx* represents the model number of the AP that you are converting. This number will likely be 1130, 1200, 1240, or 1250 depending on which AP you are working with. Holding down the Mode button until the AP LED turns red (approximately 20–30 seconds) makes the AP default to a 10.0.0.1 IP address and then searches for a TFTP server in that range to download an IOS image from. Once this is complete, the AP should reboot in autonomous mode with a factory default configuration.

CERTIFICATION OBJECTIVE 8.02

Configuration of Cisco Unified Wireless LAN Components

Now that we have examined the configuration and operations that can be performed on a standalone access point, we can discuss deployment of the Cisco Unified Wireless LAN. The following section will cover configuration and monitoring of a centralized/unified wireless network with a Cisco Wireless LAN Controller–based architecture. After reading this section, you should have a basic understanding of the tasks that need to be accomplished to deploy a wireless network using Cisco Wireless LAN Controllers and lightweight access points.

Understanding Controller Components

Before discussing the configuration of a Cisco Wireless LAN Controller from the ground up, it is essential that you understand the key components of a controller. In this section, we will describe these components and show how they interoperate with each other to make the controller an intelligent and essential part of the Cisco Unified Wireless Network.

Ports

Ports are the physical entities that a Wireless LAN Controller uses to connect to the rest of the network. There are two different kinds of ports on a controller:

- **Distribution system ports** The ports that are used for connecting the controller to a neighbor switch and, by extension, to the rest of the network. Depending on the model of Cisco controller, there can be anywhere from 2–8 distribution system ports on a given controller for connecting to a switch via a copper or fiber connection, or via a switch backplane in the case of the Cisco Wireless Services Module (WiSM). Distribution system ports can carry multiple VLANs via an 802.1q trunk and, because there is more than one of them, they can be configured in either primary/backup port configuration, where there is a primary port for traffic and a backup port should that primary port be disabled, or link aggregation (LAG) configuration, where all of the ports on a controller are bundled into a single 802.3ad *port channel*. When a port out of that link-aggregated bundle is disabled, the connection stays up, albeit with lower overall data throughput. LAG is the default port configuration on modern controllers and is the only mode of operation on a WiSM module. We will discuss the relationship between distribution system ports and controller interfaces in the next section.
- **Service port** The physical port optionally used for out-of-band management of the controller and directly mapped to the service-port interface discussed in the "Static Interfaces" section. The service port does not support 802.1q trunks and should only be used in the event of a network failure. It does not play any role in interacting with the distribution system ports and should not

even have an IP address on the same supernet. Although it is not covered in the exam, it is good to know that the service port also acts as the controller's connection to the switch's supervisor module in the case of the WiSM module.

Static Interfaces

There are two different types of interfaces on a Wireless LAN Controller. An interface on a controller is similar in function to an interface on a Cisco router or switch except for the fact that it is a virtual/logical entity. Although there are cases where an interface maps to a specific physical port (discussed later in this section), the concept of controller interfaces remains logical. The first class of controller interfaces is *static interfaces*. Static interfaces are a default part of every controller and are absolutely necessary for a controller to operate, even if there are no access points or wireless networks associated to it. There are four specific static interfaces that are always found on every Cisco Wireless LAN Controller:

- **Management interface** Used for in-band management of the controller and serves as its primary connectivity interface to important elements such as the RADIUS server for wireless security, other Wireless LAN Controllers, and Layer 2 communication with lightweight access points. When checking for connectivity to a controller, the management interface is the address that you would normally ping via ICMP. As mentioned previously, the management interface's IP address is the one that is used for managing the controller via Telnet, SSH, or the Web-based GUI. In smaller wireless networks, the management interface (and its attached network) can be used by individual SSID/WLANs as a client VLAN.
- **AP-manager Interface** Used for all Layer 3 communications between the controller and lightweight access points once the access points have successfully joined the controller. The AP-manager IP address is used as the tunnel source for LWAPP/CAPWAPP packets from the controller to the access point and as the destination for these packets from the access point to the controller. The AP-manager interface is optimally on the same subnet as the management interface but is still fully operational even if it runs on a different subnet.

- **Virtual interface** Used to support mobility management, DHCP relay and embedded Layer 3 security such as guest Web authentication. When the DHCP Proxy feature is enabled, the virtual interface acts as the DHCP server placeholder for wireless clients that obtain their IP address from a DHCP server. When guest access Web authentication is deployed, the virtual interface IP address serves as the redirect address for the Web authentication login page.

The IP address that is typically used for the virtual interface is 1.1.1.1 but, in an environment that already has a third-party guest access server in place for wireless, this address may conflict with that product's redirection IP address. Be sure to understand the operation of any third-party guest access/Web redirect servers when deploying a controller, and take that into consideration when setting the virtual interface IP address.

- **Service-port interface** The only interface on the Wireless LAN Controller that is directly mapped to a physical port (the service port). As discussed previously, the service-port interface must have an IP address from a different supernet from any other interface or port and is strictly used for out-of-band management of the controller in the event of a network outage.

Many network administrators make the mistake of putting the service-port interface into a live subnet on the wired network because it is mapped to a conspicuous copper port. Some have even used the service-port interface IP to manage the controller remotely by routing that network. Although this is not necessarily a harmful practice, it undermines the purpose of the service-port interface, which is to provide out-of-band management in an event of a network outage. Always use the management interface for remote access and keep the service-port interface on a known but nonrouted/localized network subnet.

Dynamic Interfaces

Dynamic interfaces or *VLAN interfaces* are analogous to the VLANs on standalone access points mentioned in the "Standalone Access Point Configuration" section of this chapter. Dynamic interfaces are created specifically to tie the controller in to the VLANs that it needs to provide to specific wireless SSIDs. In addition to this, dynamic interfaces provide DHCP relay services for associated clients. A single controller can support up to 512 dynamic interfaces (VLANs).

WLANs

Wireless LANs (WLANs) are simply the configuration section where a wireless service set identifier (SSID) is tied to an interface. The terms *SSID* and *WLAN* are often used interchangeably because of this close relationship. WLANs are configured with wireless security, QoS, radio policies, and other advanced wireless network parameters. Up to 16 WLANs can be configured per controller.

Dynamic interfaces are used to provide a connection to the VLAN that is used by wireless clients who associate to a wireless SSID. WLANs tie the SSID to the dynamic interface which, in turn, ties them into the VLAN.

Controller Configuration

Now that a foundation of basic controller concepts has been established, we can move into the actual configuration of a Cisco Wireless LAN Controller. Although this information is widely accessible on Cisco Connection Online (CCO), this section provides a concise guide that will be important to understand for the purposes of the exam. This section will take you from initial controller bootup all the way to configuring a fully functional WLAN with wireless security features.

Initial Controller Configuration

When a controller is taken out of the box and is in factory default mode, you must complete an initialization process to give it basic functionality and the ability to be managed by the CLI and GUI. The GUI cannot be accessed prior to completing this process. To start the initialization process, you have to configure the controller using a DB9 connected console cable and a terminal program on your computer. After the initial bootup sequence, the controller will check for a valid configuration. If one is found, you will be prompted to enter your username and password. If one is not found, you have the option of using the CLI Configuration Wizard shown in Figure 8-8.

FIGURE 8-8 CLI Controller Configuration Wizard

```
Welcome to the Cisco Wizard Configuration Tool
Use the '-' character to backup
System Name [Cisco_be:dd:2b]:
Enter Administrative User Name (24 characters max):
Enter Administrative Password (24 characters max):
Re-enter Administrative Password :

Service Interface IP Address Configuration [none][DHCP]:

Management Interface IP Address: 192.168.1.5
Management Interface Netmask: 255.255.255.0
Management Interface Default Router: 192.168.1.1
Management Interface VLAN Identifier (0 = untagged):
Management Interface Port Num [1 to 8]: 1
Management Interface DHCP Server IP Address: 192.168.1.1

AP Transport Mode [layer2][LAYER3]:
AP Manager Interface IP Address: 192.168.1.6

AP-Manager is on Management subnet, using same values
AP Manager Interface DHCP Server (192.168.1.1):

Virtual Gateway IP Address: 1.1.1.1

Mobility/RF Group Name: MyMobilityGroup

Enable Symmetric Mobility Tunneling [yes][NO]:

Network Name (SSID): MyWirelessNetwork

Allow Static IP Addresses [YES][no]:

Configure a RADIUS Server now? [YES][no]:
Warning! The default WLAN security policy requires a RADIUS server.
Please see documentation for more details.

Enter Country Code list (enter 'help' for a list of countries) [US]:

Enable 802.11b Network [YES][no]:
Enable 802.11a Network [YES][no]:
Enable 802.11g Network [YES][no]:
Enable Auto-RF [YES][no]:

Configure a NTP server now? [YES][no]:
Configure the system time now? [YES][no]:

Warning! No AP will come up unless the time is set.
Please see documentation for more details.

Configuration correct? If yes, system will save it and reset. [yes][NO]:yes
```

In the CLI Configuration Wizard, you can configure the following:

- Controller hostname/system name
- Administrative username and password
- Service interface IP address
- Management interface IP address/network mask/default gateway
- AP-manager interface IP address
- Virtual interface IP address
- Mobility group name
- An initial WLAN/SSID with 802.1X security
- RADIUS servers for authentication
- NTP servers for time synchronization

Once you have taken a note of all of the information you have entered, the controller will reboot and prompt you with a username and password on the CLI. Although the controller runs its own software, it behaves much like IOS, and you can make configuration changes using the `config` command. In addition to this, you can perform various `show` commands and run debugs to diagnose controller behavior. These commands are described in detail in Chapter 11.

The Controller Graphical User Interface (GUI)

In order to access the controller's GUI, you will need to navigate to the controller's management IP address using a Web browser and the HTTPS command. When you access the controller's GUI, the first screen you see will look like Figure 8-9.

After clicking the Login button, you will be prompted to enter the administrative username and password you created in the CLI-based Configuration Wizard. Once that is successful, you will be brought to the controller's Summary screen shown in Figure 8-10.

From here you can see the basic menu items for the controller, some of which will be covered in more detail in the rest of this chapter:

- **Monitor** Takes you to the main controller Summary screen. The menu items in the Monitor menu are read-only and used to examine performance of various controller components, including APs and their radios, associated wireless clients and their details, and rogue access points.

FIGURE 8-9

Initial Controller GUI splash screen

FIGURE 8-10 Controller Summary screen

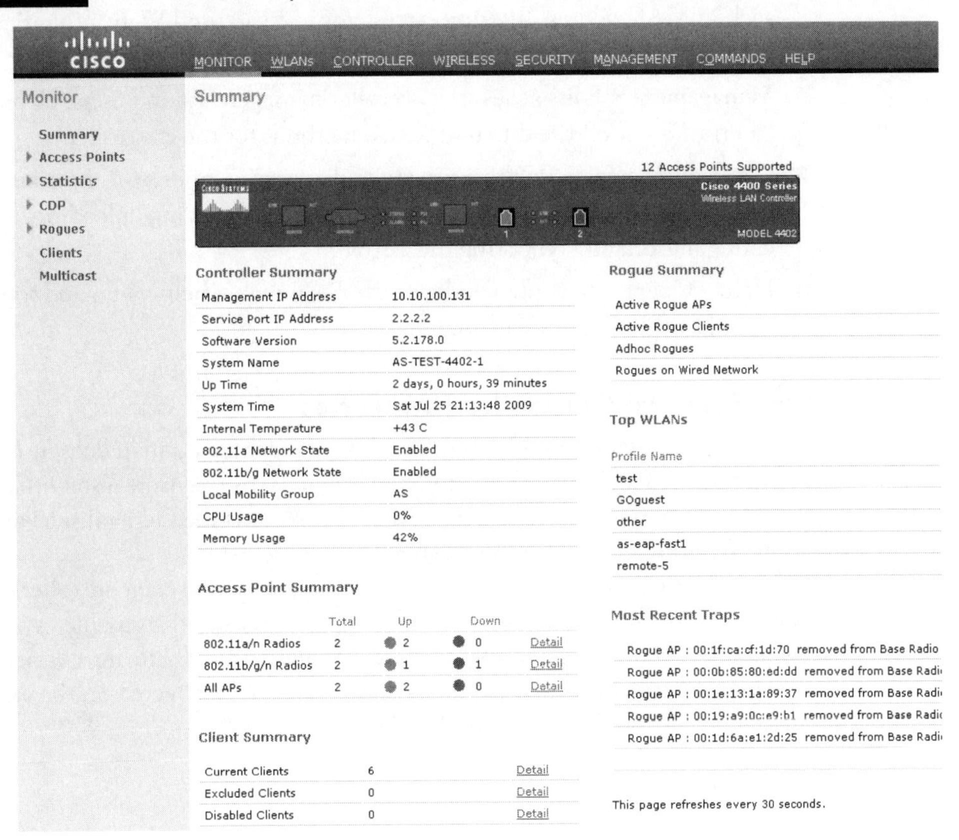

- **WLANs** Displays all the currently configured WLANs on the controller. You can add, edit, and remove these WLANs from this menu and also perform more advanced functions like creating AP groups and establishing anchor controller relationships (two concepts that are not covered on the exam).
- **Controller** Allows you to modify global settings for the controller. This is also where you go to configure and edit dynamic interfaces and configure mobility group options (which you will use in multicontroller and guest access deployments).
- **Wireless** Shows all the access points currently associated to the Wireless LAN Controller and is also the section where you can make configuration changes to APs, radios, and global radio frequency (RF) settings such as Radio Resource Management (RRM), which is covered later in this chapter.
- **Security** Specifies wireless security related items such as RADIUS authentication and accounting servers. This is also where additional items such as MAC address filtering, access control lists, and Web authentication for guest access can be configured.
- **Management** Edits access to controller management and users. Here, you can turn on or off the different access methods for the controller.
- **Commands** Takes you to a variety of diagnostic options for the controller including upgrading software, setting system time, saving/uploading configuration, and rebooting the system.
- **Help** Opens a separate window with a searchable help menu and topics sorted by the preceding menu items.

Setting General Controller Properties

Now that you have familiarized yourself with what each menu item does, it is important to first make general controller configurations to ensure a smooth deployment of the network. The first section to look at is the General sidebar selection under the Controller menu, shown in Figure 8-11.

In this particular window, you can configure items such as the controller's hostname and the mobility group and RF group name, as well as enable or disable features such as LAG mode for the controller ports. There are a variety of other important features and commands that will not be covered on the exam but are important to take note of in this display window.

FIGURE 8-11 Controller general settings

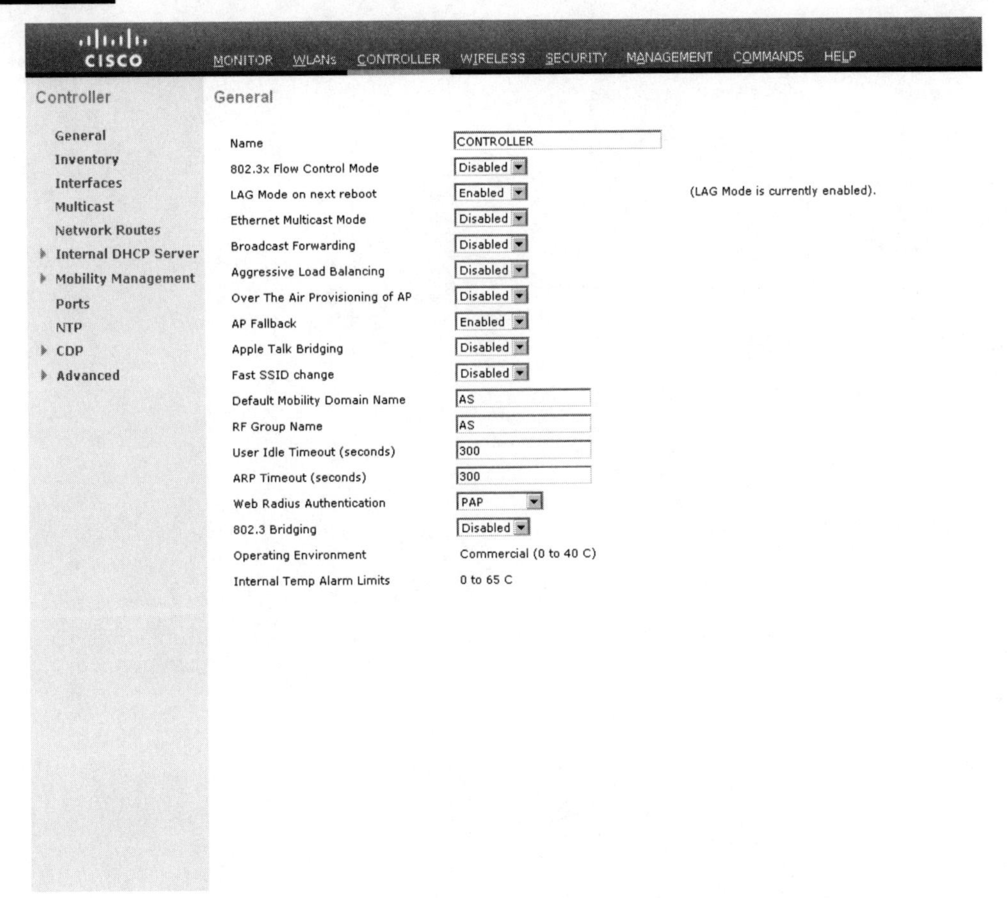

Under the same Controller menu item, you can set a very important feature that will assist you greatly in adding new access points to the controller. Under the NTP menu, you can add a Network Time Protocol server for the controller to synchronize system time with. The function, shown in Figure 8-12, is a very simple IP address for the time server and a polling interval, which is configured for all time servers. Synchronizing (or at least setting) controller time is very important, as the controllers associate with access points using certificates. These certificates are timestamp sensitive and, because of this, will not work or will seem expired because of an improperly set controller system time. To avoid this situation, setting the NTP is one of the first tasks that is usually performed when bringing a new controller online.

FIGURE 8-12 Adding an NTP server to the controller

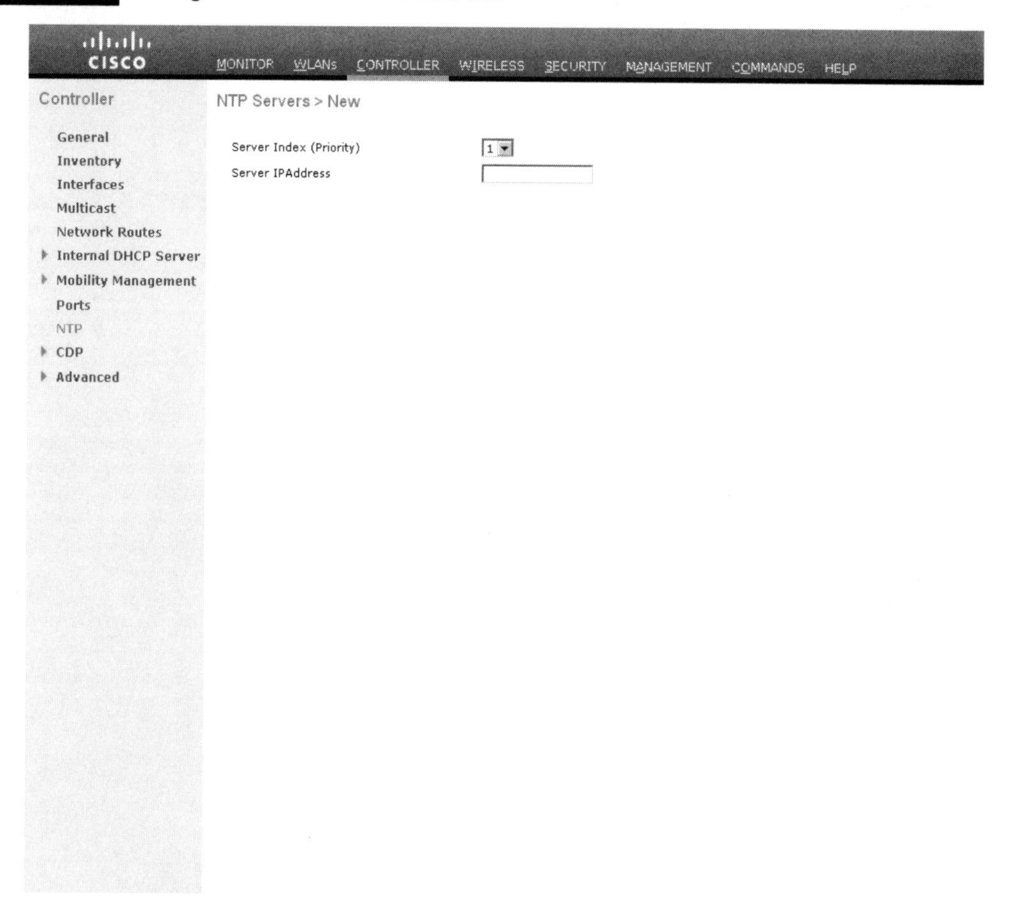

Configure Dynamic Interfaces

Before you can even create a wireless LAN (WLAN) for your wireless clients to connect to, it is necessary to establish the networks and VLANs they will be accessing. In order to create proper "hooks" to the network for these WLANs to attach to, a controller needs to have dynamic interfaces configured. Follow these steps to accomplish this:

1. Under the Controller menu item, select Interfaces. On this screen, you can view and edit all of the currently configured interfaces on the controller, including the required static interfaces. Click the New button to create a dynamic interface; the window shown in Figure 8-13 appears.

FIGURE 8-13 Creating a new interface

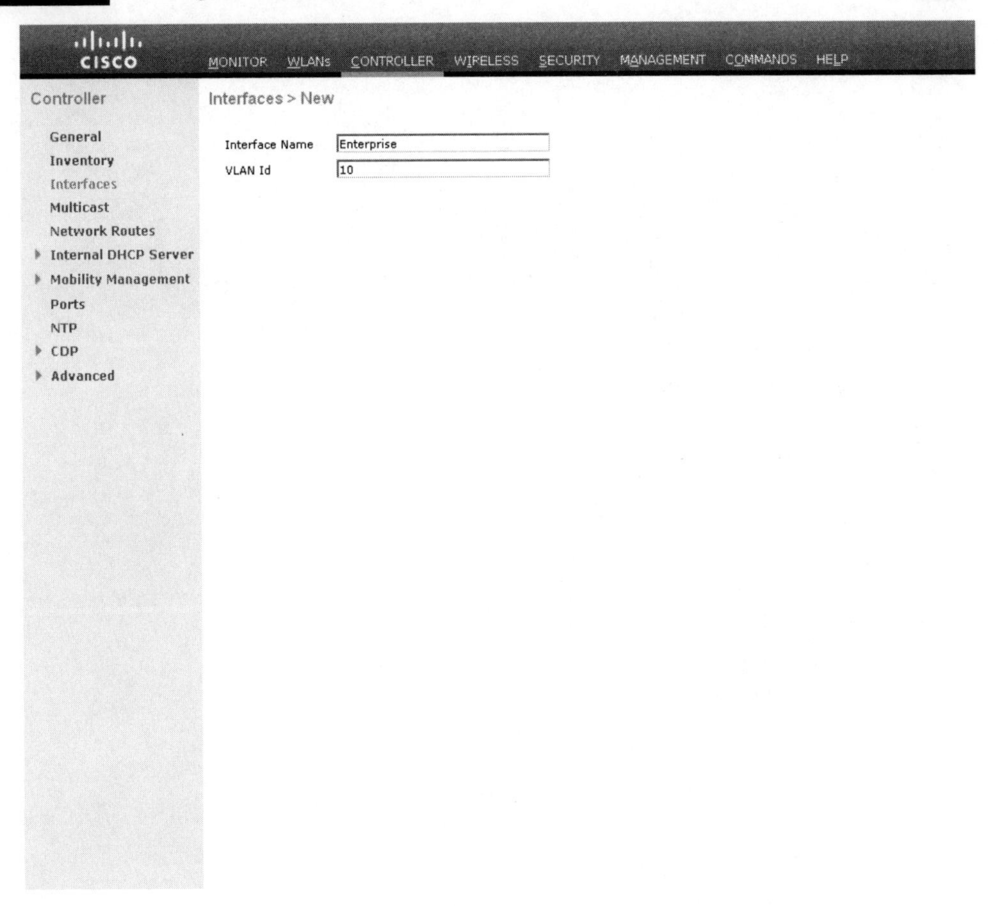

2. Create a meaningful interface name and applicable VLAN number (in an 802.1q trunked link, this is almost always the VLAN tag that is configured on the connected switch) and click Apply. The controller will take you to the Edit window for that interface (Figure 8-14). This is the same window that you will see when editing an existing interface as well.
3. Note that, within a specific network range or VLAN, each controller uses a single IP address for its dynamic interface within that VLAN. This is important when planning IP addressing space and setting DHCP address ranges, especially in networks where there will be multiple controllers providing access to the same VLAN. In the Interface Edit window, you can set the IP address, network mask, and default gateway for the dynamic interface

FIGURE 8-14 Interface Edit window

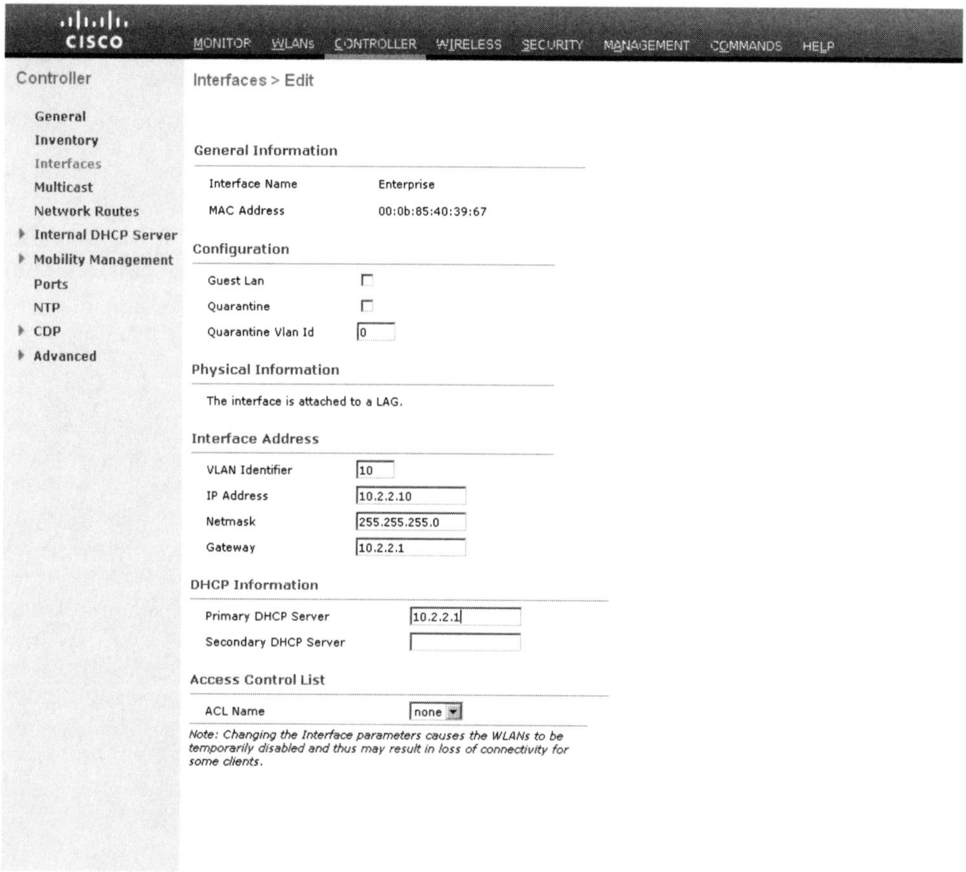

4. In addition to basic IP addressing information for the interface, you can optionally set primary and secondary DHCP servers that are used by the controller's DHCP proxy feature for wireless client IP address requests.

5. After you have verified all of your settings, click Apply at the top of the screen. The controller will take you back to the Controller Interfaces window (Figure 8-15), where you should be able to see the newly created dynamic interface. Make a note of this interface for the section on WLAN configuration.

FIGURE 8-15 Controller Interfaces window

MONITOR WLANs CONTROLLER WIRELESS SECURITY MANAGEMENT COMMANDS HELP

Controller

- General
- Inventory
- Interfaces
- Multicast
- Network Routes
- Internal DHCP Server
- Mobility Management
- Ports
- NTP
- CDP
- Advanced

Interfaces

Interface Name	VLAN Identifier	IP Address	Interface Type	Dynamic AP Management
ap-manager	untagged	10.10.100.132	Static	Enabled
autotrack	802	10.74.54.130	Dynamic	Disabled
enterprise	10	10.2.2.10	Dynamic	Disabled
guest	15	5.1.1.10	Dynamic	Disabled
management	untagged	10.10.100.131	Static	Not Supported
oasis	801	10.74.54.66	Dynamic	Disabled
service-port	N/A	2.2.2.2	Static	Not Supported
virtual	N/A	1.1.1.1	Static	Not Supported

It is rare that the management interface of a Wireless LAN Controller is in the same network as the wireless clients. In an enterprise environment, it is a Cisco best practice to separate control plane traffic and data plane traffic for proper traffic segmentation. By segmenting management traffic and client traffic, we shield the management network from end users and ensure that the management interface is free from unnecessary traffic and network overhead often observed in a LAN environment, such as trivial broadcast traffic.

Configure Security

In this particular scenario, we are building a secured enterprise WLAN with WPA2 and 802.1X security. In order to accomplish this, it is necessary to make sure that there is, at the very minimum, a single RADIUS server that we can use for 802.1X EAP authentications. When you refer back to Chapter 5, you will recall that the Wireless LAN Controller plays the role of authenticator. In this role, the Wireless LAN Controller needs to have access to both the client that is associating as well as the authentication server (RADIUS). Follow these steps to establish the relationship between the authenticator (controller) and the server (your administered AAA RADIUS server):

1. Under the Security menu item, select RADIUS to display a set of submenu items. Click Authentication.
2. Click the New button at the top right hand side of the screen. It will bring you to a window similar to the one seen in Figure 8-16.

FIGURE 8-16 Creating a new RADIUS authentication server

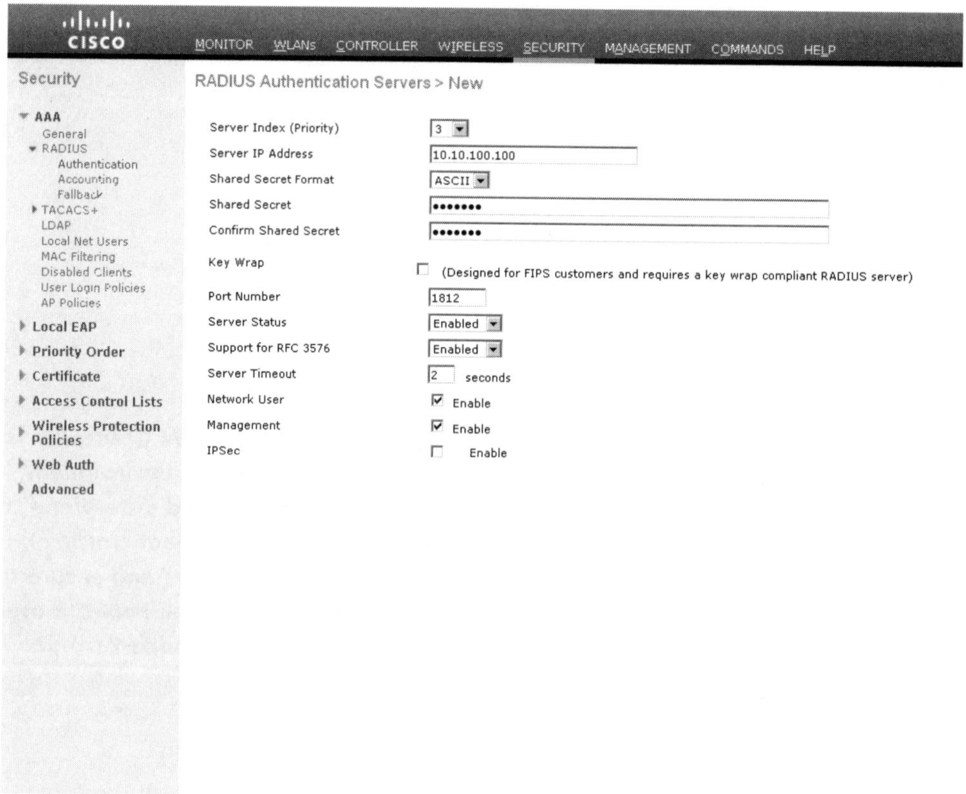

3. Within this display you can configure basic settings for the RADIUS authentication server including its IP address, shared secret, and port number (1812 by default). It is important that you or a security administrator have properly set the controller's management IP address as an AAA client on the server side with a matching shared secret. To ensure that this RADIUS server can be queried by authenticating clients, be sure that the Network User box is checked.
4. After you have verified all of your settings, click Apply at the top of the screen. The controller will take you back to the RADIUS Authentication Servers window (Figure 8-17), where you should be able to see your newly created RADIUS authentication server.

FIGURE 8-17 RADIUS Authentication Servers window

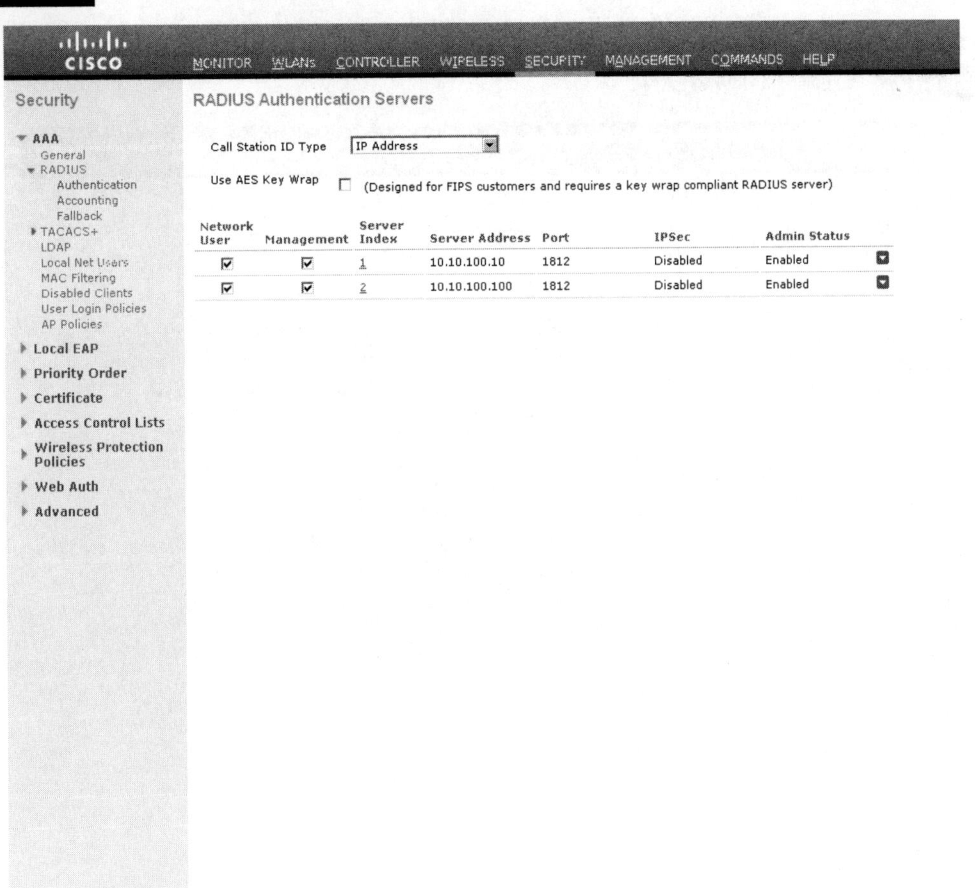

5. In most enterprise deployments, there is more than one RADIUS server that can be queried for wireless authentication for the pure purpose of redundancy. To configure additional RADIUS authentication servers, simply repeat steps 1–4.

Configure Wireless LANs (WLANs)

With a dynamic interface built and RADIUS authentication servers configured on the controller, you now have all of the primary pieces needed to make the WLAN functional for wireless clients. Follow these steps to connect a newly created WLAN what you already have configured.

1. The WLANs menu item, as mentioned earlier, displays all the WLANs configured on the controller. On the upper right hand corner, select Create New from the drop-down box. Click the Go button next to that to begin creating a new WLAN. The window shown in Figure 8-18 will appear.

FIGURE 8-18 Creating a new WLAN

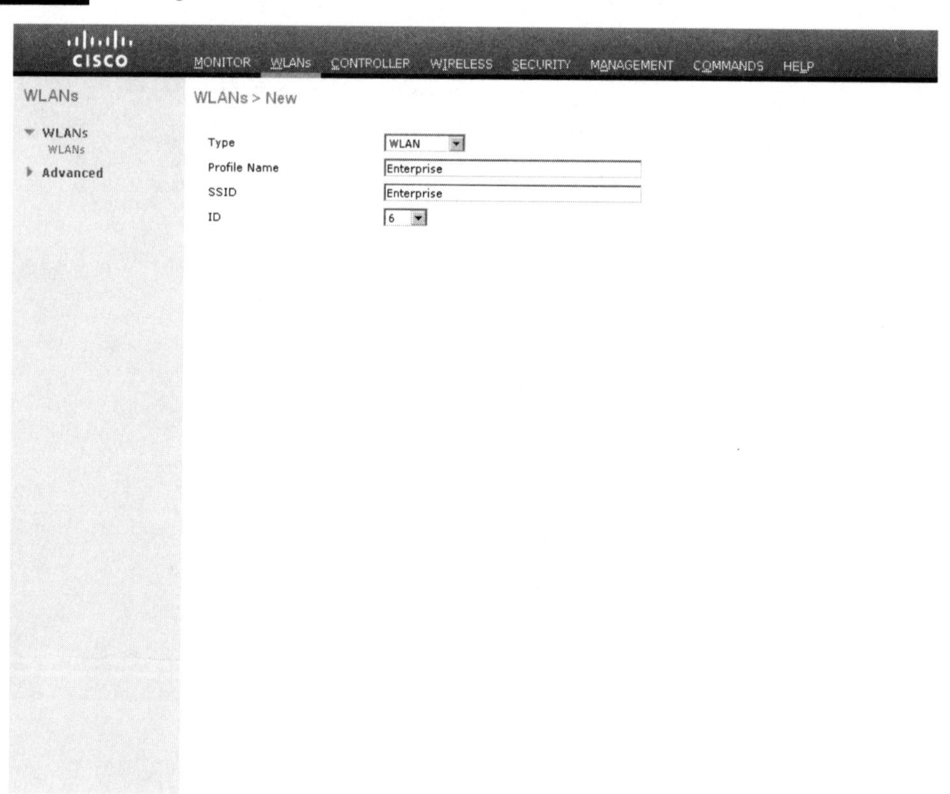

2. Create a meaningful WLAN name and applicable SSID (for the sake of simplicity, simply call this new SSID Enterprise) and click Apply. The controller will take you to the Edit window for that WLAN (Figure 8-19). This is the same window that you will see when editing an existing WLAN.
3. The WLAN Edit window is broken up into tabs for configuring different elements of the WLAN. The General tab allows you to set the interface for the WLAN (or the network where clients associated to this WLAN will go). In this case, you will use the Enterprise dynamic interface that you previously configured. In addition to mapping the WLAN to a dynamic interface, the General tab allows you to enable or disable the WLAN, utilize the Radio Policy to dictate which AP radios will use the WLAN on, and enable or

FIGURE 8-19 WLAN Edit window

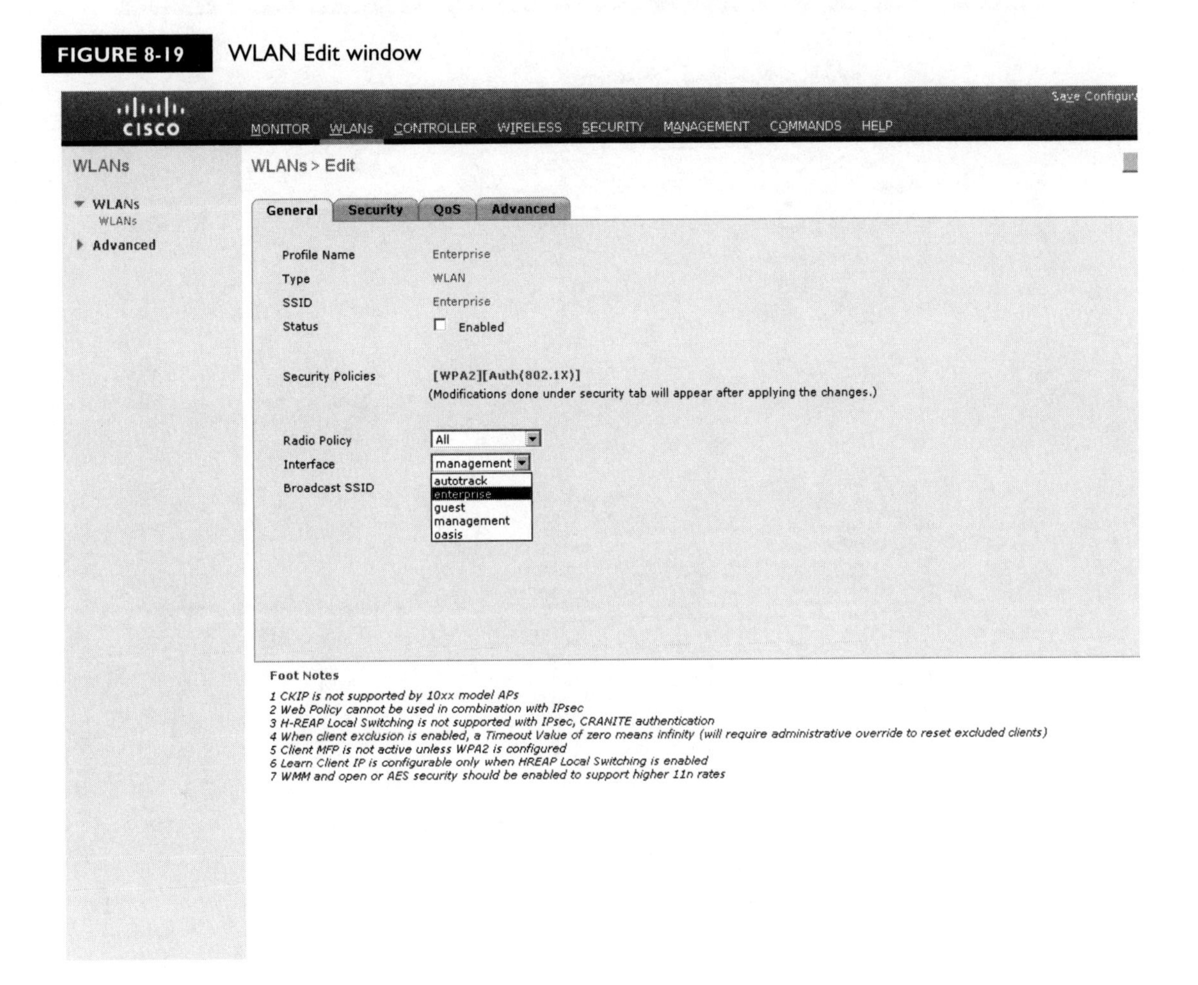

disable the Broadcast SSID feature. In most cases it is recommended that the Broadcast SSID feature be disabled, except in the case of a pervasive guest access network. At this point, we will proceed to configuring security options on the WLAN, which brings you to the Security tab shown in Figure 8-20.

4. The Security tab is also broken up into a subsection of tabs. For the scope of this book, we will only cover what is needed to configure a wireless network to WPA2 802.1X standards. The first tab is the Layer 2 security tab, where the majority of authentication and encryption configuration on the

FIGURE 8-20 WLAN Edit Security Layer 2 window

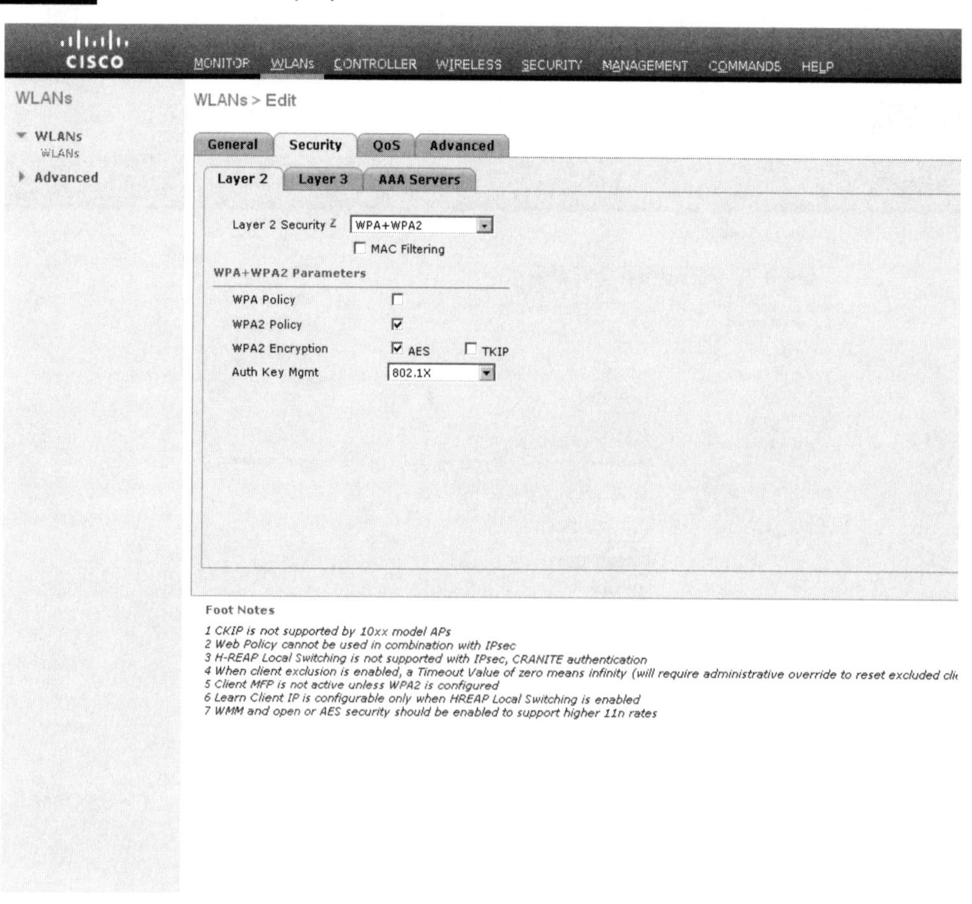

controller is completed. On newer controller code versions, WPA+WPA2 for Layer 2 Security and 802.1X for Auth Key Mgmt are the defaults. In addition, the WPA+WPA2 parameters allow you to choose which WPA versions you want to allow as well as the encryption method that is used. In this scenario, we will stick with the defaults of WPA2 with AES encryption. The Layer 3 tab, which is not covered in the exam, allows you to configure additional Layer 3 security such as VPN and Web authentication (for guest users). The AAA Servers tab is shown in Figure 8-21.

FIGURE 8-21 WLAN Edit Security AAA Servers window

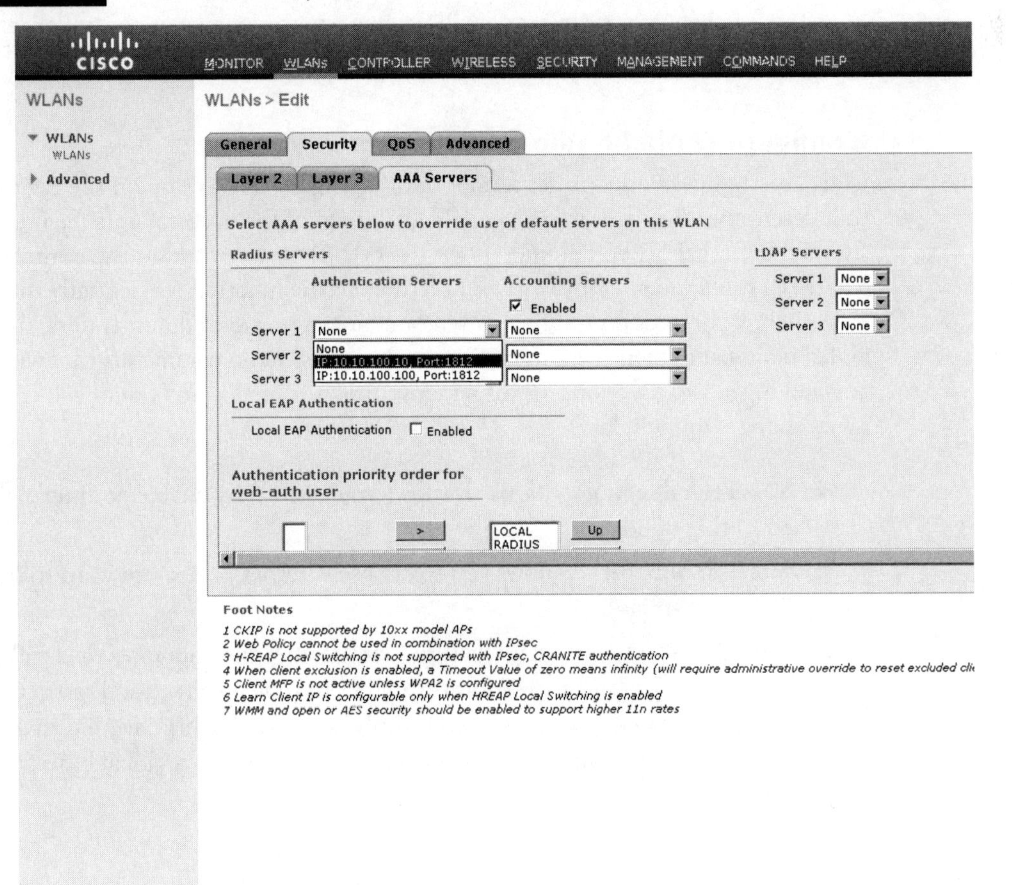

5. In this window, you can specify which AAA servers (configured in the security section) will be used by this VLAN for 802.1X wireless authentication. By clicking the drop-down boxes under Authentication, the servers that you configured in the previous section will automatically appear. When you have selected these servers and verified your settings in all of the tabs in the WLAN section, you can click Apply in the upper right hand corner of the window.
6. With a fully functioning WLAN that is connected to a dynamic/VLAN interface and querying against a known AAA RADIUS server, you have all of the basic elements necessary to support wireless clients. To ensure that these changes are saved and will survive a power outage or controller reboot, click the Save Configuration link at the top of the window.

Configure Global Radio Parameters

In this section, we will look at an important configuration section on the controller that determines the radio frequency (RF) behavior of the access points. Figure 8-22 shows the Global Radio Parameters for the 802.11b/g network on the controller. There is a similar menu for the 802.11a radio, but the function is essentially the same.

Although this section is rarely edited or changed by most administrators, the global radio parameters work in concert with the AP radio parameters that will be covered in the next section. In this window, the most important values you can make changes to include:

- **802.11b/g Network Status** Allows you to globally enable or disable all 802.11b/g radios.
- **802.11g Support** Allows you to choose whether or not you want to support 802.11g standard.
- **Data Rates** A very useful feature that specifies the data rates that will be supported by the access points in order to tune coverage and restrict undesirable clients (such as 802.11b-only devices). At this time, limiting the data rates on the wireless network can only be done on a global basis from this window.

FIGURE 8-22 802.11b/g Global radio parameters

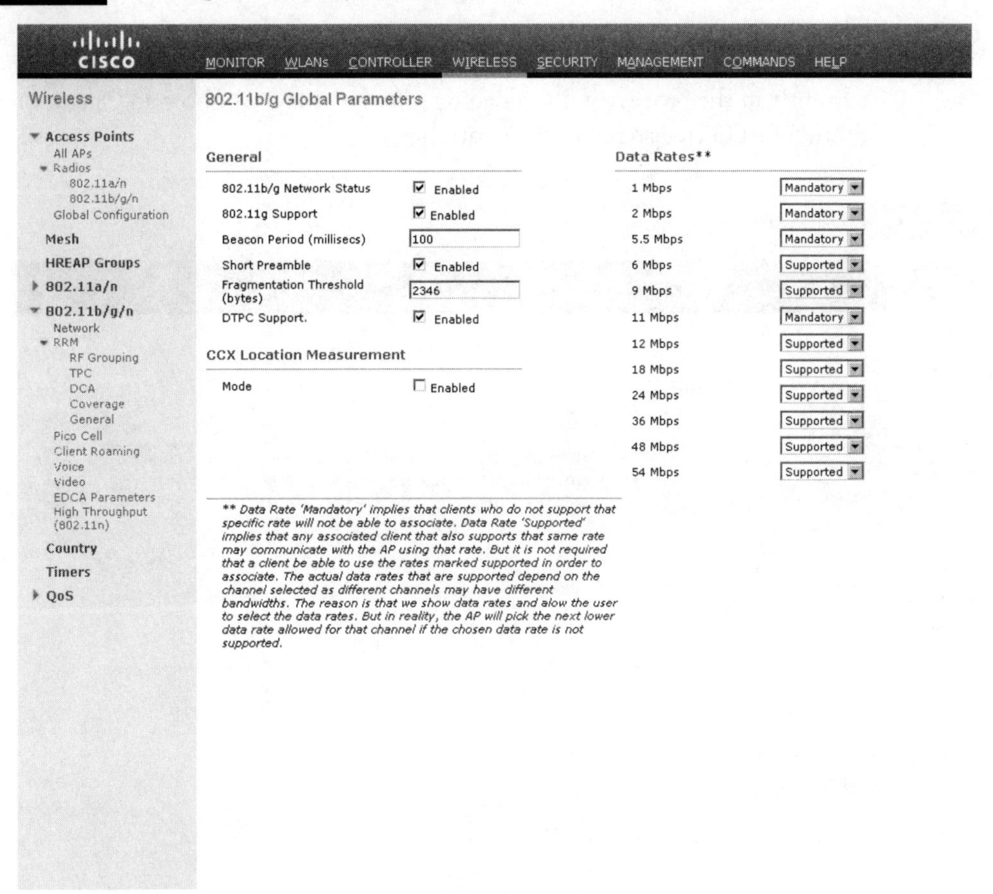

Configuring Access Points Using the Controller

Although the logic of a centralized/unified wireless solution results in lightweight access points, this does not mean that lightweight access points are free of configuration. There are still essential configuration steps for optimizing an access point for management and wireless network performance. This section will cover the most important tasks in AP configuration using the Wireless LAN Controller.

General Access Point Configuration

The Wireless menu item takes you to a full view of all of the access points that are currently associated to the controller (Figure 8-23). Unless the access points are currently in the process of downloading software, they will show an Operational Status of REG (registered to the controller).

FIGURE 8-23 All APs window

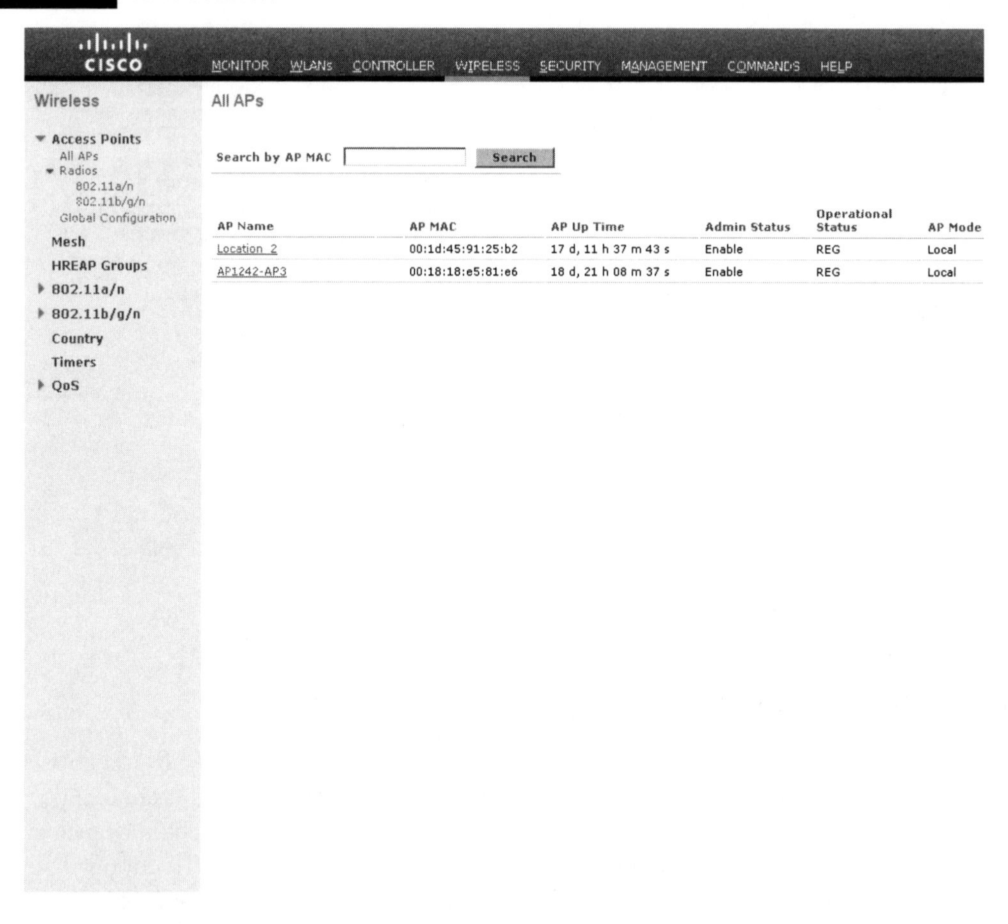

In this display, you can also see information like the AP's MAC address, up time, and the mode it is running in. In addition to this, the Wireless display has a search function so you can troubleshoot individual APs based on their MAC address. To edit the general configuration of an access point, click the AP name, which should be a blue hyperlink. The Details window for that specific AP will appear as shown in Figure 8-24.

FIGURE 8-24 AP Details window

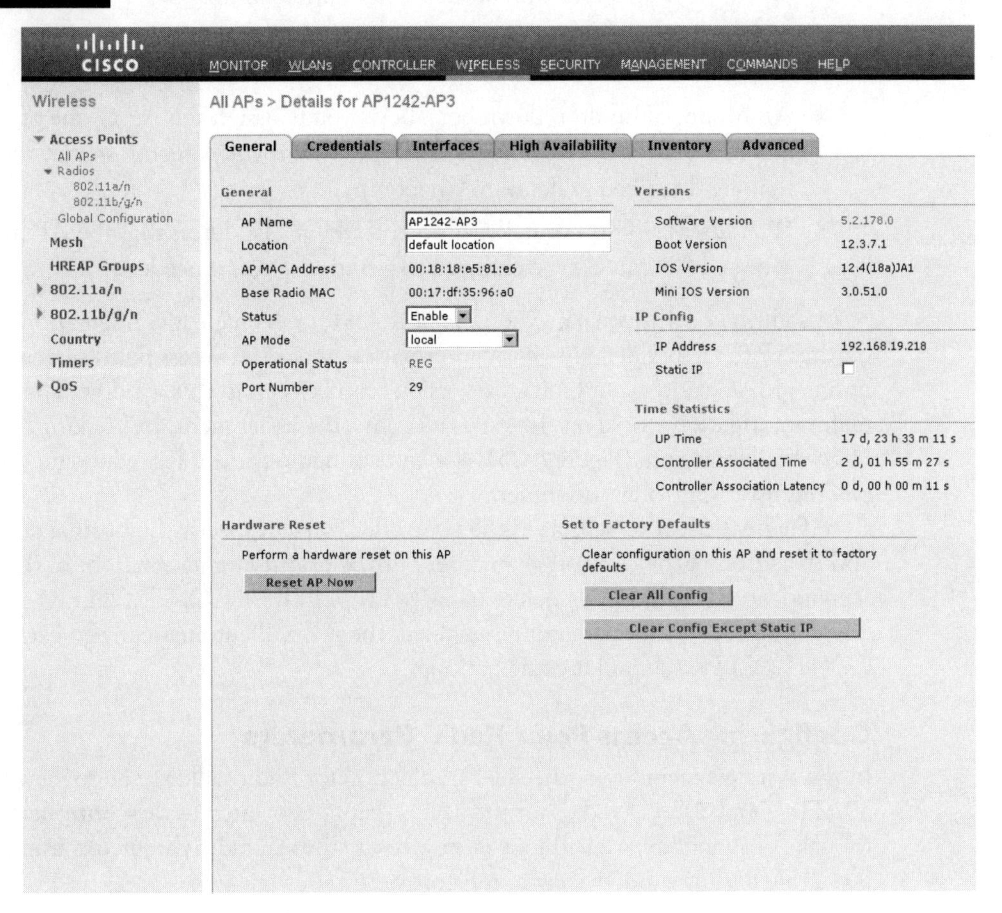

As with the WLANs Edit window, there are a series of tabs for different information and configuration. On the General tab, you can make changes to the following properties of the AP:

- **AP Name** The hostname of the AP. Typically something that helps identify its location and/or role in the network.
- **Location Description** Along with the hostname, this can be used to provide a brief description of the AP for administrators to reference.
- **Status** The AP can be administratively enabled or disabled with this drop-down box.
- **AP Mode** This drop-down box allows you to assign the AP to one of several different AP modes (local, H-REAP, monitor, rogue detector, sniffer, bridge) that are described in detail in Chapter 6.
- **IP Address** This normally shows the DHCP IP address that the AP has acquired but can also be changed to a static address if needed.

In addition to this you can reset/reboot the AP, or even clear configuration on the AP to factory defaults. Basic information about the access point such as uptime, MAC address, and software versions can be seen in this window. The other major information stored on the AP is its controller assignment. By clicking the High Availability tab (Figure 8-25), you can configure the AP for redundancy and deterministic controller assignment.

In Figure 8-25, you see that in the controller you can specify the hostname and IP address of the controllers in order of their priority for association. In this scenario, an AP will only associate to CONTROLLER2 if CONTROLLER1 goes down. If the AP Fallback feature is enabled, the AP will automatically go back to CONTROLLER1 should it come back up.

Configuring Access Point Radio Parameters

In the Wireless menu item, the sidebar selection for Radios allows you to view the 802.11b/g and 802.11a radios for the access points associated to the controller. For the sake of simplicity, we will look at how to modify AP radio parameters using the 802.11b/g Radios window shown in Figure 8-26.

FIGURE 8-25 AP High Availability window

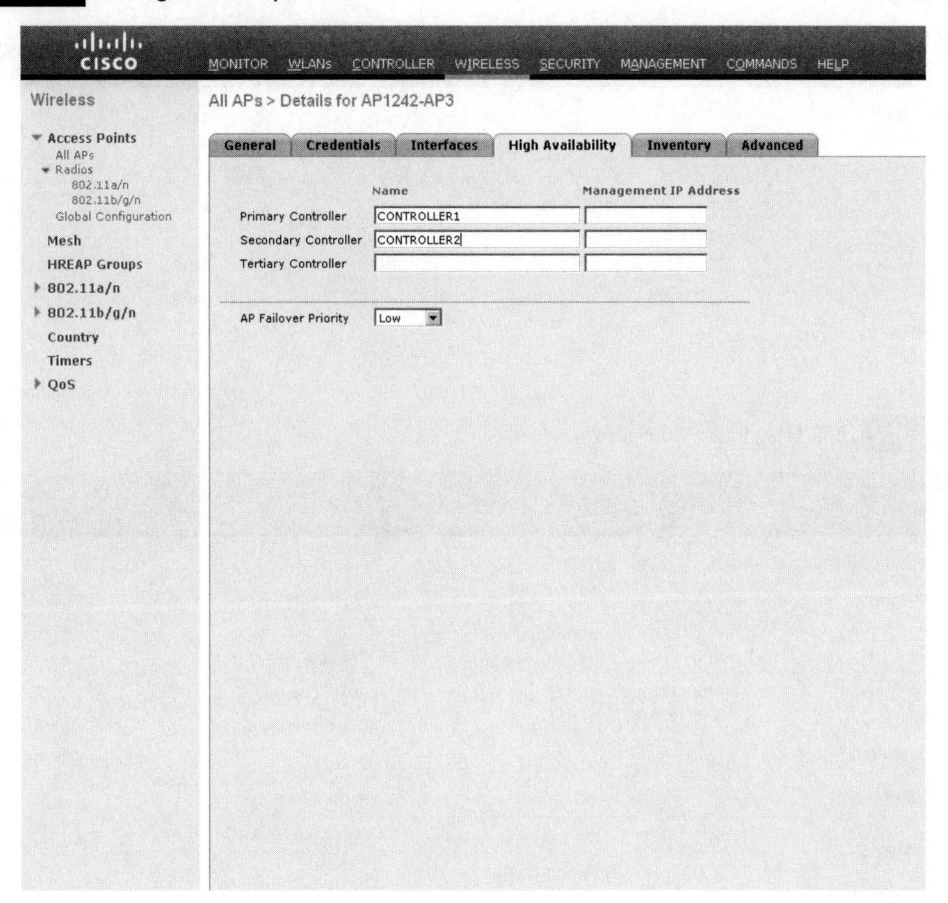

In this window, you can view radio-specific information such as the up/down status of the radio and RF-related information such as transmit power and channel settings. Similar to the All APs window, you can configure individual AP radios by clicking the blue drop-down arrow and selecting Configure to open up the configuration window, shown in Figure 8-27.

FIGURE 8-26 Wireless 802.11b/g Radios window

MONITOR WLANs CONTROLLER WIRELESS SECURITY MANAGEMENT COMMANDS HELP
Save Configuration

802.11b/g/n Radios

AP Name	Radio Slot#	Base Radio MAC	Admin Status	Operational Status	Channel	Power Level	Antenna
AP1242-AP3	0	00:17:df:35:96:a0	Enable	UP	1 *	3	External
Location_2	0	00:17:df:a7:66:80	Disable	DOWN	1 *	1 *	External

* *global assignment*

FIGURE 8-27 802.11b/g Radio Configure window

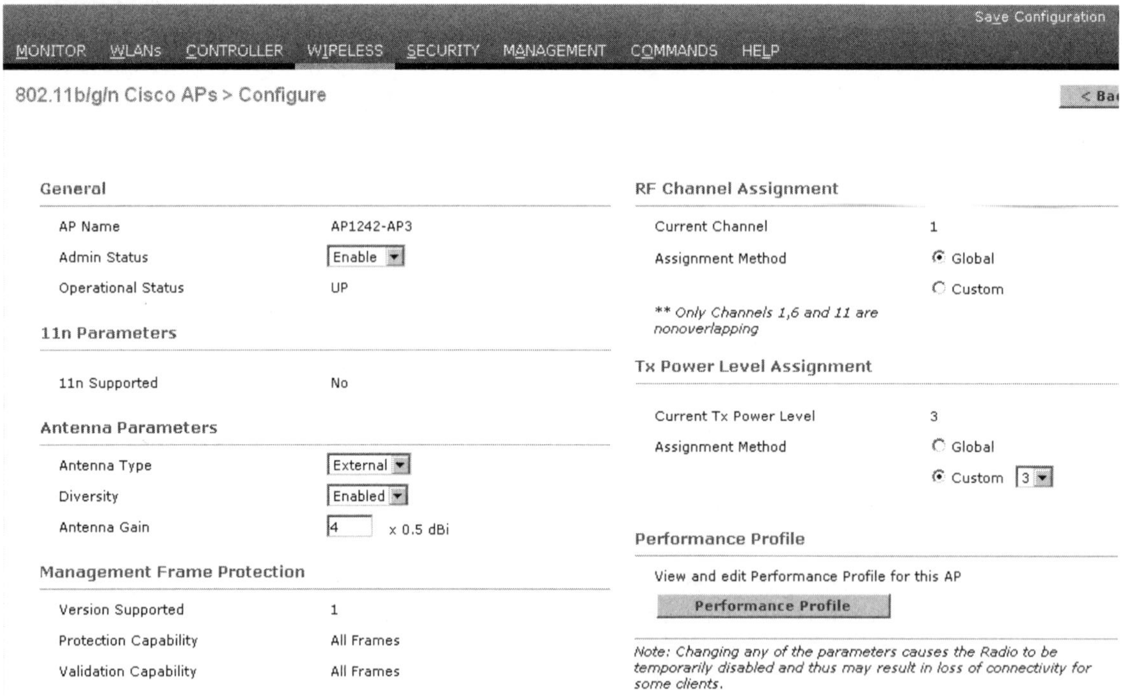

MONITOR WLANs CONTROLLER WIRELESS SECURITY MANAGEMENT COMMANDS HELP
Save Configuration

802.11b/g/n Cisco APs > Configure

< Back

General

AP Name	AP1242-AP3
Admin Status	Enable
Operational Status	UP

11n Parameters

11n Supported	No

Antenna Parameters

Antenna Type	External
Diversity	Enabled
Antenna Gain	4 x 0.5 dBi

Management Frame Protection

Version Supported	1
Protection Capability	All Frames
Validation Capability	All Frames

RF Channel Assignment

Current Channel	1
Assignment Method	(•) Global
	( ) Custom

*** Only Channels 1,6 and 11 are nonoverlapping*

Tx Power Level Assignment

Current Tx Power Level	3
Assignment Method	( ) Global
	(•) Custom 3

Performance Profile

View and edit Performance Profile for this AP

Performance Profile

Note: Changing any of the parameters causes the Radio to be temporarily disabled and thus may result in loss of connectivity for some clients.

In each AP's radio (802.11b/g or 802.11a), you can perform the following key tasks:

- **Admin Status** Allows you to administratively enable or disable this radio on the access point.
- **RF Channel Assignment** Assigning channels is typically reserved for the controller's Radio Resource Management (RRM) feature described in the next section; by default the Global radio button does just that. You can, however, manually set the AP's channel by selecting the Custom radio button and selecting a channel from the drop-down box.
- **Tx Power Assignment** Similar to channel assignment, assigning transmit power is another task typically performed by the RRM feature. By default, the Global radio button defers to the controller's setting, but the Custom radio button allows a similar drop-down box to the channel assignment menu, from which you can select transmit powers. In the transmit power numbering scheme, a power level of 1 indicates the highest power level with higher numbers, indicating lower overall transmit powers. The milliwatt (mW) and dBm values for these vary by country.

CERTIFICATION OBJECTIVE 8.03

Radio Resource Management (RRM) Concepts

Radio Resource Management Overview

Radio Resource Management (RRM), a key feature on Cisco's Wireless LAN Controllers, analyzes the existing RF environment and makes various adjustments on AP power levels and channel configurations in order to improve overall coverage and reduce co-channel interference. RRM is composed of three different algorithms, each of which performs a different task. These algorithms are discussed in the next three sections.

Dynamic Channel Assignment (DCA) Algorithm

The *Dynamic Channel Assignment (DCA)* algorithm's purpose is to optimize AP channel settings based on what it sees from neighboring APs. The primary purpose of DCA is to mitigate co-channel interference by setting APs that are adjacent to each other on nonoverlapping channels within the shared frequency spectrum. Because the process of a channel change on an AP can be a disruptive event for clients associated to that AP, there is a system by which certain APs are more likely to have their channels adjusted. Channel adjustments are performed locally in order to prevent disruptive larger scale channel changes and APs with higher utilization (that is, more users and more traffic) are also less likely to have their channel changed under DCA. In this way, DCA performs its task of AP channel optimization with as little disruption to the overall wireless network as possible.

Dynamic Transmit Power Control (TPC) Algorithm

The *Dynamic Transmit Power Control (TPC)* algorithm runs every 600 seconds by default and is primarily responsible for limiting overcoverage by access points by turning power levels down. Like DCA, this is intended to minimize co-channel interference and optimize overall coverage. In the TPC algorithm, each AP keeps a list of APs sorted by strongest RSSI (Received Signal Strength Indicator) value. TPC will lower AP transmit powers until the AP sees its third strongest neighbor at the preconfigured RSSI level (–70 dBm by default). It is important to note that in order for TPC to work, you will need to have at least four APs in the environment and those APs should technically be transmitting at full power to start for TPC to have its full effect.

Coverage Hole Detection and Correction Algorithm

The *Coverage Hole Detection and Correction* algorithm is similar to TPC in that it adjusts the transmit power settings on the AP radios. Instead of decreasing power, Coverage Hole Detection increases the transmit power of APs based on client signal strength statistics. If a wireless client's signal-to-noise ratio (SNR) falls below a certain point, indicating a disabled or missing AP, the APs in the surrounding areas will increase their transmit power to make up for the coverage loss.

INSIDE THE EXAM

Standalone Access Point Configuration

Because standalone AP configuration does not have a lab requirement on the exam, you will only need to understand the basics of Cisco standalone access points. Be sure to understand the different methods for acquiring the IP address of the AP and the different methods for accessing it. Familiarize yourself with the Express Setup and Express Security screens and know on a basic level how to configure an SSID with security.

Understand how to upgrade access point IOS software and also be familiar with the process of converting an autonomous/standalone AP to LWAPP mode, and vice versa. Know the difference between standalone and lightweight access points.

Configuration of Cisco Unified Wireless LAN Components

The "Configuration of Cisco Unified Wireless LAN Components" section is the most important part of the chapter. It is essential to understand the components of the controller and understand the concepts of ports and interfaces. It is also important to be able to distinguish static interfaces from dynamic interfaces. Understand the relationship between WLANs and dynamic interfaces.

The controller configuration section should give valuable insight into the process of configuring a controller to service wireless clients from the ground up. You should have a complete understanding of the initial steps of building a controller using the CLI Configuration Wizard. Be comfortable with the menu structure of the controller GUI and tasks that are performed under each menu item. Be very familiar with the concepts of building dynamic interfaces and tying them in to WLANs that you build for wireless client access. Understand that for a secure, functional wireless network to exist, there must be a valid VLAN (dynamic interface) and security servers (AAA RADIUS) for the WLAN to attach to in order to provide proper network and authentication services. The step by step process for configuring all of these in order is shown in this section of this chapter.

In addition to this, be aware of the difference between configuring controller properties and settings versus using the controller to configure associated access points. The access point configuration guidelines will show you what can be configured on a per access point basis, and you can see the different settings that can be performed on an access point to select redundant controllers and override RRM settings if need be.

INSIDE THE EXAM

Radio Resource Management (RRM) Concepts

The primary thing you need to pay attention to on the Radio Resource Management (RRM) section is the different algorithms for performing channel assignment and alterations to transmit power on the APs. Be familiar with what triggers each of the algorithms to run and the basic requirements for each. DCA is exclusive to channel assignment, TPC lowers transmit power settings, and Coverage Hole Detection and Correction raises transmit power settings based on client information.

CERTIFICATION SUMMARY

When considering the configuration of a wireless network, it is important to recognize the importance of standalone APs as well as the controller-based unified wireless solution. Cisco standalone APs run IOS and can be configured via console, Telnet, or GUI. After you have gone through the process of discovering the AP IP address—using the console, the DHCP server or the IP Setup Utility (IPSU)—you can access the GUI of the AP. When configuring a Cisco standalone (or autonomous) AP using the GUI, there are several key displays that will allow you to quickly perform configuration on the AP to make it operational. The Express Setup window allows you to configure basic settings such as the APs hostname and IP address and the Express Security window allows you to configure the AP's wireless properties such as SSID and wireless security settings. If needed, autonomous access points can have their software upgraded via the GUI or can be converted to lightweight LWAPP APs using the WCS, the AP Conversion tool, or the same GUI. In addition to this, lightweight APs can also be converted to autonomous APs using a Wireless LAN Controller or TFTP server.

As the need for more robust and centralized wireless solutions arose, the deployment and use of the Cisco unified wireless network increased as well. Also known as a controller-based architecture, this model is widely adopted by large enterprises all over the world. It is important to understand the various concepts behind a controller. Ports are the controller's physical connection to the network (usually through a neighboring switch). Interfaces are virtual/logical entities that specify VLANs and enable the controller to talk to specific parts of the network using ports. WLANs are the pieces of controller configuration that tie wireless SSIDs to dynamic interface VLANs. It is important to first configure dynamic interfaces and security information such as RADIUS servers before building the WLANs that will tie into them. Controllers also contain global radio parameters that affect the RF performance of all associated APs.

AP configuration, although accomplished with a controller, is separate from controller configuration. The information on an AP that is not stored in the controller includes the AP Name; IP address; primary, secondary, and tertiary controller assignments; and the administrative enable/disable state. AP configuration also involves the ability to tune radio settings on the radios of the AP. In most cases, RRM will handle items such as transmit power and channel settings but these values can be overridden on an AP with custom configurations for each.

The Radio Resource Management feature on the Wireless LAN Controllers dynamically makes changes to AP transmit power settings and channel assignments to optimize coverage and performance of the wireless network. It accomplishes this using three algorithms that perform different tasks. Dynamic Channel Assignment (DCA) handles channel assignments to prevent co-channel interference. Dynamic Transmit Power Control (TPC) exists to lower transmit power when more than three neighboring APs are seen at a particular RSSI (default –70 dBm). Coverage Hole Detection and Correction is used to raise transmit power when an AP in a particular coverage area has been disabled.

TWO-MINUTE DRILL

Standalone Access Point Configuration

- Cisco standalone APs run IOS, which can be configured with Telnet, a console CLI, or a GUI.
- A standalone AP's IP address can be discovered by using the console, viewing the DHCP server, or using the IP Setup Utility (IPSU).
- The Express Setup window in a standalone Cisco IOS AP allows you to configure basic settings such as the AP's hostname and IP address.
- The Express Security window sets SSID and wireless security settings.
- Standalone and autonomous access points can have their software upgraded via the GUI, or they can be converted to lightweight LWAPP APs using the WCS, the AP Conversion tool, or the same GUI. Lightweight APs can also be converted to autonomous APs using a Wireless LAN Controller or TFTP server.

Configuration of Cisco Unified Wireless LAN Components

- It is necessary to configure NTP settings to set the time on controllers to ensure proper association of APs.
- Ports are the controller's physical connection to the network.
- Interfaces are virtual or logical entities that specify VLANs and enable the controller to talk to specific parts of the network using ports.
- WLANs are the wireless SSIDs to dynamic interface VLANs.
- Dynamic interfaces and security information such as RADIUS servers are referenced by WLANs and should typically be configured first when network and security information is known.
- Controllers contain global radio parameters that tune the overall wireless network, the most important of which is data rate adjustments.
- AP configuration is separate from controller configuration.
- Configuration items such as AP Name; IP address; primary, secondary, and tertiary controller assignments; and the administrative enable/disable state are all configured and stored directly on the access point.

- AP configuration allows tuning of radio settings. Values such as channel and transmit power can be set to override those set with the controller's RRM feature.

Radio Resource Management (RRM) Concepts

- Radio Resource Management is a controller function that automatically makes changes to radio properties to optimize wireless network performance and coverage. It is made up of three algorithms that make changes to AP settings to accomplish this.
  - **Dynamic Channel Assignment (DCA)** Adjusts channel settings to prevent co-channel interference.
  - **Dynamic Transmit Power Control (TPC)** Lowers transmit power settings to prevent co-channel interference and excessive coverage.
  - **Coverage Hole Detection and Correction** Fills coverage holes and deficient coverage areas by allowing surrounding APs to increase their transmit power settings.

SELF TEST

The following self-test questions will help you measure your understanding of the material presented in this chapter. Read all the choices carefully, as there may be more than one correct answer. Choose all correct answers for each question.

Standalone Access Point Configuration

1. Which of the following things about standalone access points is not true?
   A. Standalone APs run on IOS software.
   B. Standalone APs require static IP addresses while Lightweight LWAPP APs do not.
   C. Most standalone APs can be converted to LWAPP mode and back.
   D. Standalone APs are still used today.
2. Which of the following is not a method for managing a standalone AP?
   A. Telnet
   B. Console port
   C. Wireless controller
   D. Web GUI

Configuration of Cisco Unified Wireless LAN Components

3. Which of the following is something you cannot specify on the controller using the CLI Configuration Wizard?
   A. A dynamic interface for enterprise users
   B. The management interface
   C. A RADIUS server for authentication
   D. A WLAN/SSID for corporate users
4. Which of the following is not true of static and dynamic interfaces?
   A. Both static and dynamic interfaces can be mapped to WLANs.
   B. Static and dynamic interfaces are configured with IP addresses.
   C. Static interfaces and dynamic interfaces can be freely added and removed.
   D. Like dynamic interfaces, static interfaces are logical, not physical, entities.
5. When configuring a dynamic interface for WLAN use, which of the following is not true?
   A. The dynamic interface contains the network information needed by the clients on a WLAN.
   B. You can assign more than one WLAN to a dynamic interface.

C. A dynamic interface uses an IP address within the range that it is created for.

D. Once a WLAN has mapped to a dynamic interface, it can be removed from the controller.

6. When configuring a WLAN, which of the following is not something you can edit?

A. Interface

B. Layer 2 security

C. Authentication servers

D. DHCP address pool

7. When configuring an access point's general properties, which of the following functions will affect the manner in which the access point behaves on the overall wireless network?

A. AP Name

B. Location

C. AP Mode

D. IP address

8. When comparing the access point's radio settings with controller radio settings, which of the following parameters can not be altered on the access points?

A. Allowed data rates

B. Channel selection

C. Transmit power

D. Radio enable/disable

Radio Resource Management (RRM) Concepts

9. When configuring access point radios, which of the following is not true in regard to Radio Resource Management (RRM)?

A. An access point radio set to Custom power or channel settings overrides RRM.

B. An access point radio set to Global power or channel settings follows RRM.

C. An access point is set to use Custom power settings by default.

D. An access point is set to use Custom power settings by default.

10. Which of the following RRM algorithms does not adjust the transmit power of the access points?

A. Dynamic Channel Assignment (DCA)

B. Coverage Hole Detection

C. Dynamic Transmit Power Control (TPC)

D. All of the above adjust the transmit power of the access points.

SELF TEST ANSWERS

Standalone Access Point Configuration

1. Which of the following things about standalone access points is not true?

A. Standalone APs run on IOS software.

B. Standalone APs require static IP addresses while Lightweight LWAPP APs do not.

C. Most standalone APs can be converted to LWAPP mode and back.

D. Standalone APs are still used today.

☑ **B.** Although it is beneficial to have static IPs on standalone APs, it is not a hard requirement for them to function properly.

☒ **A, C,** and **D** are incorrect because they all represent characteristics of standalone APs.

2. Which of the following is not a method for managing a standalone AP?

A. Telnet

B. Console port

C. Wireless controller

D. Web GUI

☑ **C.** A wireless controller is used only to manage lightweight LWAPP APs. Standalone APs do not use a controller at all.

☒ **B, C,** and **D** are incorrect because they are all valid methods for managing a standalone AP.

Configuration of Cisco Unified Wireless LAN Components

3. Which of the following is something you cannot specify on the controller using the CLI Configuration Wizard?

A. A dynamic interface for enterprise users

B. The management interface

C. A RADIUS server for authentication

D. A WLAN/SSID for corporate users

☑ **A.** Unfortunately, dynamic interfaces are not one of the interfaces that can be configured in the CLI Configuration Wizard on Wireless LAN Controllers.

☒ **B, C,** and **D** are incorrect because they are all settings that can be configured using the CLI Configuration Wizard.

4. Which of the following is not true of static and dynamic interfaces?

A. Both static and dynamic interfaces can be mapped to WLANs.

B. Static and dynamic interfaces are configured with IP addresses.

C. Static interfaces and dynamic interfaces can be freely added and removed.

D. Like dynamic interfaces, static interfaces are logical, not physical, entities.

☑ **C**. Static interfaces cannot be removed at all, and dynamic interfaces cannot be removed if they have a WLAN mapped to them.

☒ **A**, **B**, and **D** are incorrect because their statements are all true of static and dynamic interfaces.

5. When configuring a dynamic interface for WLAN use, which of the following is not true?

A. The dynamic interface contains the network information needed by the clients on a WLAN.

B. You can assign more than one WLAN to a dynamic interface.

C. A dynamic interface uses an IP address within the range that it is created for.

D. Once a WLAN has mapped to a dynamic interface, it can be removed from the controller.

☑ **D**. A dynamic interface cannot be removed from a controller unless it is no longer bound to any WLANs. In addition to this, removing a dynamic interface would remove the necessary network settings that the WLAN and its clients depend on.

☒ **A**, **B**, and **C** are incorrect because they describe valid properties and behaviors of dynamic interfaces in relation to WLANs and the controller.

6. When configuring a WLAN, which of the following is not something you can edit?

A. Interface

B. Layer 2 security

C. Authentication servers

D. DHCP address pool

☑ **D**. DHCP addressing configuration is strictly a function of the dynamic interface. The WLAN configuration plays no role in that or IP addressing for clients, other than passing them to the assigned interface.

☒ **A**, **B**, and **C** are incorrect because they are all editable configurations on a WLAN.

7. When configuring an access point's general properties, which of the following functions will affect the manner in which the access point behaves on the overall wireless network?
   A. AP Name
   B. Location
   C. AP Mode
   D. IP address

☑ C. The WPA2 standard requires AES/CCM and does not support the use of TKIP.

☒ **A**, **B**, and **D** are incorrect because they are all valid according to the WPA and WPA2 standards.

8. When comparing the access point's radio settings with controller radio settings, which of the following parameters can not be altered on the access points?
   A. Allowed data rates
   B. Channel selection
   C. Transmit power
   D. Radio enable/disable

☑ A. Allowed Data Rates are a property that can currently be configured only on the controller level on a global basis.

☒ **B**, **C**, and **D** are incorrect because they represent options that can be configured on the Access Point Configuration window.

Radio Resource Management (RRM) Concepts

9. When configuring access point radios, which of the following is not true in regard to Radio Resource Management (RRM)?
   A. An access point radio set to Custom power or channel settings overrides RRM.
   B. An access point radio set to Global power or channel settings follows RRM.
   C. An access point is set to use Custom power settings by default.
   D. An access point is set to use Custom power settings by default.

☑ D. Access points are *not* set to custom power settings by default.

☒ **A**, **B**, and **C** are incorrect because they are all true statements about AP settings in relation to Radio Resource Management.

10. Which of the following RRM algorithms does not adjust the transmit power of the access points?

A. Dynamic Channel Assignment (DCA)

B. Coverage Hole Detection

C. Dynamic Transmit Power Control (TPC)

D. All of the above adjust the transmit power of the access points.

☑ **A**. Dynamic Channel Assignment (DCA) is the only RRM algorithm that does not adjust transmit control power. Its only role is to adjust channel assignments on APs.

☒ **B** and **C** are incorrect because both algorithms that adjust transmit control power: Coverage Hole Detection raises transmit powers and TPC lowers it.

9

Understanding and Deploying the Wireless Control System

CERTIFICATION OBJECTIVES

9.01	The Wireless Control System (WCS) and Navigator
9.02	Installing and Configuring the Wireless Control System (WCS)
9.03	Monitoring with the Wireless Control System
✓	Two-Minute Drill
Q&A	Self Test

As the size of wireless networks has grown, the centralized/unified wireless model has had to evolve to respond to this growth. As a result, the Cisco Wireless Control System (WCS) was introduced as a GUI-based management platform for the Cisco Unified Wireless Network (CUWN). In this chapter we will look at where the WCS fits into the overall scheme of a wireless LAN deployment; we will also describe how to use WCS for centralized wireless configuration and monitoring.

CERTIFICATION OBJECTIVE 9.01

The Wireless Control System (WCS) and Navigator

In order to understand how best to use the WCS, it is necessary to provide information about the WCS and the WCS Navigator products before discussing their day-to-day use. The WCS is a very powerful tool for maintaining an enterprise wireless network and, in this section, we will give an overview of its role and function.

The Wireless Control System (WCS)

The Cisco Wireless Control System (WCS) is a server-based product that can run on Microsoft Windows Server, Red Hat Linux, or VMware. The WCS is typically deployed in large enterprise environments and is intended to simplify management of larger wireless LAN environments by centralizing management tasks over multiple wireless LAN controllers. Like any network management device, the WCS is often deployed in a centralized location to configure and monitor controllers and access points throughout the corporate infrastructure.

WCS Requirements

The Cisco Wireless Control System is supported on the following operating systems at the time of this writing:

- Windows 2003 SP1 or later
- Red Hat Linux AS/ES v4 (release 4.2 and later) and Red Hat Linux AS/ES v5 (releases 4.2.x or 5 or later)
- VMware ESX Server 3.0.1 or later

The Cisco Wireless Control System has the following minimum server requirements at the time of this writing:

- **Cisco WCS High-End Server** (3000 lightweight access points, 1250 standalone access points, 750 wireless LAN controllers) Intel Xeon Quad Core CPUs; 3.16 GHz, 8GB RAM, 200GB DD (free space)
- **Cisco WCS Standard Server** (2000 lightweight access points, 1000 standalone access points, 450 wireless LAN controllers) Intel Dual Core CPU; 3.2 GHz, 4GB RAM, 80GB HDD (free space)
- **Cisco WCS Low-End Server** (500 lightweight access points, 200 standalone access points, 125 wireless LAN controllers) Intel CPU; 3.06 GHz, 2GB RAM, 50GB HDD (free space)

WCS Versions and Licensing

The WCS can be ordered in two different versions:

- **WCS Base** Supports standard Cisco WCS capabilities.
- **WCS PLUS** A more feature-rich version that includes all WCS Base features plus the ability to track the location of a single Wi-Fi device; perform N:1 High Availability (HA) for WCS servers; and integrate with the Cisco 3300 Series Mobility Services Engine (MSE), which can expand location capabilities by adding Cisco Context-Aware Software to simultaneously track up to 18,000 wireless devices.

In addition to these requirements, the Cisco Wireless Control System has a variety of licensing options depending on the size and need of the deployment. These options are broken down into two categories:

- **Single Server License** For single server deployments, these licenses can be procured in increments of 50, 100, or 500 APs supported by each WCS. This license type is available with WCS PLUS.

- **Enterprise Server License** For single or multiple server deployments, these licenses can be procured in increments of 1000, 2500, 10,000, or 50,000 APs and supported over multiple WCS platforms.

There are a wide variety of licensing options for current customers and customers who are transitioning from the Wireless LAN Solution Engine (WLSE) that are detailed in the *Cisco Wireless Control System (WCS) Licensing and Ordering Guide* on Cisco Connection Online (CCO) at www.cisco.com/en/US/prod/collateral/wireless/ps5755/ps6301/ps6305/product_data_sheet0900aecd804b4646.html.

WCS Connectivity/Ports

The WCS uses the ports shown in Table 9-1 for proper operation of standard and optional features. The primary communication medium used by WCS to monitor and configure controllers and access points is the Simple Network Management Protocol (SNMP).

TABLE 9-1 WCS Ports

Port	Function
21	FTP
25	SMTP (e-mail alerts)
69	TFTP
80	HTTP Web Access
UDP 161-162	SNMP Traps
UDP 169	SNMP Trap Receiver
443	HTTPS Secure Web Access
1299	WCS RMI
1315	Database
6789	Web Listening Port
8005	Web Container (Java)
8009	Web Container (Java)
8456	HTTP Connector
8457	HTTP Connector Redirect

WCS Features

The WCS has the following features, most of which are possible because of its use of the Simple Network Management Protocol (SNMP):

- **Controller and AP Configuration** As mentioned before, the WCS uses SNMP to verify and change configurations on controllers and access points. WCS can simplify configuration of multiple devices through the use of templates.
- **Fault and Alarm Monitoring** The controllers send SNMP traps to the WCS, which gathers them in a central repository for operations and network administrators to view and respond to.
- **Wireless Network Planning with Maps and Context-Aware/Location Services** Using uploaded maps and floor plans, network administrators can get a graphical view of access point configurations and locations. These maps also tie directly into Context-Aware/Location Services.
- **Wireless LAN Troubleshooting Tools** The WCS has various connectivity and troubleshooting tools for access points, controllers, and clients, which can aid in problem resolution.

The WCS can also integrate with Cisco products such as the Cisco 2700 Series Location Server and the newer-generation Cisco 3300 Series Mobility Services Engine (MSE) to provide the following features:

- **Context Aware/Location Services** As opposed to standard location services on the WCS PLUS, the MSE and Location Server allow multiple simultaneous wireless devices to be tracked and monitored on WCS maps.
- **Active Wireless Intrusion Protection System (wIPS)** An advanced security feature available only on the MSE, this feature also requires a monitor mode access point to perform detection and mitigation of various security threats on a wireless network.

The WCS Navigator

The *WCS Navigator* is a device that sits directly above the WCS in the wireless network management hierarchy. The WCS Navigator is designed to manage multiple Cisco Wireless Control Systems to provide a unified view of the network. Using SOAP/XML over HTTPS to communicate with individual Wireless Control Systems, Navigator is able to monitor and generate reports on multiple Wireless Control Systems. The WCS Navigator is typically used in large-scale wireless networks that have geographically diverse locations.

WCS Navigator Requirements

The following server hardware and software is required to support Cisco WCS Navigator for Windows or Linux:

- 3.2 GHz Intel dual core processor with 4GB of RAM and an 80GB hard drive
- 40GB minimum available disk space on your hard drive

WCS Navigator Features

The Cisco WCS Navigator provides the following features:

- Management of multiple Cisco Wireless Control Systems
- Support for up to 20 Cisco WCS management platforms
- Management of up to 30,000 Cisco Aironet lightweight access points from a single management console
- Overview and multilevel monitoring of wireless networks
- The ability to run reports for all Wireless Control Systems
- Network-wide searches

CERTIFICATION OBJECTIVE 9.02

Installing and Configuring the Wireless Control System (WCS)

The installation and configuration of WCS is an important process to understand. This section will give specific guidance on the proper installation and deployment of the WCS platform. In this book, and for the purposes of the exam, we will cover WCS from the Windows Server perspective.

Installing/Upgrading WCS

Once you have built or acquired a server that meets or exceeds the WCS minimum requirements, the process for installing is much like that of any other Windows application. After you run the WCS installation executable file, a series of windows with the following prompts will appear:

- **License Agreement** Click the radio button for acceptance of the terms and then click the Next button.
- **Install/Upgrade (If Applicable)** In servers where there is already a WCS or previous version of WCS installed, you will receive a prompt to upgrade. In nearly all instances of a WCS upgrade, the best process is to back up the current WCS database and then uninstall the older version before attempting to install the new version.
- **Check Ports** This window prompts you to change the system port information for items like HTTP and HTTPS. Typically, these are left at the default values.
- **Set Root Password** The root is the primary administrative account for the WCS. This password will grant you initial administrative access. Because of the importance of this account, it is strongly urged that you set a strong password for the WCS root.

- **FTP/TFTP Setup** This window allows you to select folders on the local server hard drive for the FTP and TFTP servers.
- **Choose Interfaces** This window allows you to select the roles of each network interface card (NIC) in the server in relation to the WCS.
- **Installation Folder** This window lets you specify the directory where you want the WCS program files to be installed.
- **Installation Summary/Verification** If you want to proceed with installation, click Next or Finish where applicable and the installation should complete.

When installation is complete, a shortcut folder under the Start menu of the Windows Server should appear for the WCS. From there you can perform functions like backing up and restoring the WCS database, as well as stopping and starting the WCS service. Note that once WCS is successfully installed on a Windows server, it is treated as a service and not as a standard application.

Initial WCS Login

To perform the initial login to the WCS, direct your browser to the IP address of the WCS using the https:// prefix. A window similar to the one shown in Figure 9-1 should appear.

FIGURE 9-1 WCS Login screen

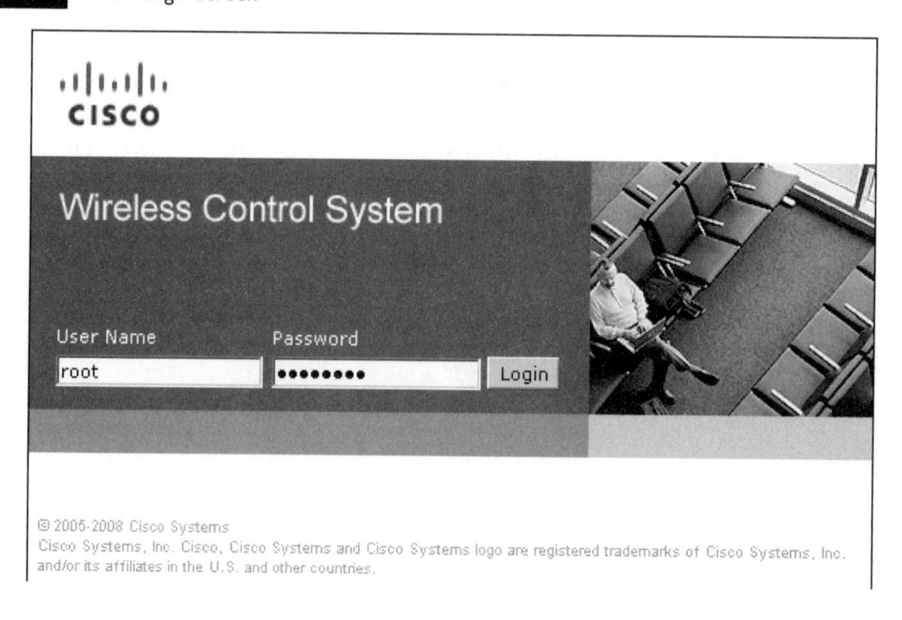

Use the root username with the root password that was configured during installation to gain administrative access to the WCS. When login is successful, you will see a window similar to the one in Figure 9-2.

In this display there is a series of tabs that allows you to monitor various pieces of the WCS:

- **General** The initial tab displayed upon WCS logon. It displays information such as a historical client count graph, the status of the controllers, radios and location servers managed by the WCS, and general coverage information, largely predicated on having maps uploaded into the WCS.

FIGURE 9-2 WCS Home display

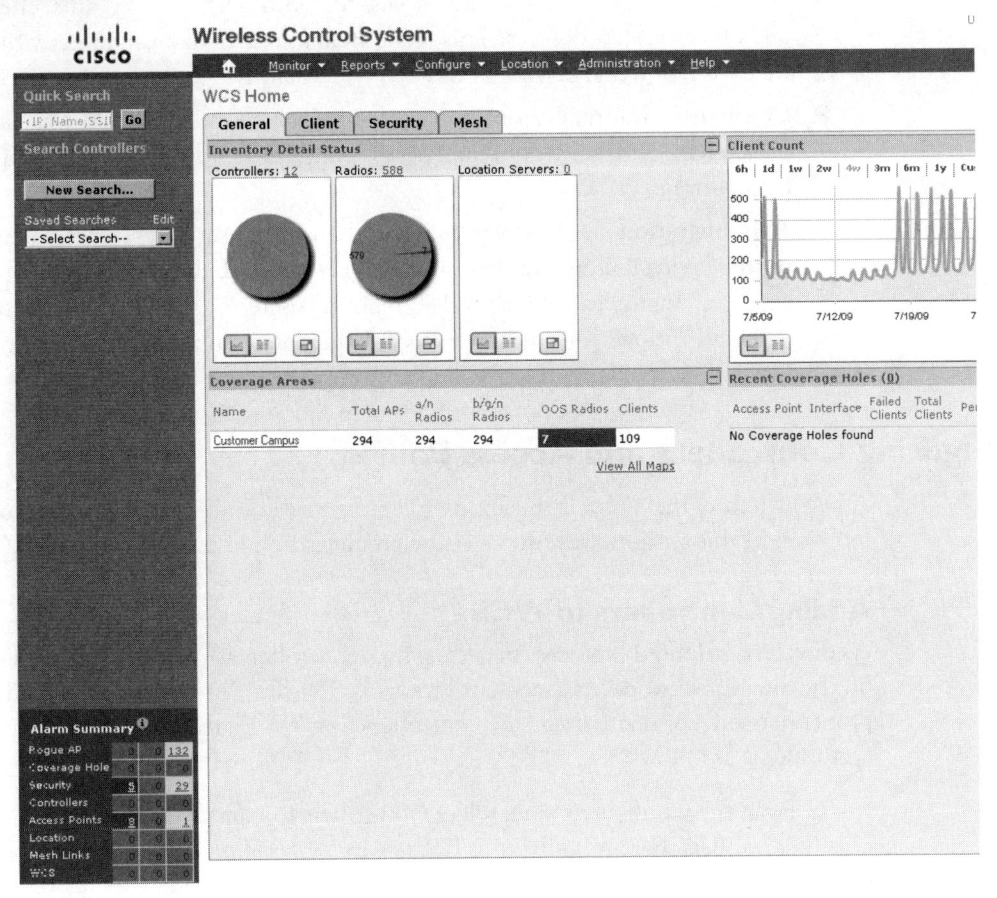

- **Client** The first step in client troubleshooting. This tab provides a list of the most active APs by client count, a list of client-related alarms, and a historical graph of client traffic.
- **Security** A concise display of security alerts occurring on the controller, including rogue AP detections, detected attacks, and general security alarms (typically related to authentication).
- **Mesh** A way to track wireless mesh statistics in networks where the Cisco Lightweight Mesh APs are being utilized. For indoor enterprise networks, this tab is not often utilized.

In addition to the tabs on the Home display, there are three menu items that will be the focus of this chapter:

- **Monitor** Allows you read-only access to controllers, access points, clients, and other wireless devices. This menu item also will take you directly to the Maps function and allow you to view WCS alarms.
- **Configure** Informally seen as a read-write selection. From here you can make changes to controllers and access points either directly or via template configuration.
- **Administration** Allows you to make changes to user access rules, including setting up local users, configuring the WCS for RADIUS or TACACS+ access, changing passwords, and configuring user groups. The Administration menu also allows you to change WCS behaviors such as changing polling intervals for background tasks and custom settings for logging.

Configuring Controllers and Access Points

A key benefit of the WCS is the ability to configure multiple controllers from one location. In this section, we will cover the configuration of controllers using WCS.

Adding Controllers to WCS

As discussed before, the process of adding new controllers to WCS depends heavily on the Simple Network Management Protocol (SNMP). Assuming that there is proper network connectivity to the controllers from WCS, the process of adding a controller is as follows:

1. From the Configure menu, select Controllers to view and edit all the controllers that are added to WCS. A window similar to Figure 9-3 will appear.

FIGURE 9-3 All Controllers window

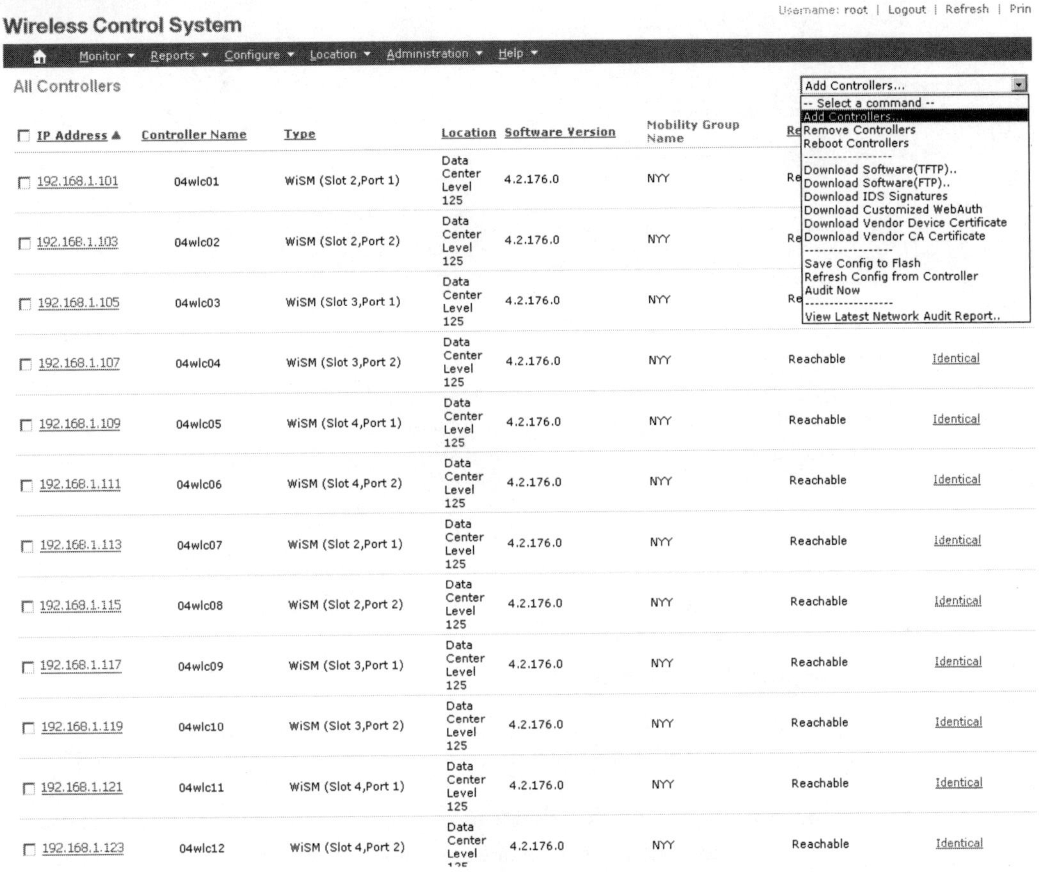

Username: root | Logout | Refresh | Prin

Wireless Control System

Monitor ▾ Reports ▾ Configure ▾ Location ▾ Administration ▾ Help ▾

All Controllers

Add Controllers...

-- Select a command --
Add Controllers...
Remove Controllers
Reboot Controllers

Download Software(TFTP)..
Download Software(FTP)..
Download IDS Signatures
Download Customized WebAuth
Download Vendor Device Certificate
Download Vendor CA Certificate

Save Config to Flash
Refresh Config from Controller
Audit Now

View Latest Network Audit Report..

IP Address ▲	Controller Name	Type	Location	Software Version	Mobility Group Name	Re	
192.168.1.101	04wlc01	WiSM (Slot 2,Port 1)	Data Center Level 125	4.2.176.0	NYY	Re	
192.168.1.103	04wlc02	WiSM (Slot 2,Port 2)	Data Center Level 125	4.2.176.0	NYY	Re	
192.168.1.105	04wlc03	WiSM (Slot 3,Port 1)	Data Center Level 125	4.2.176.0	NYY	Re	
192.168.1.107	04wlc04	WiSM (Slot 3,Port 2)	Data Center Level 125	4.2.176.0	NYY	Reachable	Identical
192.168.1.109	04wlc05	WiSM (Slot 4,Port 1)	Data Center Level 125	4.2.176.0	NYY	Reachable	Identical
192.168.1.111	04wlc06	WiSM (Slot 4,Port 2)	Data Center Level 125	4.2.176.0	NYY	Reachable	Identical
192.168.1.113	04wlc07	WiSM (Slot 2,Port 1)	Data Center Level 125	4.2.176.0	NYY	Reachable	Identical
192.168.1.115	04wlc08	WiSM (Slot 2,Port 2)	Data Center Level 125	4.2.176.0	NYY	Reachable	Identical
192.168.1.117	04wlc09	WiSM (Slot 3,Port 1)	Data Center Level 125	4.2.176.0	NYY	Reachable	Identical
192.168.1.119	04wlc10	WiSM (Slot 3,Port 2)	Data Center Level 125	4.2.176.0	NYY	Reachable	Identical
192.168.1.121	04wlc11	WiSM (Slot 4,Port 1)	Data Center Level 125	4.2.176.0	NYY	Reachable	Identical
192.168.1.123	04wlc12	WiSM (Slot 4,Port 2)	Data Center Level 125	4.2.176.0	NYY	Reachable	Identical

2. If this is a new WCS, there will be no controllers listed in this window. In the drop-down box on the top right hand side, select Add Controllers and click the Go button. A window similar to Figure 9-4 will appear.
3. In this window, you will be able to enter in the IP address and SNMP information (version, read-write community string, timeouts) for the controller to manage it. Optionally, you can upload a CSV file with a list of controllers and SNMP strings.
4. Click OK to add the controller. A success message should appear and the controller will be added to the list of controllers on the WCS.

FIGURE 9-4 Add Controllers window

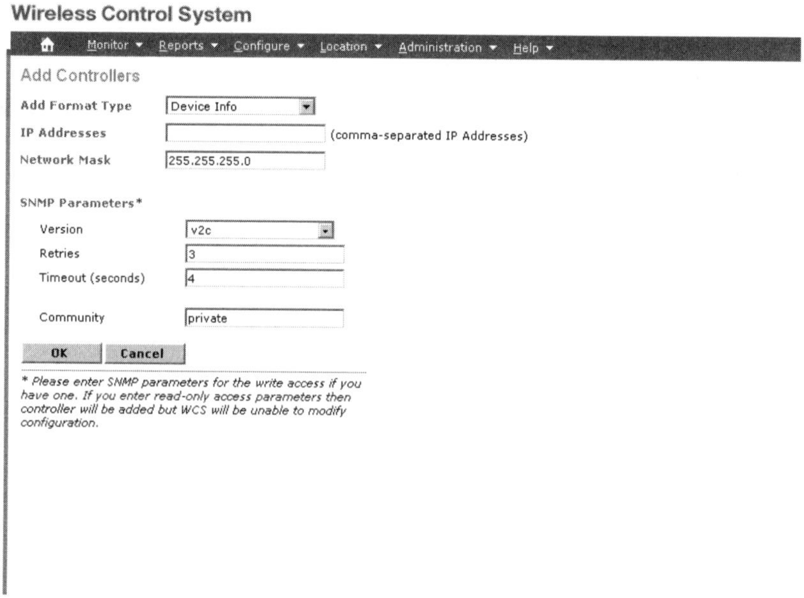

Important Commands

In the drop-down box to add controllers there are several important commands managing the controllers in the network:

- **Save Config to Flash** Performs the Save Configuration command on all checked controllers to make sure that configuration or configuration template changes survive a controller restart. It is essential that you issue this command after making any direct or template-based controller configuration change.
- **Refresh Config from Controller** Allows you to quickly update the WCS with the latest information from the controller, especially when changes have been made directly on controllers that WCS may not be aware of.
- **Audit Now** Somewhat related to the Refresh Config from Controller command but does not actually revert the WCS to the controller's configuration. Instead, this command will compare what is known on the WCS to what is configured on the controller and give you the option to refresh config from controller or push the WCS values to the controller.

- **Download Software** When we discussed upgrading or changing controller software code on WCS in Chapter 8, this was the command on WCS we referred to. With this command, you can use a TFTP or FTP server (on the WCS or elsewhere) to push new controller software out to one or several controllers that are managed by the WCS.

Configuring Controllers Using WCS

After adding the controller to the WCS, you can directly click the name of the controller you want to configure; you will see a window similar to Figure 9-5.

FIGURE 9-5 Controller Properties window

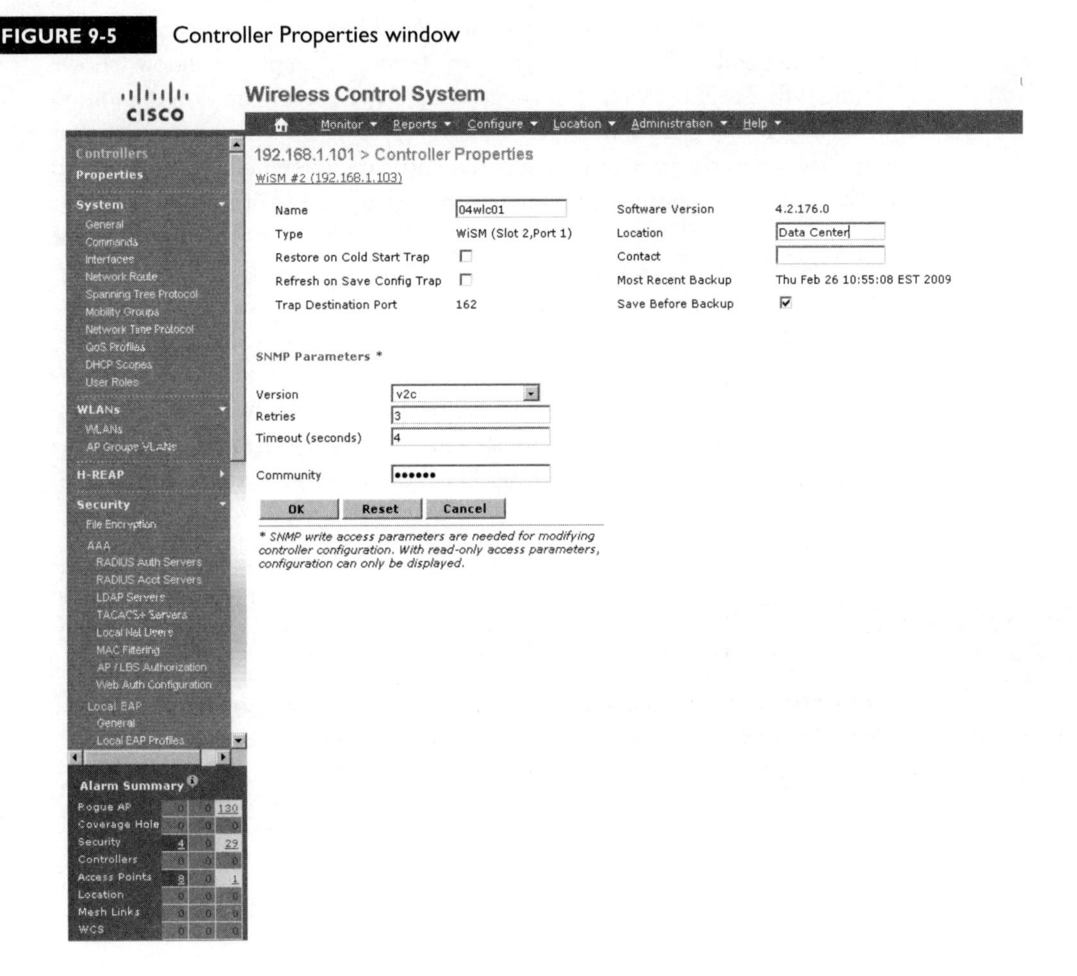

As you can see from the Controller Properties window, the menu selections and configurations are very similar to the ones covered in Chapter 8. Under the System sidebar menu, you can see the General and Interfaces configuration that we covered in Chapter 8; you can also see the configuration sections for WLANs and RADIUS Authentication Servers under the WLANs and Security sections. Configuring a controller using the WCS is very much like configuring the controller directly, except that it is being performed via SNMP instead of directly on the controller GUI itself.

Understanding Controller Templates

Controller templates are one of the features of WCS that makes it an excellent tool for managing multiple wireless LAN controllers. Under the Configure menu, you can select Controller Templates to go to the General Template window (shown in Figure 9-6). From here you can create templates for configuration of multiple controllers. At first glance, this screen looks very similar to the Controller Properties window shown in the previous section. The primary difference between a controller template and direct controller configuration is that instead of going into individual controllers to make changes to WLANs and RADIUS settings, you actually build templates for individual configuration pieces. These templates can then be applied to one or more controllers. Templates can also be removed from the controllers. In this way, a WCS controller template is not just a push function.

Note that unique controller settings such as dynamic interfaces, IP addresses, and hostnames cannot have templates due to specificity of the information. Controller templates are best used for shared items such as RADIUS authentication servers, WLANs, and universal settings like mobility groups and more general configurations.

exam watch

Controller configuration on WCS is done holistically on a per-controller basis. Controller templates are built and applied on a per-configuration element basis and can be applied to multiple controllers simultaneously.

FIGURE 9-6 General Template window

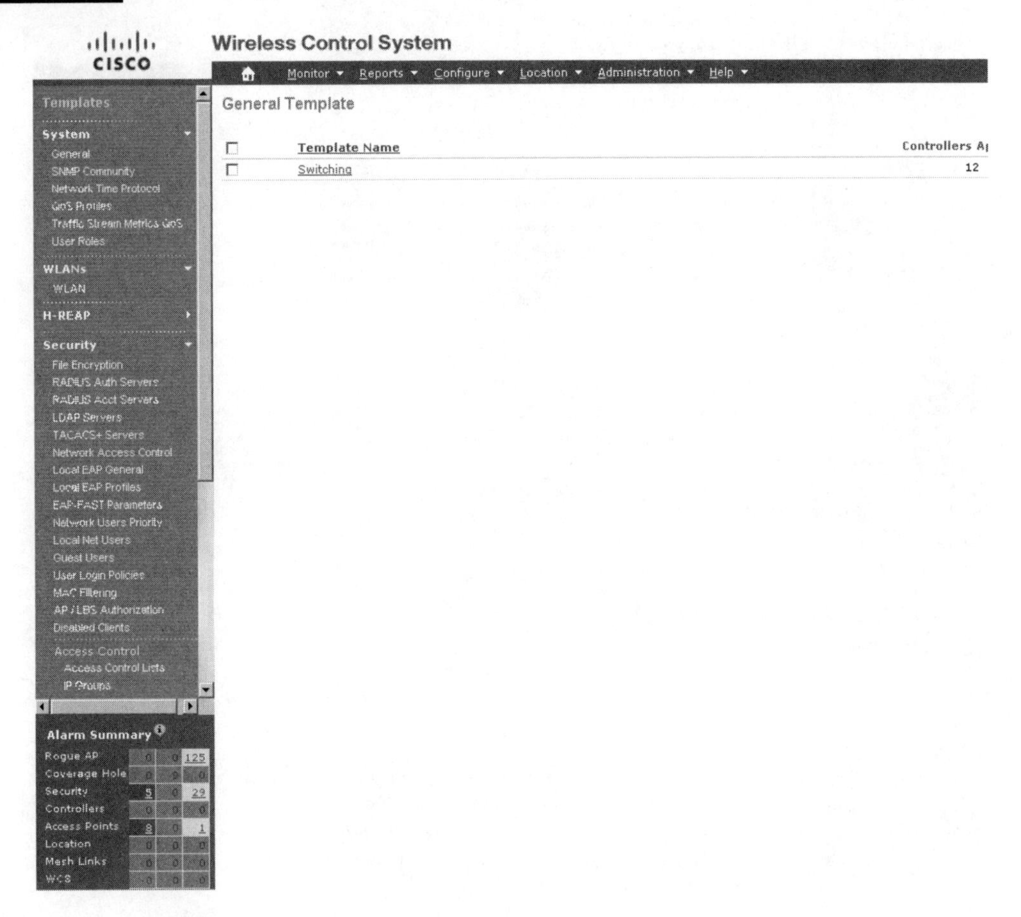

Configuring Access Points Using WCS

Like configuring controllers, configuring access points is a straightforward process with WCS. Under the Configure menu, select Access Points to see a list of access points from all of the controllers that have been added to WCS. Clicking the link to any of these access points will bring you to a window that is similar to the one shown in Figure 9-7.

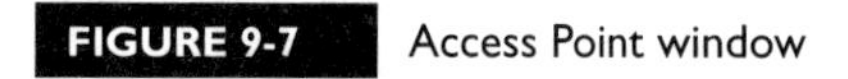

FIGURE 9-7 Access Point window

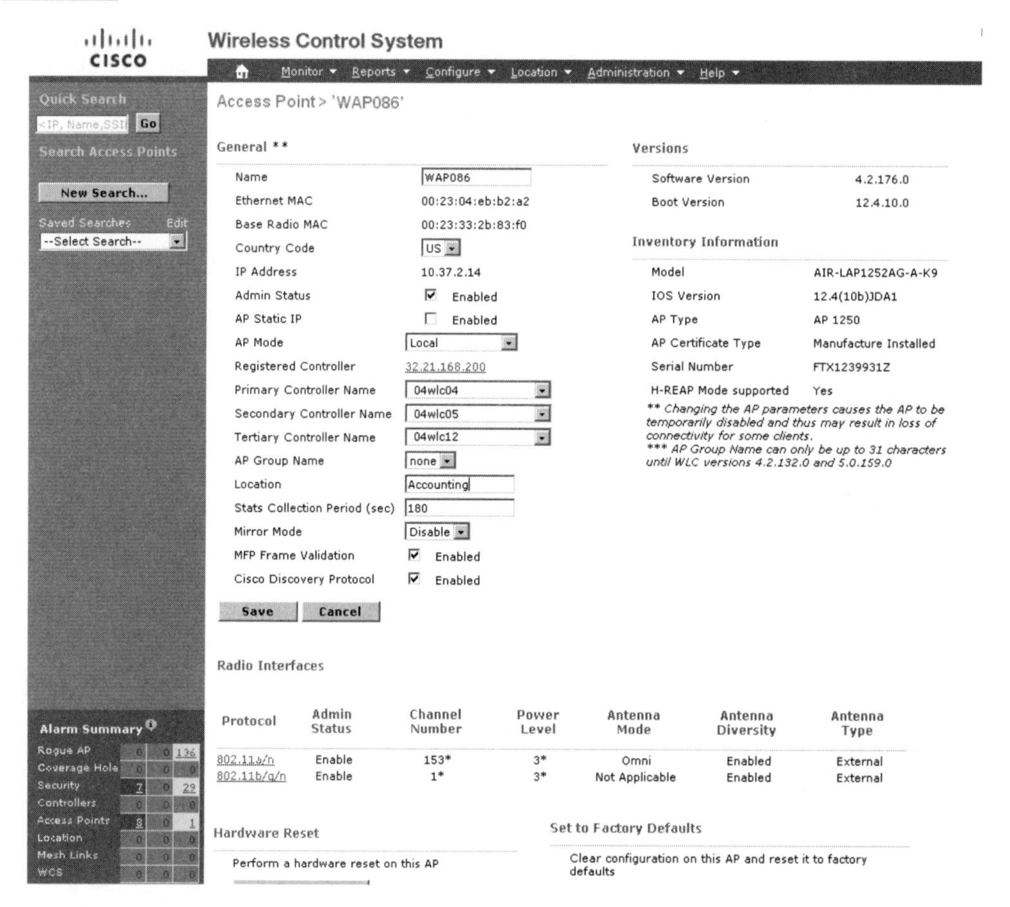

Configuring an access point on WCS is also much like configuring it on a wireless LAN controller. As seen in Figure 9-7, the same AP-specific configurations can be performed using WCS via an SNMP push as through the wireless LAN controller. An added benefit of the WCS is the relative ease with which you can configure controller assignments to individual access points. If a controller you need to associate to an access point is on the WCS, you can simply utilize the drop-down box under the Primary, Secondary, or Tertiary Controller setting and choose any of the controllers that the WCS has relationships with.

Understanding Access Point Templates

Access point templates work a little differently from controller templates because they're based on a Java application. Under the Configure menu, select Access Point Templates to see a list of various access point templates, often named for the tasks that they are built to accomplish. You can edit any of these existing templates by clicking them or using the drop-down box on the upper right hand side of the screen and clicking the Go button. In either case, you will see a window similar to the one in Figure 9-8.

FIGURE 9-8 AP/Radio Templates window

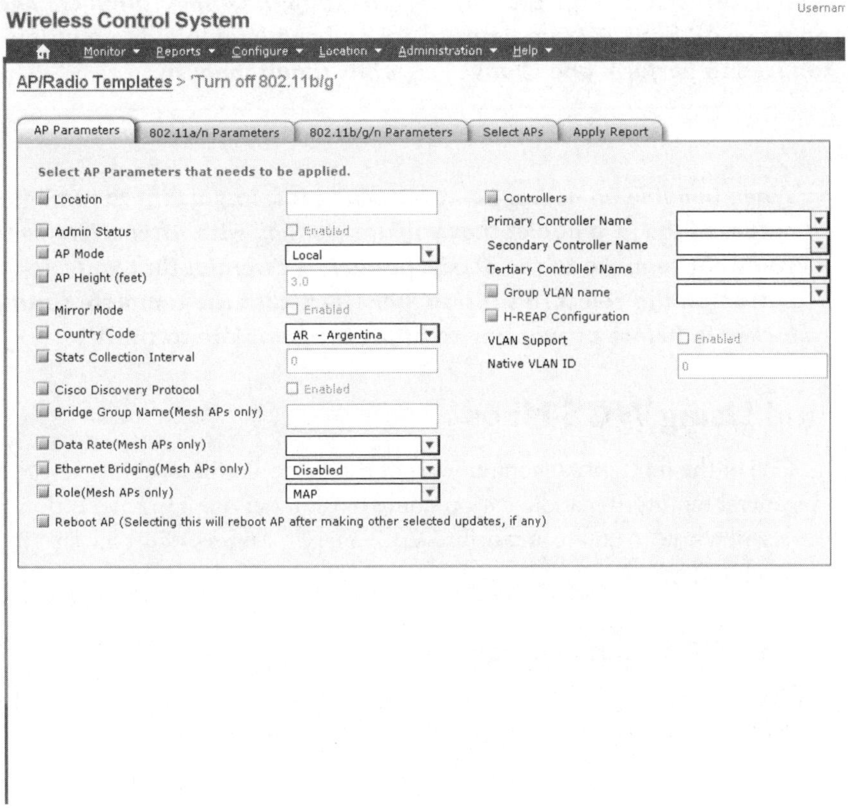

On access point templates, there are multiple tabs that you can select for making changes to AP or AP radio settings. A single access point template can make changes to as many or as few functions on the AP as you desire—just select the check box for the configuration item you wish to apply. On the Select APs tab you can select one or several APs by various criteria (such as controller, location, and so on) and apply the settings in the template to those access points. The primary benefit of the AP template, as with controller templates, is the ability to make changes to multiple APs without having to manually configure each AP individually.

Access point configuration on WCS is on a per-AP basis. AP templates can be customized to perform one change or a group of configuration changes on an AP and can be applied to multiple access points simultaneously.

When building an access point template, it is important to remember the nature of the Java applet that you are working with. After selecting the changes you want to make to the access point, it is essential that you click the Save button on the Select APs tab to make sure that the template settings are locked in before attempting to apply the template to the AP.

Creating and Using WCS Maps

One of the most visual components of the WCS lies in its ability to use maps for general monitoring and location/context aware services. In this section we will describe how to create and utilize WCS maps for more robust monitoring of the wireless network.

The WCS Maps Hierarchy

In order to properly add and build maps, it is important to understand the difference between the classes and categories of maps on the WCS:

- **Campus** In the WCS maps hierarchy, the campus is the largest overall entity. A campus is an optional category and is typically used to help organize locations for very large wireless networks. Examples of campuses are a hospital location with multiple buildings or a cluster of buildings in a single geographical area such as an office park. In some implementations you will find that outdoor maps and even outdoor access points are placed at the campus level.

- **Building** A building is a single structure within a campus. In some WCS implementations a building without a campus is sufficient from an organizational perspective. Buildings can be single or multiple floors.
- **Floor** Within each building are multiple floors. The individual floors are usually where the bulk of WCS maps are uploaded. Although maps and coverage for individual floors can be uploaded to WCS as part of a building, there is currently no feature for relating map information between floors.

Adding and Editing WCS Maps

When you know what kind of environment you are working with, creating campuses, buildings, and floors should be a relatively basic task, as you will see in this section.

1. Click the Monitor menu item and then click Maps to be brought to a list of maps similar to what you see in Figure 9-9.

FIGURE 9-9 WCS Maps window

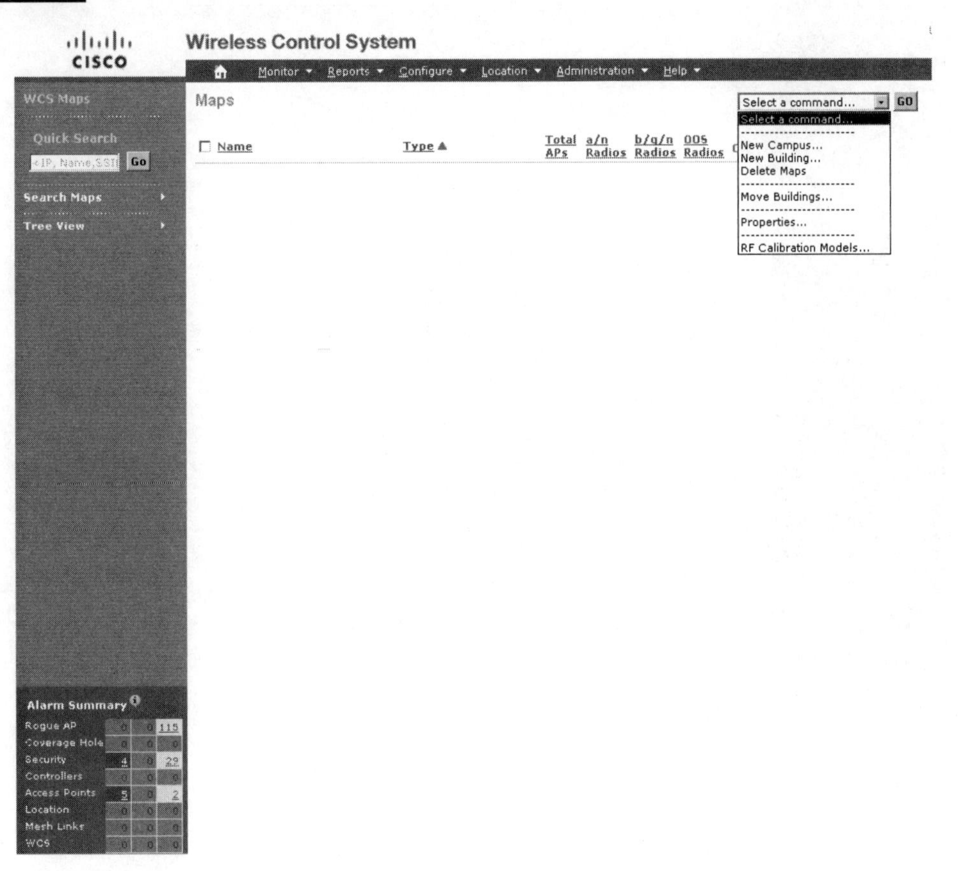

2. You can start by building a new campus or building. For the sake of instruction, we will build a new campus first. From the drop-down menu, select New Campus to see the window shown in Figure 9-10.
3. After entering in the name and image (if applicable) of the campus, click the Next button to verify the appearance of the campus and its dimensions before clicking OK to confirm. In the WCS Maps window, the campus you have created will appear as a new entry.

FIGURE 9-10 New Campus window

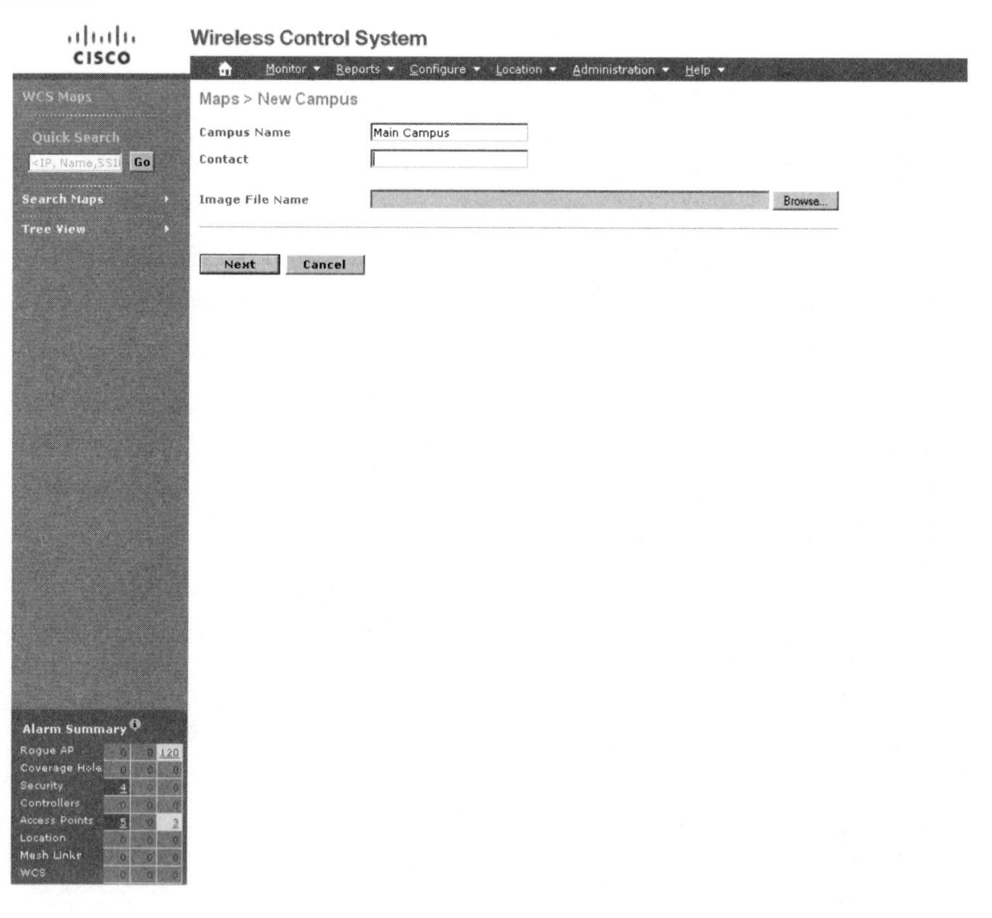

4. You can click back into the campus you have created to create a building within the campus, or you can independently create a building from the same WCS Maps drop-down box. Once you have started the process of adding a building from the WCS Maps window, the window in Figure 9-11 will appear.
5. In this view, you can add information about the building such as the number of floors and the dimensions. Do not worry about being accurate if you are not sure about the information, as this can always be changed later. Click the OK button. A window similar to Figure 9-12 will appear.

FIGURE 9-11 New Building window

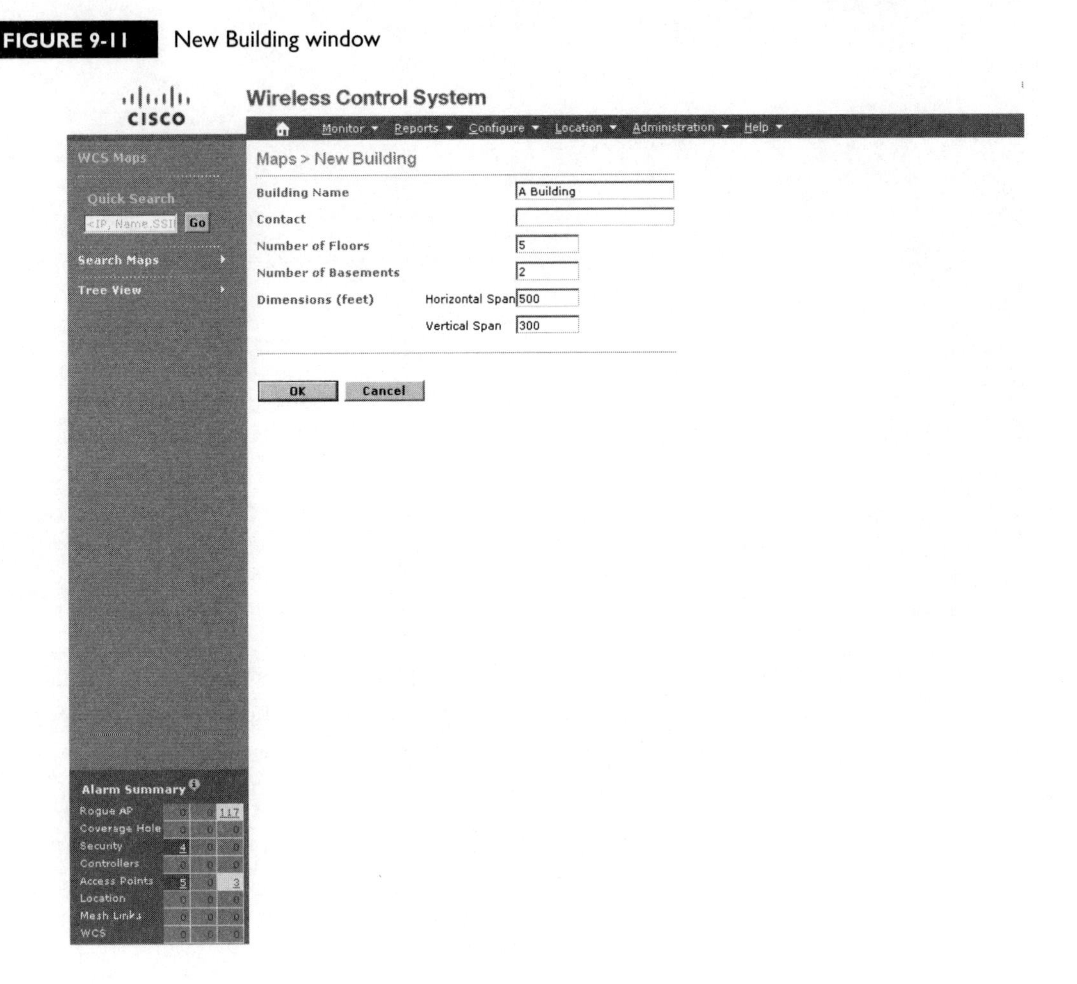

FIGURE 9-12 A Building window

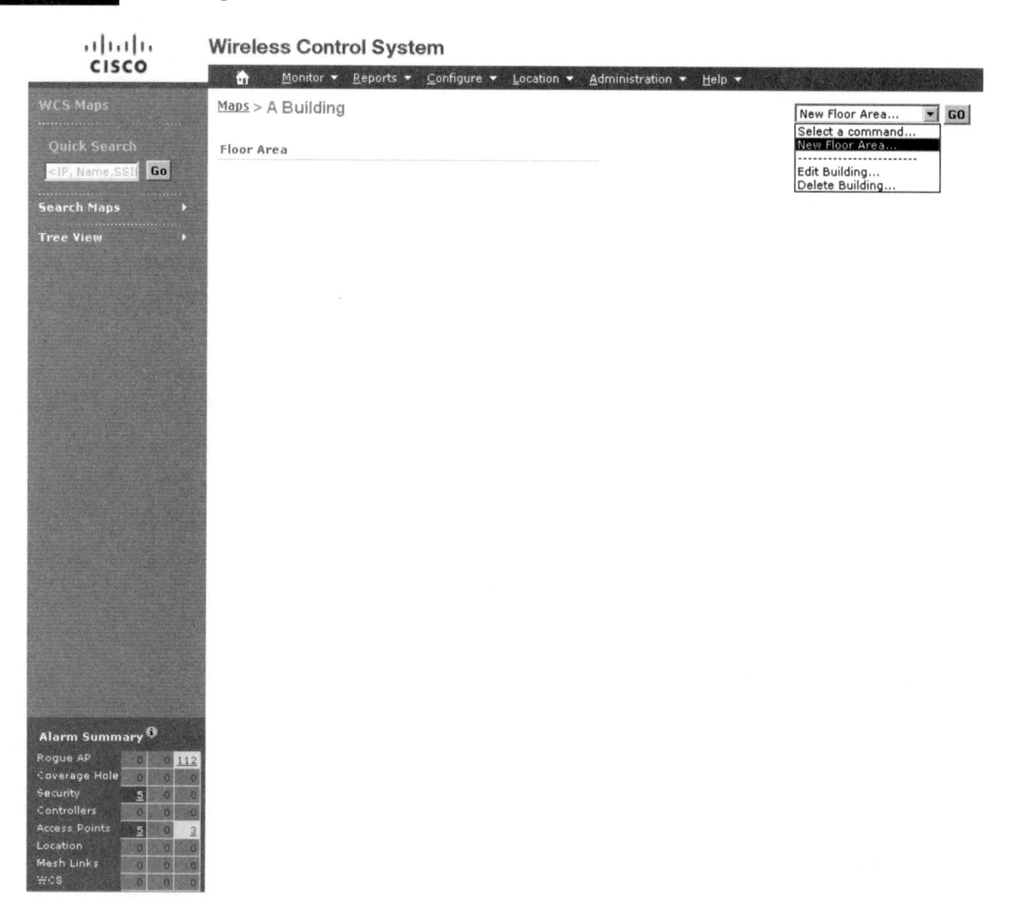

6. From here, you can begin the process of adding a new floor into the building. Select New Floor Area from the drop-down menu, and the New Floor Area window shown in Figure 9-13 will appear.

FIGURE 9-13 New Floor Area window

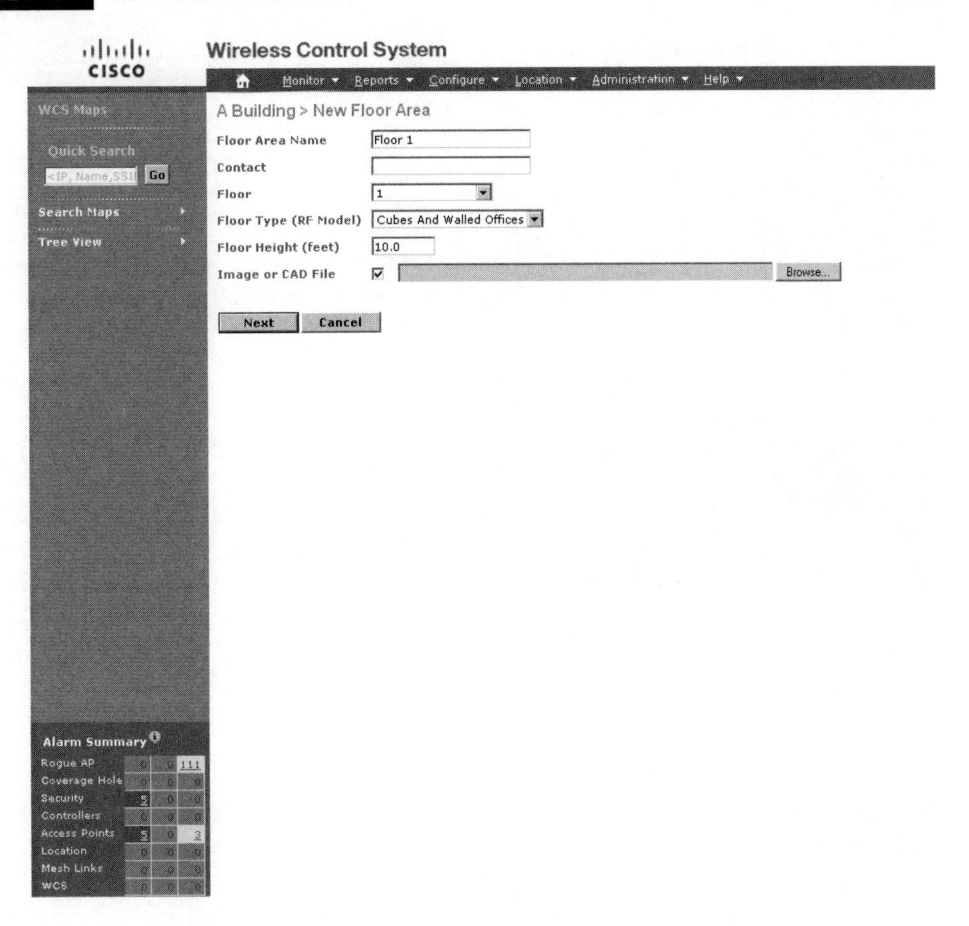

7. As with the Campus and Building windows, you can enter information about the floor, as well as pick which floor of the building this new floor will be. In the Image or CAD File field you can add a CAD file or image in JPEG, GIF, or PNG format. Click the Next button to see a window similar to Figure 9-14.

FIGURE 9-14 New Floor Area window with uploaded map

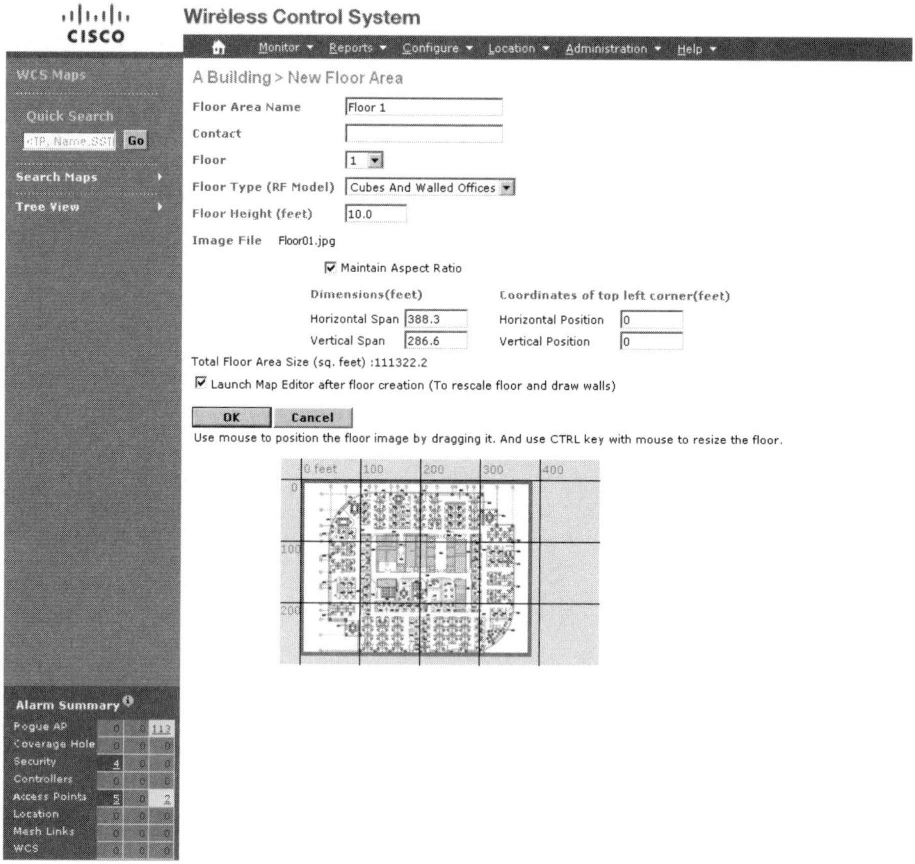

cisco

Wireless Control System

Monitor ▾ Reports ▾ Configure ▾ Location ▾ Administration ▾ Help ▾

WCS Maps

Quick Search

<IP, Name, SSID> Go

Search Maps

Tree View

A Building > New Floor Area

Floor Area Name: Floor 1

Contact:

Floor: 1

Floor Type (RF Model): Cubes And Walled Offices

Floor Height (feet): 10.0

Image File: Floor01.jpg

Maintain Aspect Ratio

Dimensions(feet)

Horizontal Span: 388.3

Vertical Span: 286.6

Coordinates of top left corner(feet)

Horizontal Position: 0

Vertical Position: 0

Total Floor Area Size (sq. feet) :111322.2

Launch Map Editor after floor creation (To rescale floor and draw walls)

OK Cancel

Use mouse to position the floor image by dragging it. And use CTRL key with mouse to resize the floor.

0 feet 100 200 300 400

0

100

200

Alarm Summary

Rogue AP	0	0	113
Coverage Hole	0	0	0
Security	4	0	0
Controllers	0	0	0
Access Points	5	0	2
Location	0	0	0
Mesh Links	0	0	0
WCS	0	0	0

8. A small version of your uploaded map will appear and give you the ability to set the dimensions of the map area. You also have the option to run the Map Editor from this window before you complete the floor area creation process. (You can access this window again in a created map by selecting the Edit Floor Area option.) Check the Launch Map Editor box and click OK. A Java window will pop up (Figure 9-15).

FIGURE 9-15 Map Editor window

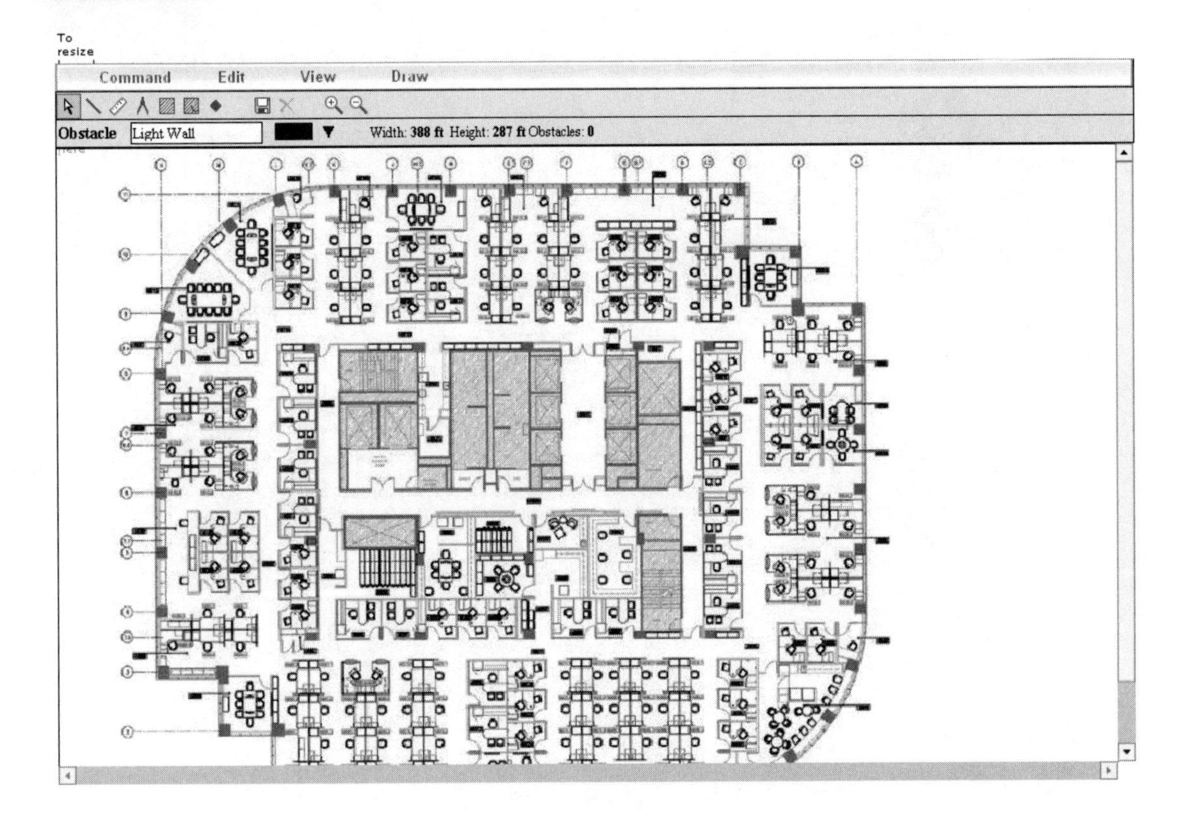

9. In the Map Editor you can scale the map, add walls and objects to aid in RF prediction, and perform various measurement functions. Close the Map Editor window and you will be brought back to the window in Figure 9-16.

10. Now that you have fully uploaded the floor map, you can use the drop-down menu to start the process of adding a new access point to the map.

FIGURE 9-16 Building/Floor window

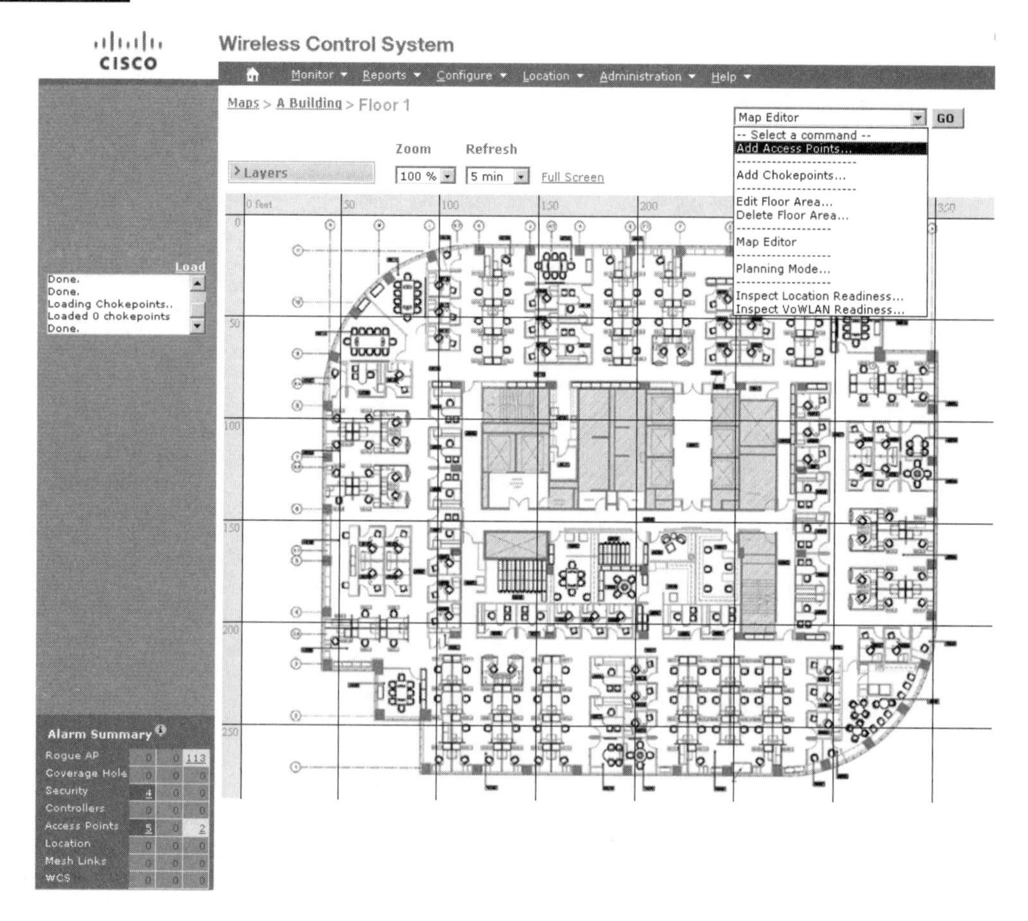

Adding and Positioning Access Points

1. Clicking the Add Access Points option in the drop-down box (shown in Figure 9-16) to see the window in Figure 9-17, which will show you all of the access points in the WCS that have not yet been added to maps.

FIGURE 9-17 Add Access Points window

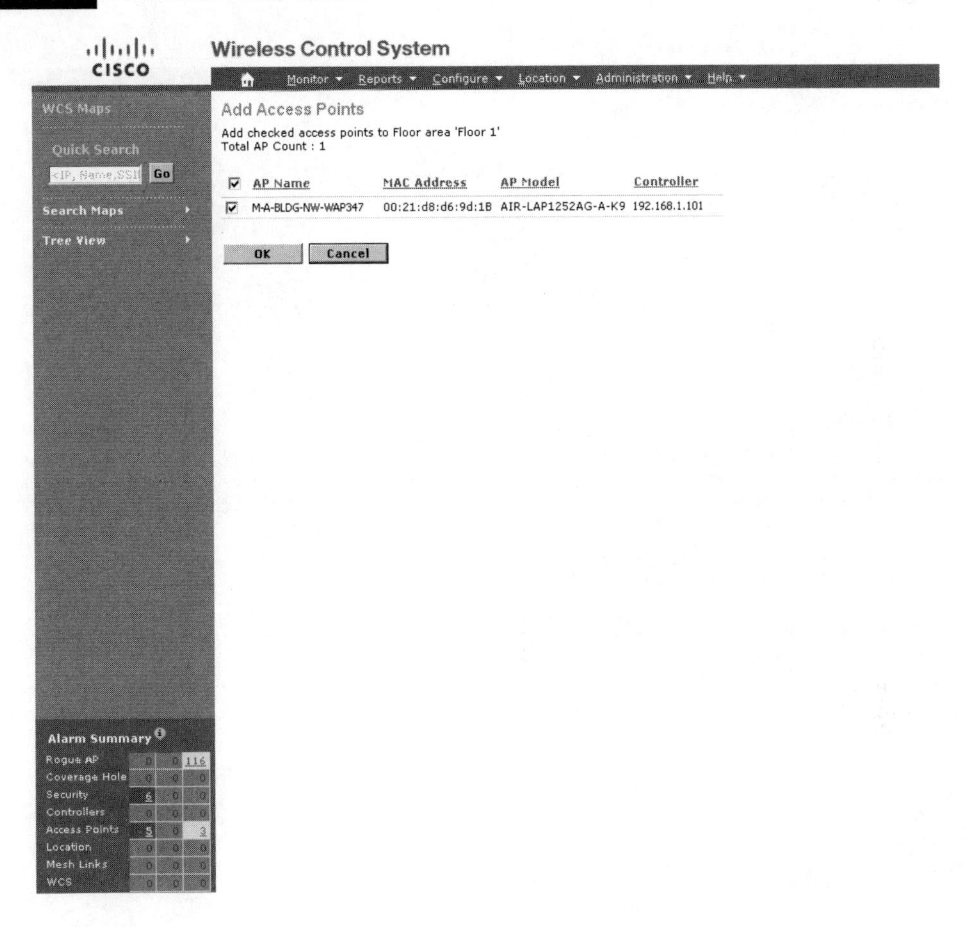

2. Check the access points you want to add and then click OK. The map you are adding the access point to will appear, and you will be able to position the access point where you please. Be aware that the placement of the access point on the map is only as accurate as the administrator who is placing it. When the access point has been placed in its desired location, click the Save button. The map with the newly placed access point will appear in Figure 9-18.

FIGURE 9-18 Building/Floor window with map of access points and coverage

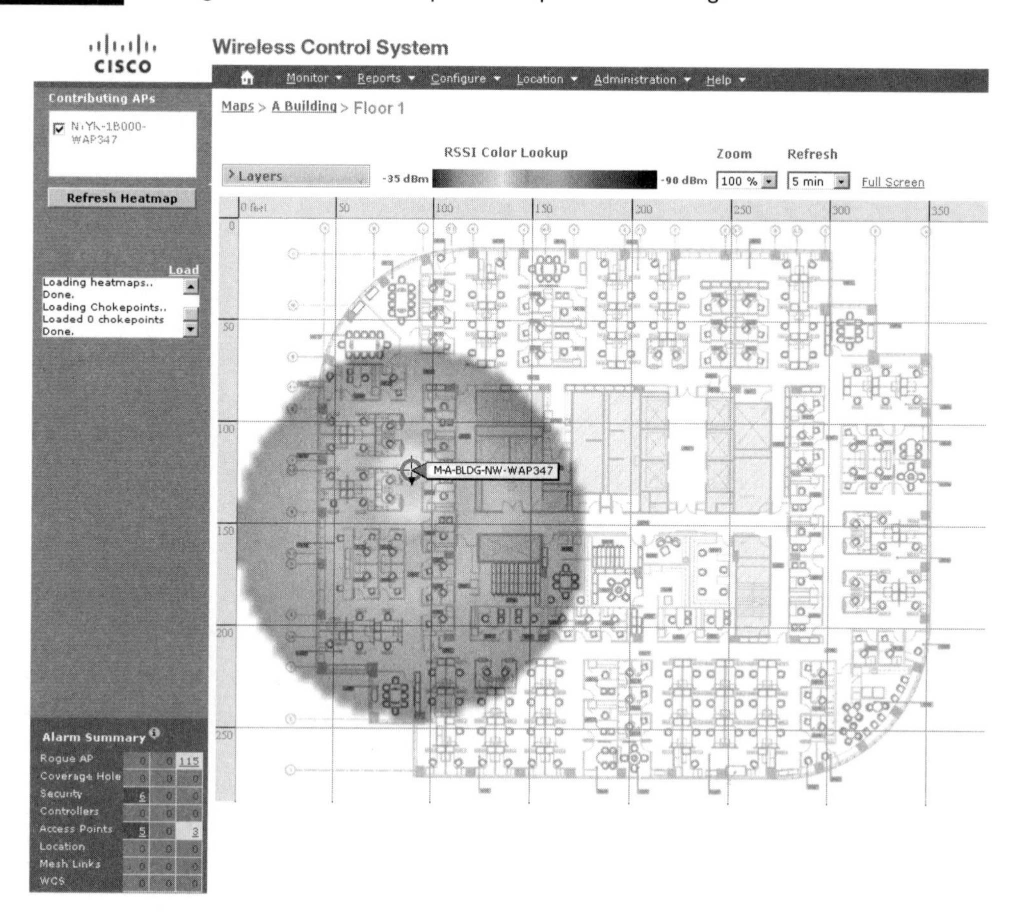

3. If you move your mouse over the access point, a window will pop up and give you the ability to monitor radio and general AP information. In environments where you have a large number of APs, this feature can be very useful. At the top left hand side of the map, click the Maps link. Figure 9-19 shows the original Maps display with a successfully added campus, building, and access point.

FIGURE 9-19 Completed WCS Maps window

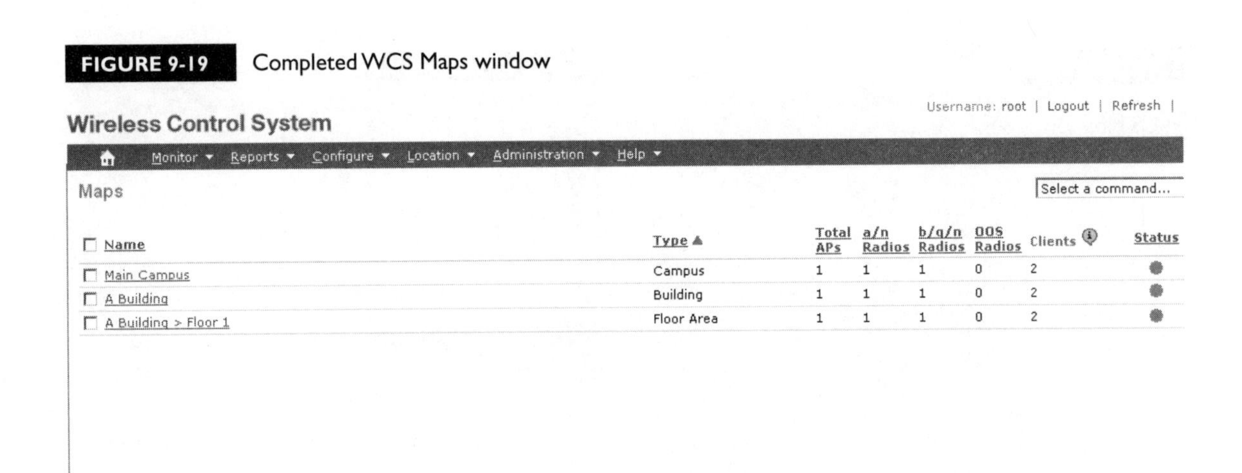

CERTIFICATION OBJECTIVE 9.03

Monitoring with the Wireless Control System

An important role of the WCS is in its monitoring of the wireless network. In the section on WCS maps, we examined one of several different methods for monitoring wireless devices using the WCS. In this section we will discuss some of the other monitoring tools that are available.

Monitoring Wireless Components

When you click the Monitor menu, a variety of read-only choices appear. These choices are shown in Figure 9-20.

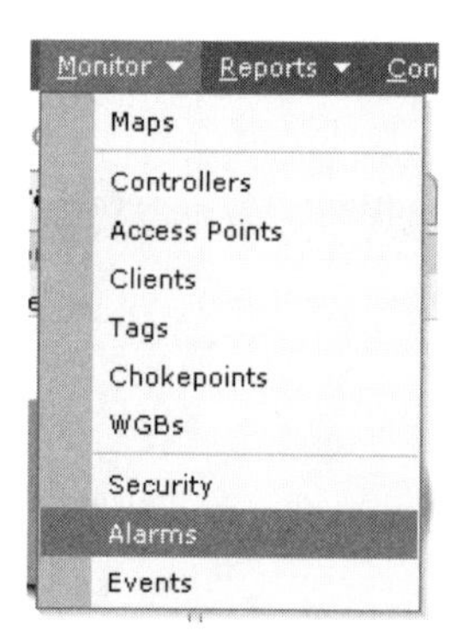

FIGURE 9-20

WCS Monitor menu

The WCS Monitor menu allows you to monitor the various wireless components; the most important of which are the following:

- **Controllers** Allows you to monitor individual controllers by viewing information such as port status, connectivity tests, and configured parameters.
- **Access Points** Gives access point information such as port status, CDP neighbors, and radio information.
- **Clients** Lets you track client usage on a variety of filters and use the powerful client troubleshooting tool, which assists IT administrators with diagnosing wireless client connectivity problems without their having to touch the client device.
- **Alarms** Allows you to view all the current alarms on the system and filter them by severity and type.

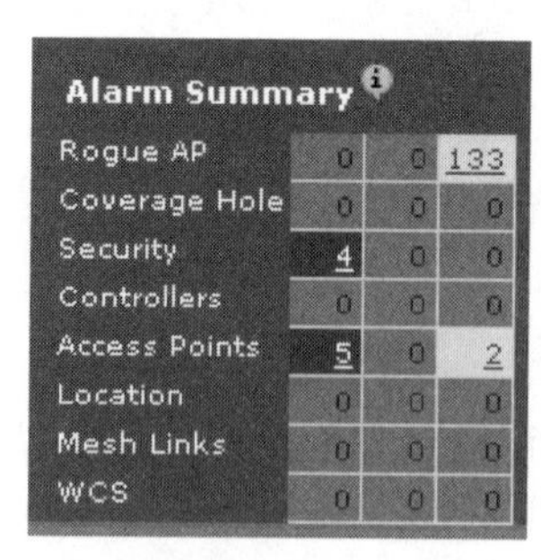

FIGURE 9-21

WCS Alarm Summary window

WCS Alarm Management

The most convenient alarm information feature on the WCS is the Alarm Summary box located in the lower left hand side of the WCS interface. This window (Figure 9-21) sorts alarms by type and severity (minor, major, and critical).

The major types of alarms are controller, access point, and security alarms. There are also useful alarms for rogue AP detection, location services, and general WCS status.

INSIDE THE EXAM

The Wireless Control System (WCS) and Navigator

Be sure to understand the role of the WCS in a unified wireless network and know that it manages wireless LAN controllers and APs. Familiarize yourself with the versions, ports, and features of the WCS. The WCS is also able to integrate with the Location Server and the Mobility Services Engine to provide extra services. Be familiar with the purpose of the WCS Navigator. Remember that the Navigator manages multiple Wireless Control Systems and is used primarily as a centralized management solution by corporations with geographically disparate locations.

Installing and Configuring the Wireless Control System (WCS)

This section is an important one to understand. Be sure to familiarize yourself with the WCS graphical user interface and know the process of building a WCS from initial server installation all the way to entering the GUI. Know that upgrading a WCS involves uninstalling existing software, backing up the database, installing new software, and restoring from the old database. Have a good understanding of the WCS ports and understand the major features of WCS and where it fits in an overall unified wireless network.

Be familiar with the important commands within the Controller Configuration drop-down box. Configuration of controller and access points using WCS is very similar to the process of configuring them directly. Be able to distinguish regular configuration of controllers and access points from template-based configuration. Most importantly, be able to identify the displays and screenshots of various controller and access point configuration windows. The chapter includes step-by-step procedures for adding and editing WCS maps. Understand the hierarchy of WCS maps and be familiar with the procedure for uploading CAD and image files to create a map. Be familiar with the screens for placing and moving access points within the map.

Monitoring with the Wireless Control System

The primary takeaway from this section is that the WCS allows a good amount of read-only monitoring capabilities for access points, controllers, clients, and other devices. Be sure to recognize the Alarm Summary window and know what kind of information the monitor and alarm windows can provide. Know that the Monitor menu item contains most, if not all, of the information that an administrator needs to examine and troubleshoot the wireless network.

CERTIFICATION SUMMARY

With larger-scale wireless networks, the need for a centralized management solution has arisen to manage multicontroller environments with hundreds, if not thousands, of access points. The Wireless Control System (WCS) is designed to provide a centralized management tool to accomplish that goal. The WCS has specific server requirements, depending on the size and scope of the wireless network and also has licensing guidelines that vary by the implementation. In addition, there are specific ports that WCS needs to operate and perform its tasks successfully. The WCS is, first and foremost, a wireless management tool that enables ease of configuration and centralized monitoring of wireless devices. The WCS can also be integrated with the Cisco Location Server or Mobility Services Engine to provide Context-Aware/ Location services and advanced security features such as Active Wireless Intrusion Protection (wIPS). The WCS Navigator is an overlay solution that allows you to manage multiple WCS platforms and have a centralized repository for alarms and monitoring.

Configuring controllers and access points can be accomplished in one of two ways, either directly with the WCS or with templates. The bulk of this chapter concentrates on configuring controllers and access points and on the principles of WCS maps. WCS maps allow you to monitor access points and are also used to integrate with Location Server and the Mobility Services Engine to provide location tracking. The step-by-step process for building WCS campuses, buildings, floors, and maps is detailed, as well as placing access points on those maps.

The monitoring function of the Wireless Control System enables you to view WCS maps and controllers, access points, clients, and alarms. The accompanying Alarms Status window on the WCS also provides detailed tracking for controllers, access points, and security alarms in addition to rogue AP detection and general WCS status.

TWO-MINUTE DRILL

The Wireless Control System (WCS) and Navigator

- The WCS manages controllers, access points, and clients and integrates with Location Server and the Mobility Services Engine.
- Depending on the size of the wireless deployment, the WCS has different system and server requirements.
- WCS comes in two flavors: WCS Base and WCS PLUS. WCS PLUS supports more access points and also has limited location tracking capabilities.
- WCS has various ports that are needed for proper operation.
- WCS allows you to configure controllers and access points, monitor, and report, as well as check the alarm status and integrate with the Location Server and the Mobility Services engine for location tracking.
- WCS Navigator centrally manages multiple Wireless Control Systems and consolidates alarms and reporting features.

Installing and Configuring the Wireless Control System (WCS)

- WCS can perform controller and access point configuration either directly with the WCS or with templates.
- Direct configuration is very much like directly accessing the controller or access point to use the controller's GUI.
- Template-based configuration allows the WCS to implement configuration changes on several controllers and/or access points at the same time using a uniform format.
- WCS maps allow you to monitor the wireless network and provide location tracking when paired with a Location Server or Mobility Services Engine.

Monitoring with the Wireless Control System

- The monitoring function of the WCS enables you to view maps, controllers, access points, clients, and alarms.
- The WCS Alarms Status window provides detailed tracking for controllers, access points, and security alarms in addition to rogue AP detection and general WCS status.

SELF TEST

The following self-test questions will help you measure your understanding of the material presented in this chapter. Read all the choices carefully, as there may be more than one correct answer. Choose all correct answers for each question.

The Wireless Control System (WCS) and Navigator

1. The Wireless Control System (WCS) manages everything except which of the following?
   A. Wireless LAN controllers
   B. Lightweight access points
   C. The WCS Navigator
   D. Location servers
2. Which of the following products does not directly support Context-Aware/Location services?
   A. WCS Base
   B. WCS PLUS
   C. Cisco Location Server
   D. Cisco Mobility Services Engine

Installing and Configuring the Wireless Control System (WCS)

3. Which of the following communication protocols is most used when the WCS interacts with the wireless LAN controllers that it manages?
   A. TFTP
   B. Telnet
   C. FTP
   D. SNMP
4. Which of the following is not an essential command on the Configure Controllers drop-down menu?
   A. Refresh Config from Controller
   B. Save Config to Flash
   C. Change Startup Config
   D. Audit Now

5. When using controller templates, which of the following statements is not true?

A. Even after a desired controller template is applied, it is important to save the configuration on the controllers.

B. Controller templates can be used to set controller hostnames and IP addresses.

C. Controller templates contain individual controller configurations.

D. Controller templates can be applied on one or multiple controllers.

6. When configuring an access point using the WCS which of the following is not something you can edit?

A. Primary controller name

B. 802.11b/g radio status

C. Client Exclusion Policy

D. Name

7. Which of the following is not a valid file format for WCS maps?

A. CAD

B. BMP

C. GIF

D. JPEG

8. Which of the following is not a mandatory hierarchy type for WCS maps?

A. Campus

B. Building

C. Floor

D. All of the above are mandatory parts of the hierarchy of WCS maps.

9. When configuring controllers and access points with WCS, which of the following is not true?

A. Most of the configurations that you can accomplish on the controllers themselves can be completed with the WCS.

B. When making changes to controllers on WCS, the Save Config command is performed by default.

C. When making changes to access points using WCS, the changes are saved automatically to the access point once they are applied.

D. Access point and controller configuration is simplified in WCS because certain values are prepopulated in drop-down boxes where they normally would not be.

Monitoring with the Wireless Control System

10. Which of the following is not a part of the Monitor menu?

A. Maps

B. Access Points

C. Controllers

D. All of the above are part of the Monitor menu.

SELF TEST ANSWERS

The Wireless Control System (WCS) and Navigator

1. The Wireless Control System (WCS) manages everything except which of the following?

A. Wireless LAN controllers

B. Lightweight access points

C. The WCS Navigator

D. Location servers

☑ **C.** The WCS does not manage the WCS Navigator; the WCS Navigator can manage multiple Wireless Control Systems.

☒ **A, B,** and **D** are incorrect because they all represent devices that are managed by WCS.

2. Which of the following products does not directly support Context-Aware/Location services?

A. WCS Base

B. WCS PLUS

C. Cisco Location Server

D. Cisco Mobility Services Engine

☑ **A.** The WCS Base version of WCS has no location tracking capabilities.

☒ **B, C,** and **D** can all provide location tracking functionality. The Location Server and Mobility Services Engine can integrate with WCS to tack multiple devices simultaneously, and the WCS PLUS version allows you to track individual devices.

Installing and Configuring the Wireless Control System (WCS)

3. Which of the following communication protocols is most used when the WCS interacts with the wireless LAN controllers that it manages?

A. TFTP

B. Telnet

C. FTP

D. SNMP

☑ **D.** SNMP is, by far, the most utilized protocol for WCS to controller communications. In the process of adding a controller to WCS, you have to have a matching SNMP community string.

☒ **A, B,** and **C** are incorrect because although they are valid ways for interacting with the WCS, they are decidedly not the most used.

4. Which of the following is not an essential command on the Configure Controllers drop-down menu?

A. Refresh Config from Controller

B. Save Config to Flash

C. Change Startup Config

D. Audit Now

☑ C. Change Startup Config is not a command on the Configure Controllers drop-down menu.

☒ **A**, **B**, and **D** are incorrect because all of these commands are listed specifically as important commands on the Configure Controllers drop-down menu.

5. When using controller templates, which of the following statements is not true?

A. Even after a desired controller template is applied, it is important to save the configuration on the controllers.

B. Controller templates can be used to set controller hostnames and IP addresses.

C. Controller templates contain individual controller configurations.

D. Controller templates can be applied on one or multiple controllers.

☑ **B.** It is understood that controller templates cannot and should not be used to push controller-specific configurations such as hostnames and IP addresses.

☒ **A**, **C**, and **D** all describe valid properties of controller templates and their application.

6. When configuring an access point using the WCS which of the following is not something you can edit?

A. Primary controller name

B. 802.11b/g radio status

C. Client Exclusion Policy

D. Name

☑ C. Client Exclusion Policies can be configured using Controller Templates or Controller Configuration on the WCS in the WLANs section, but they are not AP-specific configurations.

☒ **A**, **B**, and **D** are incorrect because they are all editable configurations on an access point using WCS.

7. Which of the following is not a valid file format for WCS maps?

A. CAD

B. BMP

C. GIF

D. JPEG

☑ **B.** BMP Files are not currently supported by WCS as a valid uploadable map type.

☒ **A, C,** and **D** are all valid and acceptable file types for maps on WCS, provided that they are of a supported resolution.

8. Which of the following is not a mandatory hierarchy type for WCS maps?

A. Campus

B. Building

C. Floor

D. All of the above are mandatory parts of the hierarchy of WCS maps.

☑ **A.** The campus, although it is a valid piece of the WCS maps hierarchy, is not a mandatory category and is often not used for smaller wireless implementations.

☒ **B** and **C** are incorrect because they are mandatory categories in order to create a map, even if the building is a simple one-floor structure. **D** is incorrect because it includes **A** as a part of its answer.

9. When configuring controllers and access points with WCS, which of the following is not true?

A. Most of the configurations that you can accomplish on the controllers themselves can be completed with the WCS.

B. When making changes to controllers on WCS, the Save Config command is performed by default.

C. When making changes to access points using WCS, the changes are saved automatically to the access point once they are applied.

D. Access point and controller configuration is simplified in WCS because certain values are prepopulated in drop-down boxes where they normally would not be.

☑ **B.** The save config command is not performed by default and must be executed in the Controller Configuration drop-down box regardless of whether you are configuring controllers directly or applying templates.

☒ **A, C,** and **D** are incorrect because they are all true statements about controller and access point configuration on the WCS.

Monitoring with the Wireless Control System

10. Which of the following is not a part of the Monitor menu?

A. Maps

B. Access Points

C. Controllers

D. All of the above are part of the Monitor menu.

☑ D. The Monitor menu is a very powerful section in that you can access WCS maps and monitor controllers, access points, and alarms.

☒ A, B, and C are all found on the Monitor menu in WCS.

10

Understanding and Installing Wireless Clients

CERTIFICATION OBJECTIVES

The majority of this book has focused on the infrastructure and networking element of Cisco wireless networks. This chapter will cover the fundamentals and operation of wireless clients (or supplicants, as they are often known), including those of the Windows, Mac, and Linux operating systems. In addition to these, we will cover Cisco wireless client products and discuss the importance of the Cisco Compatible eXtensions (CCX) program.

CERTIFICATION OBJECTIVE 10.01

Operating System Client Configuration

For the exam, it is important to understand the operation of the wireless clients on the major operating systems, which are Microsoft Windows, Mac, and Linux. In this section we will examine the operation of wireless clients for each operating system.

Windows Client Configuration

In Microsoft Windows, the built-in operating system client/supplicant is known as Windows *Wireless Zero Configuration*, or *WZC*. WZC is included in Windows XP and later versions of Windows as a utility for managing wireless connections. It is often used when there is an absence of a client from the wireless NIC vendor (such as Intel, Cisco, or IBM). WZC is a very basic configuration tool with support for only the EAP-TLS and PEAP 802.1X authentication types. Wireless Zero Configuration owes its name to the behavior it exhibits of automatically connecting wireless clients to broadcasting networks. This allows most wireless users to associate without having to perform any configuration on the client.

In cases where security needs to be configured, understanding the configuration of the Wireless Zero Configuration client is important. The following sections will give you a general overview of manually creating a preferred network versus associating to a network that the client already is aware of on a sample Windows XP SP3 client.

on the job

In personal computers running Windows XP below Service Pack 3, the capabilities of the Wireless Zero Configuration client are noticeably limited and you may not see support for such features as WPA2. Nonetheless, it is important to make sure that wireless clients are running at least Windows XP Service Pack 2 with the necessary wireless client patches. You should evaluate your environment and determine which Microsoft Windows Service Packs and hotfix patches to install on your PC.

Manually Creating a Preferred Network

1. To access the Windows XP Wireless Client Configuration, you must be logged in as the Administrator of the client device. To configure the wireless client, right-click Wireless Network Connection under the Network Connections shortcut in the Control Panel and then select Properties (Figure 10-1).
2. Click the Wireless Networks tab to see a window similar to the one in Figure 10-2. If there are already preferred networks configured, they will appear here. The windows in the next few steps are the same whether you are adding or

FIGURE 10-1 Accessing Wireless Network Connection Properties

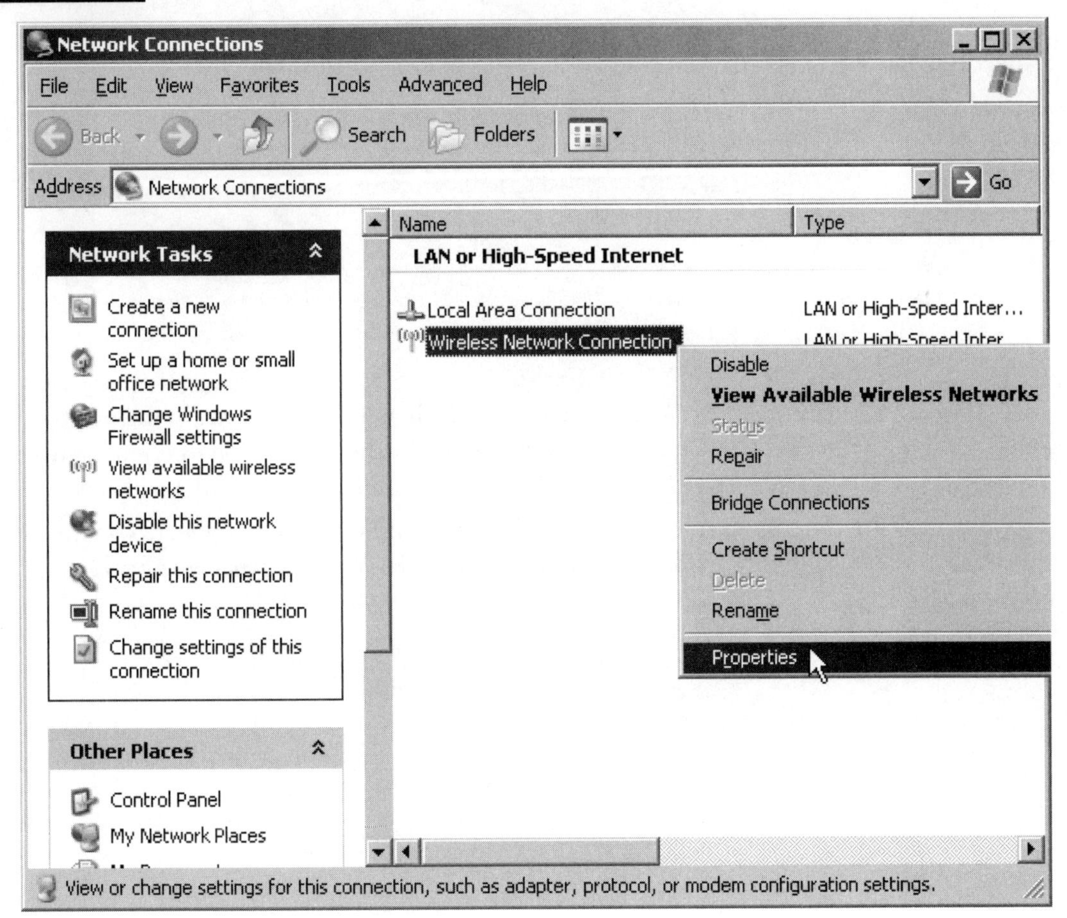

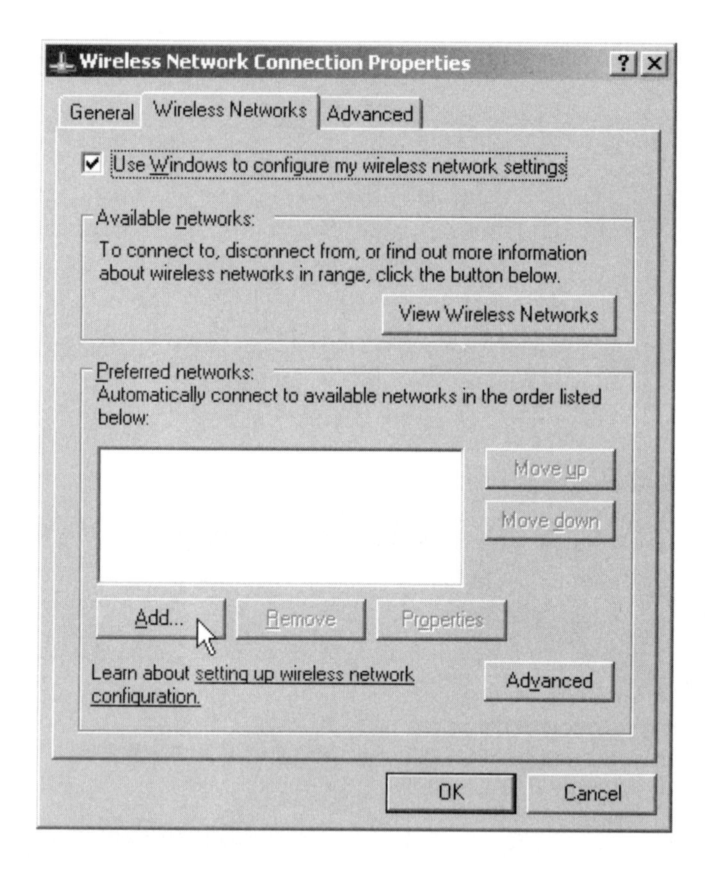

FIGURE 10-2 Wireless Network Connection Properties

editing a preferred wireless network. To adjust the priority of these networks, move them up or down in the list. Click the Add button to begin configuring the Wireless LAN settings. A window similar to Figure 10-3 will appear.

3. For this example, enter the SSID Enterprise that was configured on the wireless LAN controller in Chapter 8. For networks that are not set to

FIGURE 10-1

Wireless Network Properties Association tab

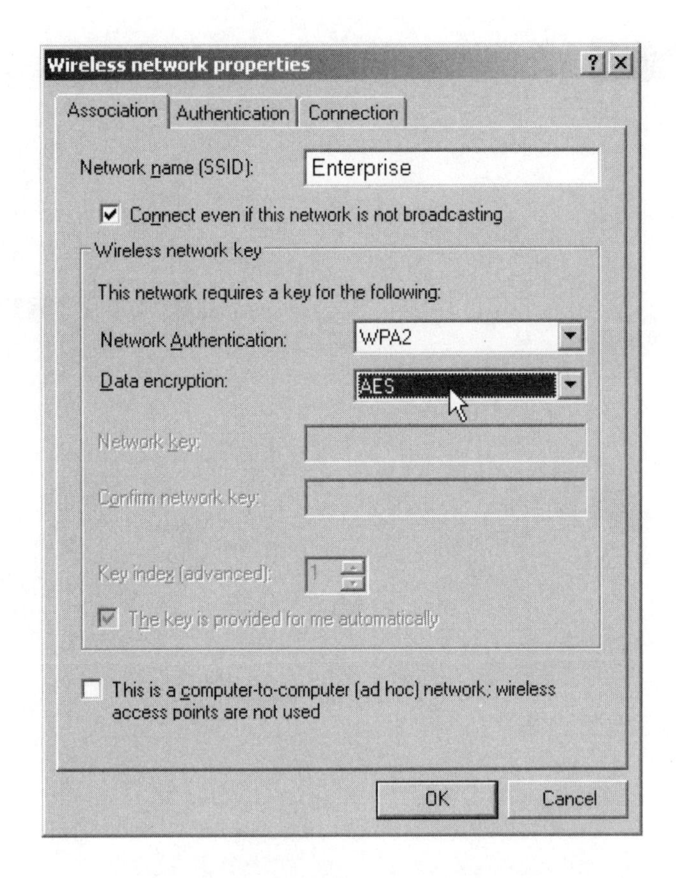

broadcast, it is important to check the Connect Even if This Network Is Not Broadcasting check box. Under Network Authentication, select WPA2 from the drop-down box, and under Data Encryption, select AES from the drop-down box (as shown in Figure 10-3). Click the Authentication tab to see the window shown in Figure 10-4.

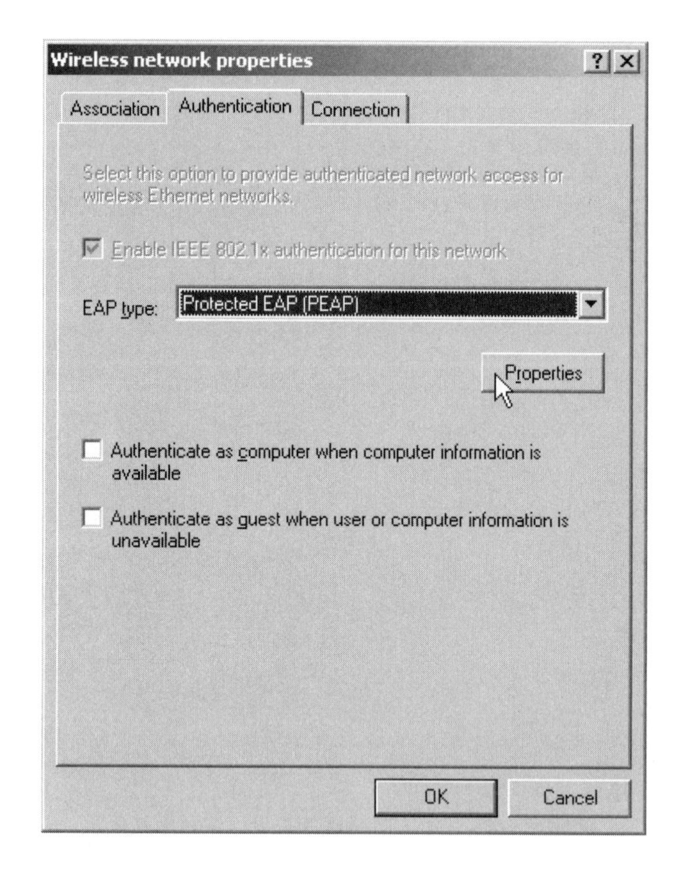

FIGURE 10-4

Wireless Network Properties Authentication tab

4. Unlike more advanced client software/supplicants, Microsoft Wireless Zero Configuration is rather limited in the 802.1X EAP security types that you are able to pick. PEAP is one of them. Click OK to save the settings for this particular network.

Connecting to an Existing/Broadcasting Network

To connect to an existing or broadcasting wireless network using the Windows Zero Configuration client, follow the below steps:

1. From the same Network Connections window, right-click Wireless Network Connection and select the View Available Wireless Networks option

FIGURE 10-5 Viewing the available networks

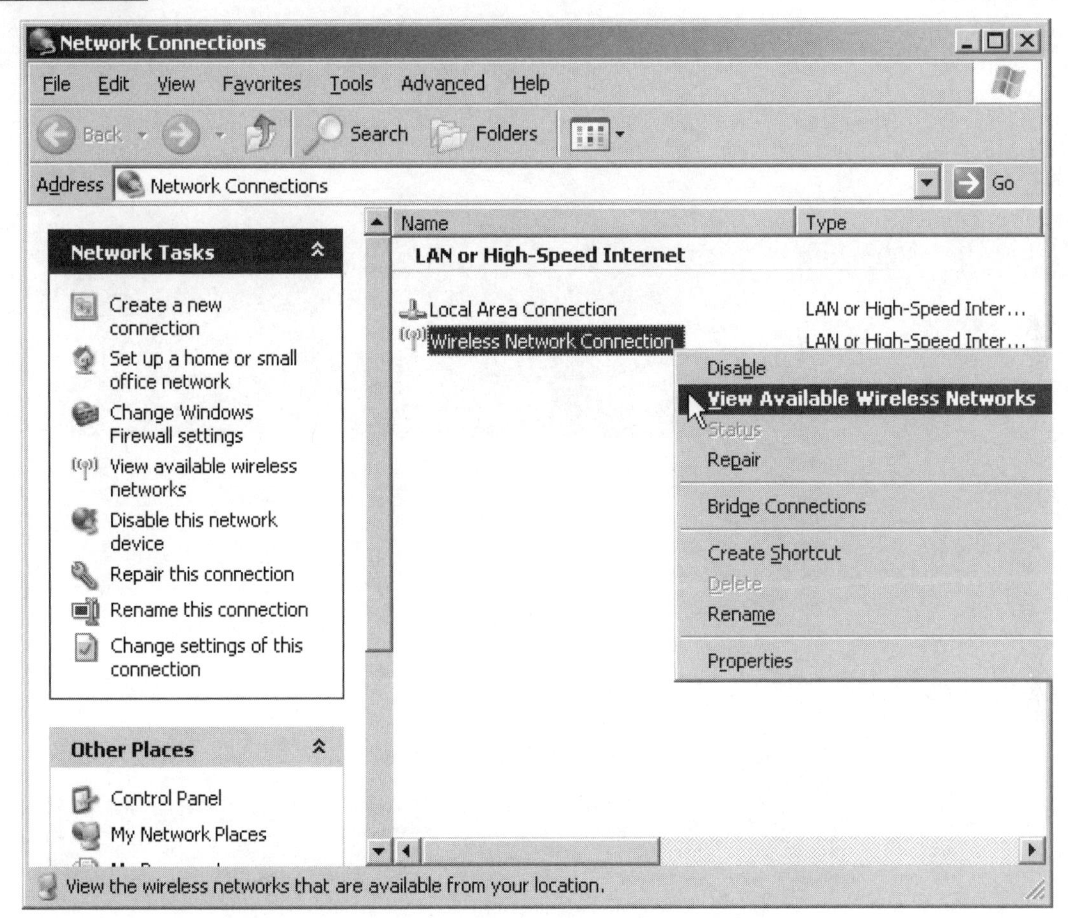

(Figure 10-5). Alternatively, you can simply double-click the Wireless Network Connection icon in the system tray (in the lower right hand side of your screen).

2. A window similar to Figure 10-6 will appear. This window will show both the preferred networks that you have configured (if available) and any broadcasting wireless networks detected by the adapter. Click the Connect button after you have highlighted your desired wireless network.

FIGURE 10-6 Choose a wireless network

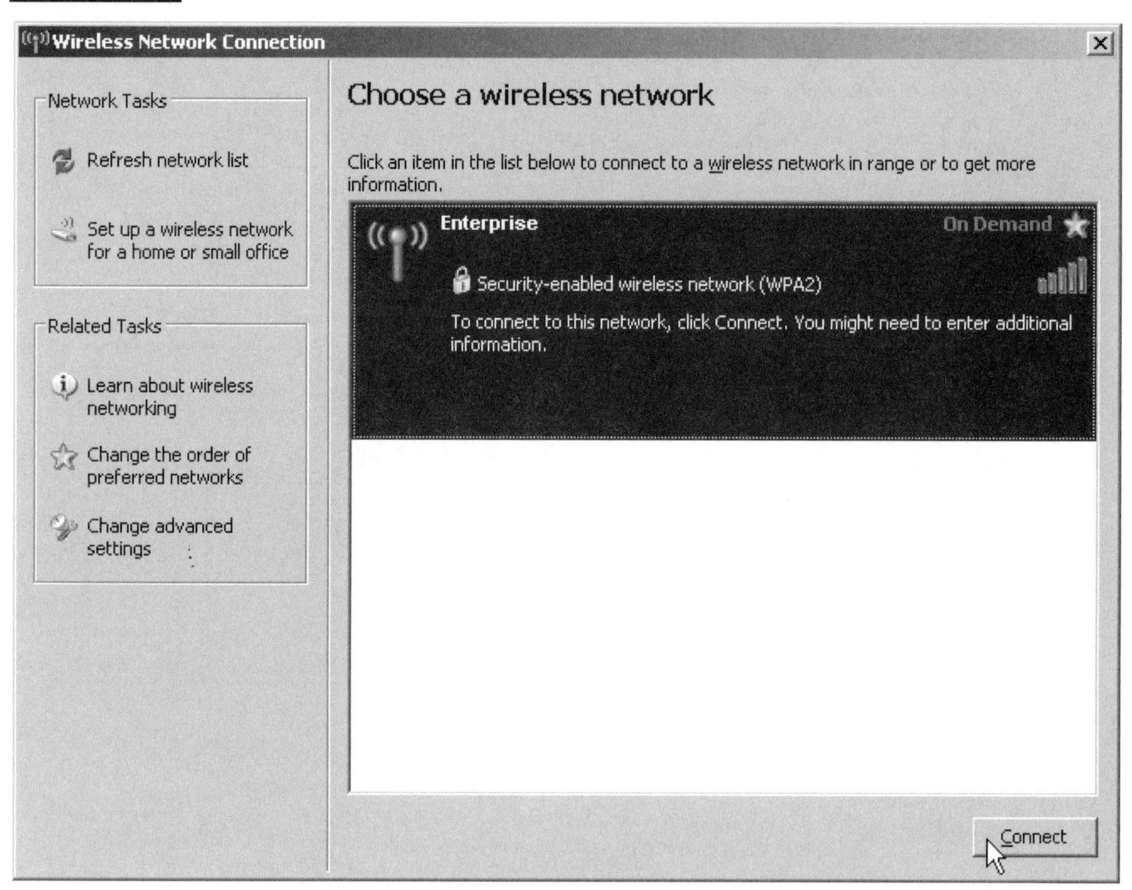

3. If there is network security such as EAP authentication configured, you may get a balloon prompt in your taskbar (Figure 10-7) and be prompted to enter a username and password (Figure 10-8).

FIGURE 10-7

Authentication balloon prompt

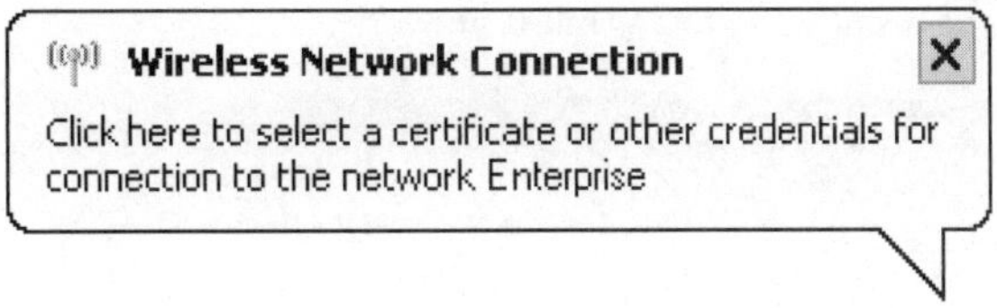

Wireless Network Connection

Click here to select a certificate or other credentials for connection to the network Enterprise

4. If the network is open with no security, you will see a successful Connected status similar to Figure 10-9; you should see this same window after your credentials have been verified.
5. The Connected status shows that you have successfully associated and authenticated to your desired wireless network.

FIGURE 10-8

Enter Credentials window

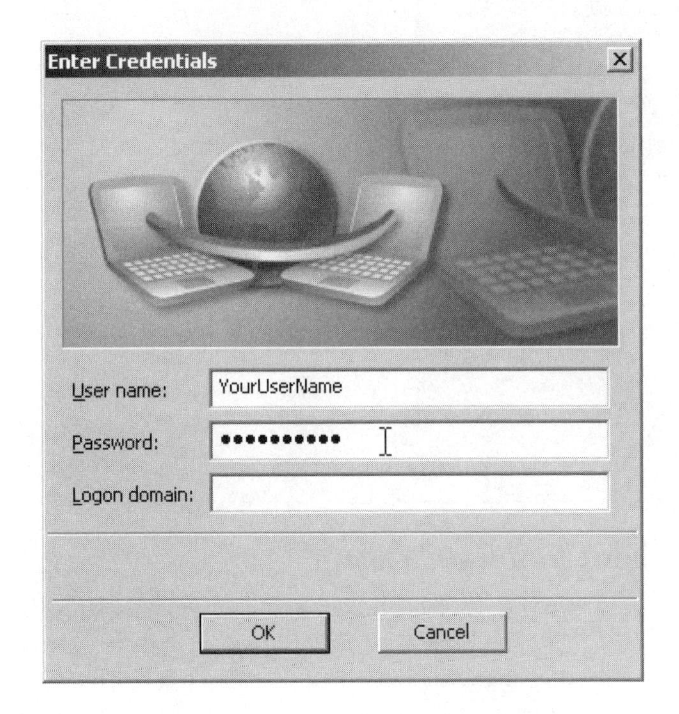

FIGURE 10-9 Connected to a wireless network

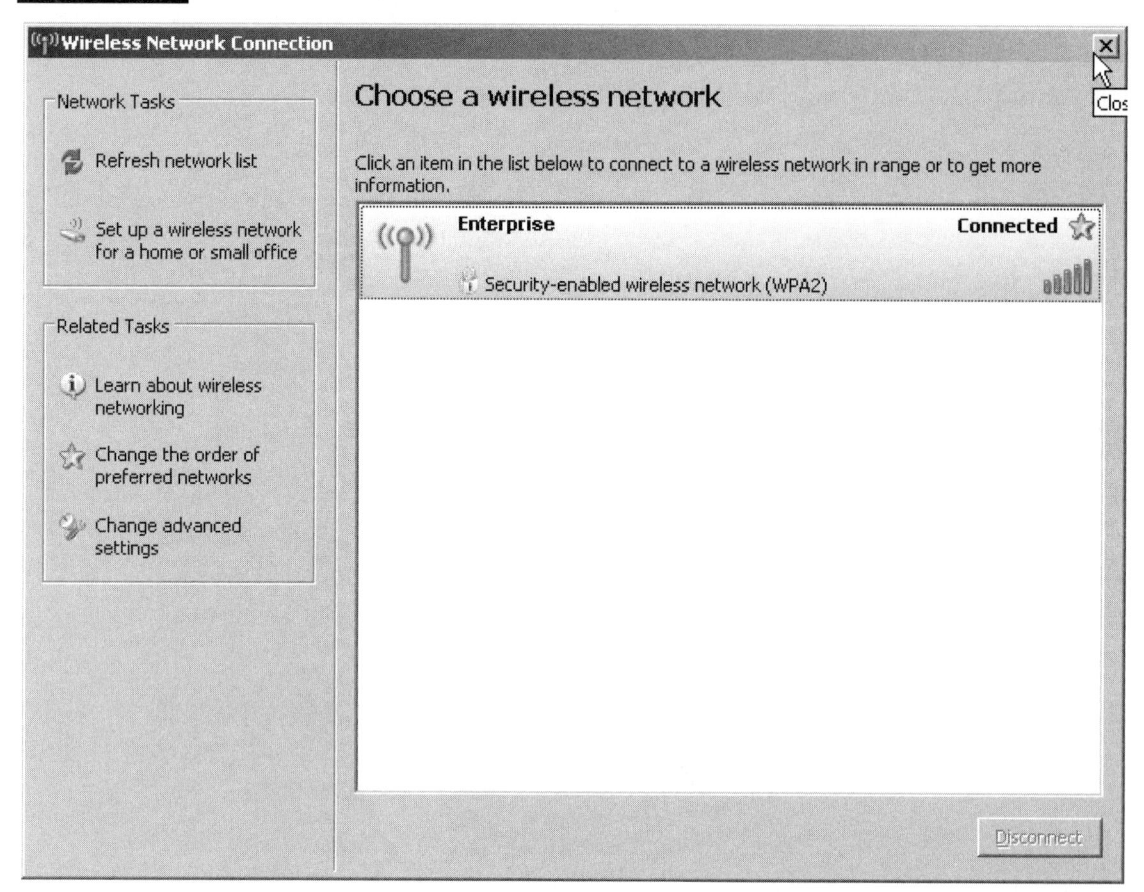

Mac Client Configuration

In Mac, the built-in operating system client/supplicant is known as *AirPort* or *AirPort Extreme*. AirPort Extreme simply refers to the 802.11g-capable version of the Apple wireless client. Because it is the most commonly used supplicant for Mac-based wireless clients, the AirPort client software supports the full range of WPA and WPA2 encryption, as well as the full range of 802.1X authentication types. This section will show you how a Mac OS is configured using a Mac OS X Leopard client.

Manually Creating a Preferred Network

As with the Windows Zero Configuration client, the Mac client allows you to manually configure a preferred wireless network. To accomplish this, follow the steps below:

FIGURE 10-10

Open Network Preferences

1. Click the wireless status icon on the top right hand side of the Mac window to see a drop-down menu (Figure 10-10); select Open Network Preferences to begin manual configuration of the wireless network.
2. Click the Turn Airport On button and then click the Advanced button (as shown in Figure 10-11), to see a tabbed AirPort window where you can configure a new preferred network (Figure 10-12).

FIGURE 10-11 Activating AirPort

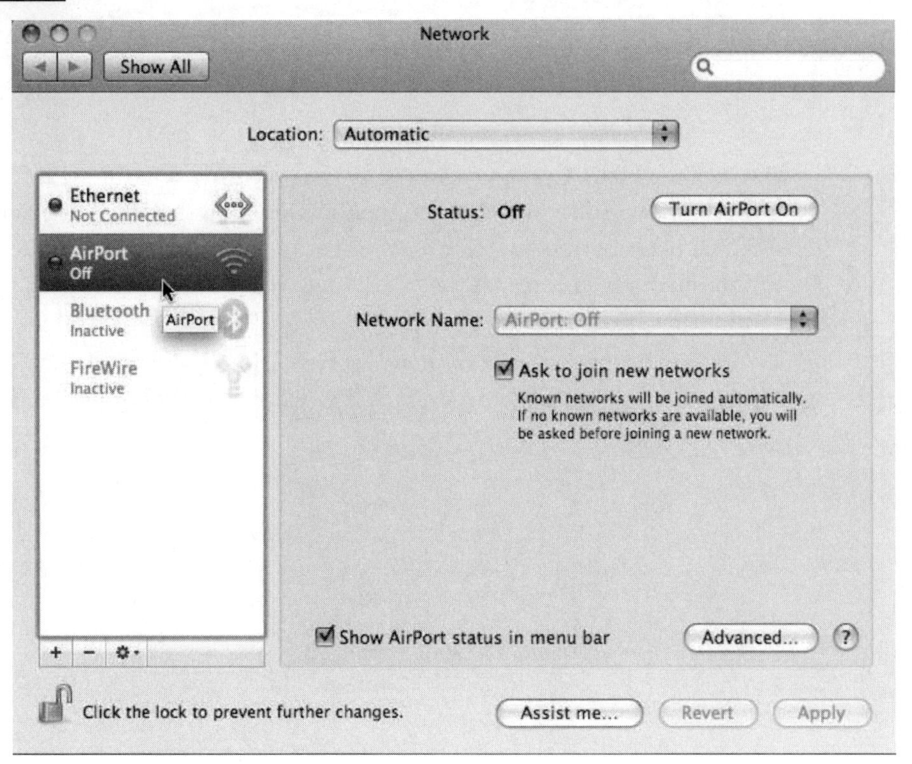

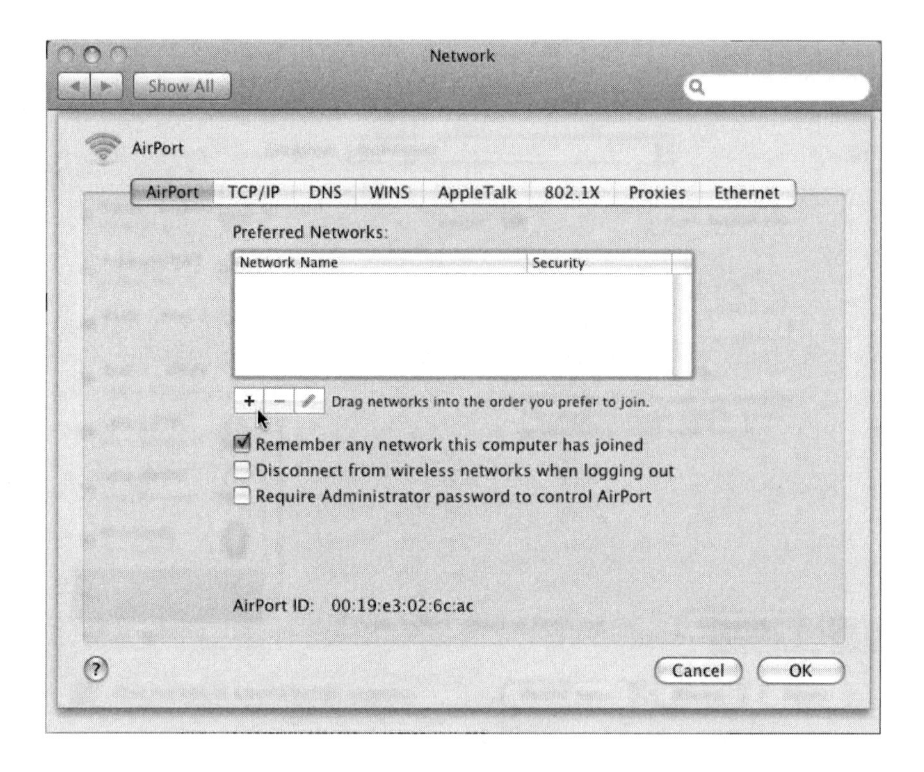

FIGURE 10-12 Add Preferred Network

3. Click the 802.1X tab to begin configuring security. We will use the same Enterprise SSID with 802.1X authentication as in the previous example. Click the + button to create a new security profile called Enterprise. The information you enter for the 802.1X tab will be similar to what is shown in Figure 10-13.
4. Go back to the AirPort tab shown in Figure 10-12 and then click the "+" button to add the Enterprise SSID. The window shown in Figure 10-14 should appear.

FIGURE 10-13

AirPort 802.1X configuration

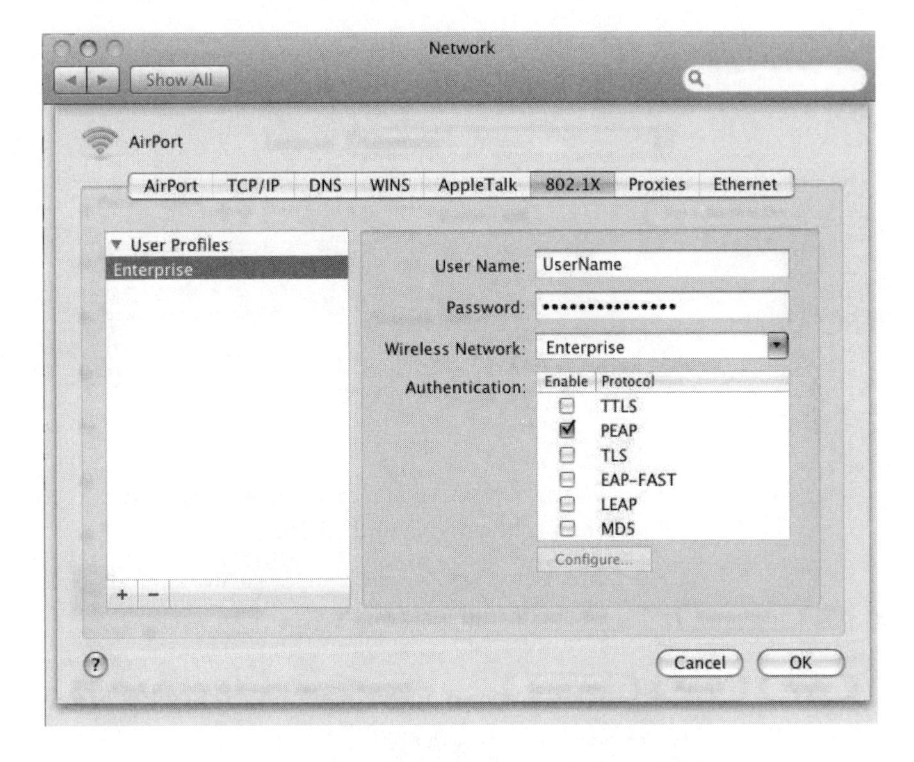

FIGURE 10-14

Enter network information

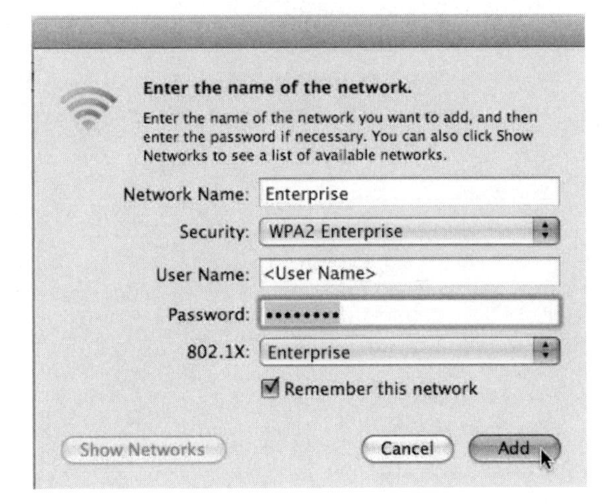

FIGURE 10-15

Enter credentials

5. When you try to connect to the wireless network, you may receive the login prompt shown in Figure 10-15.
6. After any possible certificate or password prompts, a successful wireless connection should result.

Connecting to an Existing/Broadcasting Network

To connect to an existing or broadcasting wireless network using the Mac client, follow the below steps:

1. If there is an existing or broadcast network, you can also look for those and associate to them, as seen in Figure 10-16.

FIGURE 10-16 Associating to an existing network

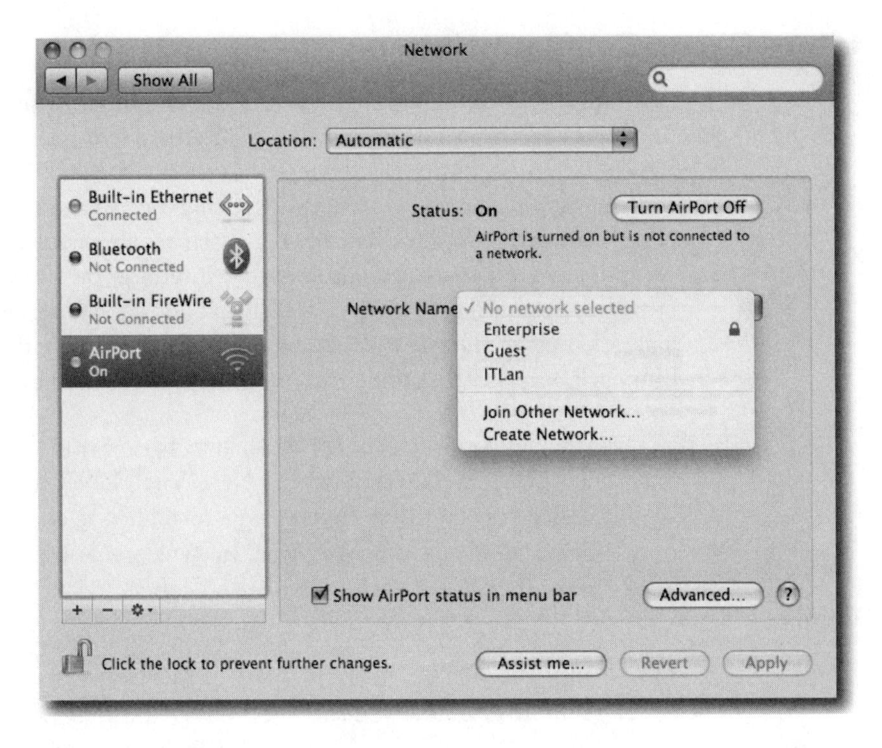

2. In this example, you can select the Enterprise, Guest, or ITLan SSIDs. You also have the Join Another Network option, which allows you to manually enter network information and the Create a Network option, which makes the wireless client an ad hoc wireless network host.
3. After selecting one of the SSIDs that are shown as available, you may get a login prompt and/or certificate prompt similar to the window shown in Figure 10-15.

Linux Client Configuration

Although the Linux operating system is known primarily for its command line interface, the exam will only cover wireless client configuration using the Linux GUI. Most Linux-based operating systems use a GUI software utility for configuring wired and wireless network settings on Unix and Linux platforms called *NetworkManager*. NetworkManager is a relatively simple software utility that serves two primary purposes: to manage network interfaces on a Linux machine and to provide a graphical user-friendly applet that allows the user or administrator to make changes to the network connections. This section will show you how a Linux wireless client is configured using the NetworkManager tool.

1. The NetworkManager applet (Figure 10-17) lists the detected wireless networks as well as the Create a New Wireless Network option (which, as in Mac, gives you the ability to create an ad hoc network) and the Connect to Other Wireless Network option. In some scenarios, you will be able to simply select the desired network name and connect.
2. To associate to the Enterprise SSID, click Connect to Other Wireless Network. A window similar to the one shown in Figure 10-18 will appear.
3. From this window, you can select any of a variety of security types from the Wireless Security drop-down box. For this example, select WPA2 Enterprise, and the window will expand to include the fields shown in Figure 10-19.

FIGURE 10-17

NetworkManager detected networks

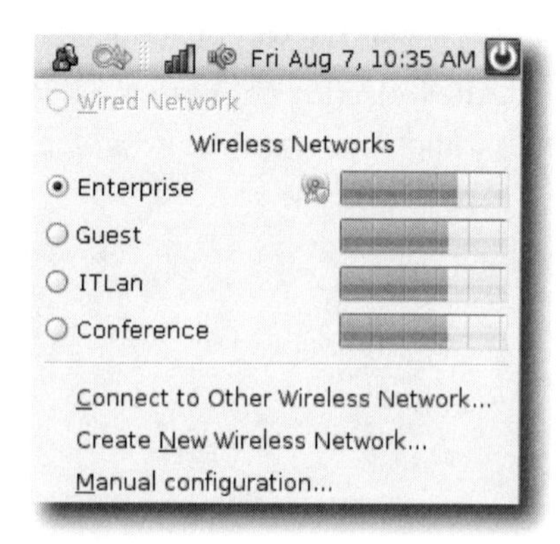

FIGURE 10-8

Connect to Other Wireless Network window

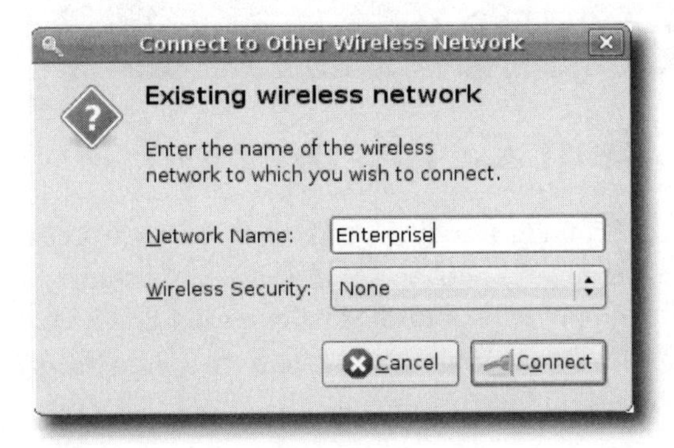

4. Here, you can configure your 802.1X authentication type (EAP Method), security credentials (username, password), and certificate file (if necessary). Once all of the necessary information for your wireless connection is entered, click the Connect button to join the desired wireless network.

FIGURE 10-19

Connect to Other Wireless Network security configuration

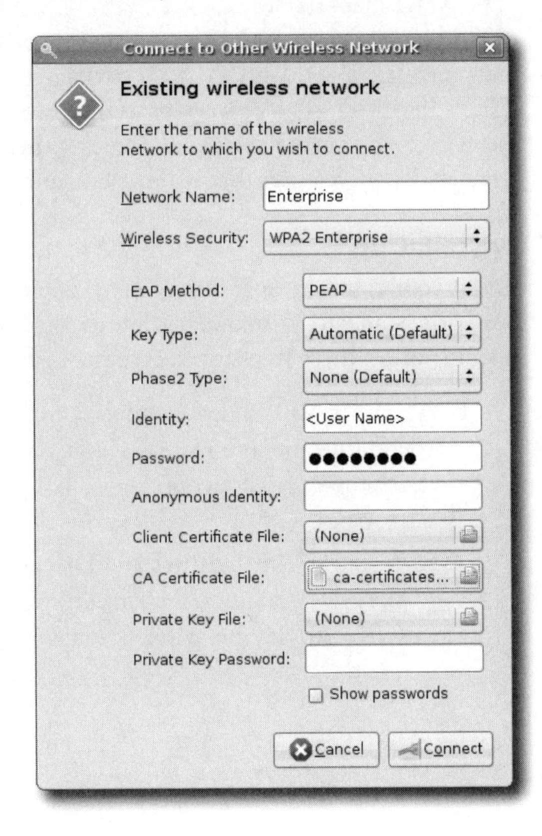

CERTIFICATION OBJECTIVE 10.02

Cisco Client Configuration

An understanding of Cisco client software/supplicants is important for both the exam and networks that utilize a substantial amount of Cisco client devices or supplicant software. The two modern Cisco wireless client products are the Aironet Desktop Utility (ADU) and the Cisco Secure Services Client; they will both be covered in this section.

Cisco Aironet Desktop Utility (ADU)

The *Cisco Aironet Desktop Utility* is a Cisco wireless client software utility that was introduced with the release of the Cisco Aironet 802.11a/b/g adapter card. The ADU supports only the PCI (PI21) and PCMCIA (CB21) versions of the 802.11a/b/g adapter and is not a supplicant that can be used with non-Cisco adapters. For this reason, ADU is packaged with the CB21/PI21 software on the Cisco download site. A successor to the Cisco Aironet Client Utility (ACU), ADU is relatively advanced compared to most other wireless client software and supports a wide variety of features that will be shown in the configuration steps of this section.

Installation

Insert your Cisco Aironet 802.11a/b/g adapter into the PC that you are working on. After the standard Windows hardware installation prompts, which you can largely ignore, follow these steps to execute the installation package:

1. As with most Windows applications, the ADU install package will prompt you for specific information about your system. The first of these prompts is the Setup Type, shown in Figure 10-20; you can choose from the following setup types:
   - **Install Client Utilities and Driver** The recommended option; it installs the ADU, additional client utilities, and the drivers needed to run the Cisco 802.11a/b/g wireless adapter.

FIGURE 10-20

Setup Type window

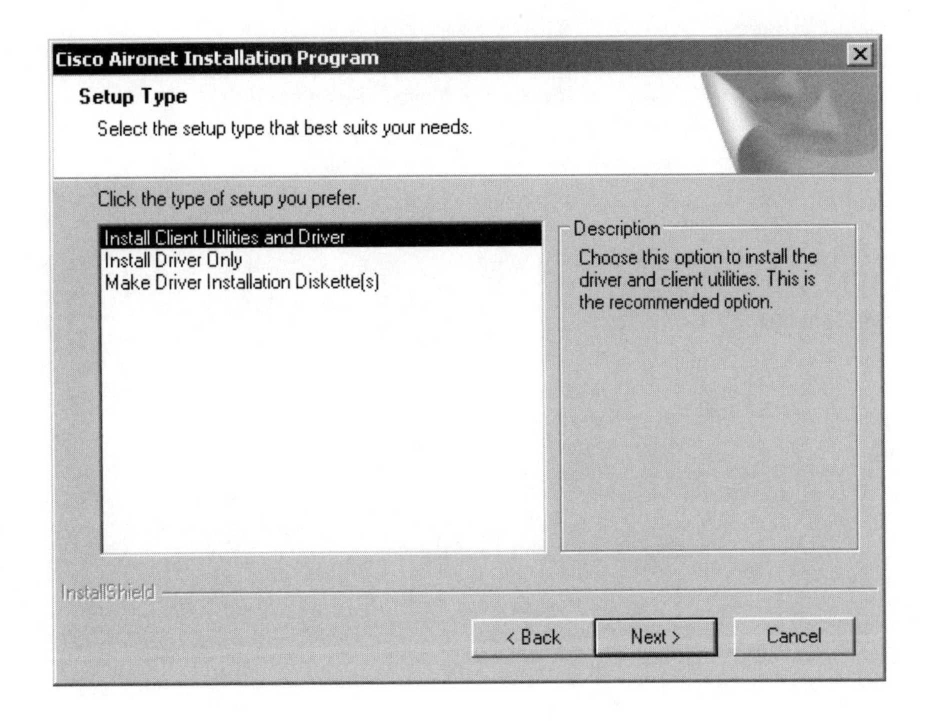

- **Install Driver Only** For users who are using Wireless Zero Configuration, CSSC (covered later in this chapter), or a third-party client software/ supplicant for their CB21/PI21 wireless adapter.
- **Make Driver Installation Diskette(s)** Allows you to extract the installation files to a CD, USB key, or other storage device for later use.

Select the Install Client Utilities and Driver option. Click the Next button; you may be prompted with a window that requires you to insert the Cisco Aironet 802.11a/b/g adapter if you have not already done so. Be sure that the adapter is installed/inserted and then click OK. A prompt like the one shown in Figure 10-21 should appear.

FIGURE 10-21

Site Survey Utility option

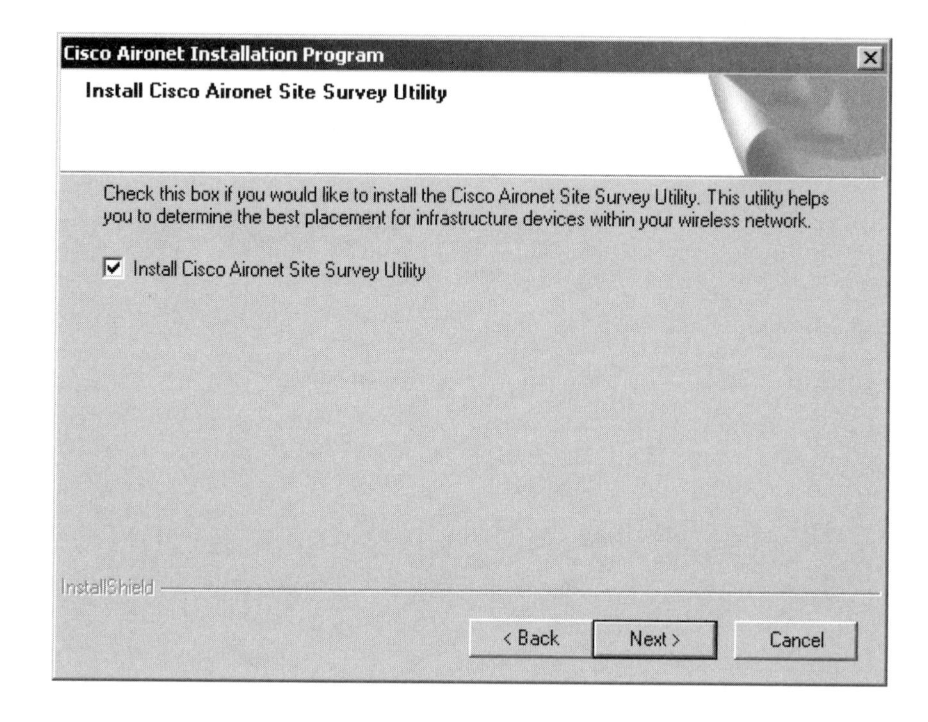

2. Click the check box to install the utility and then click Next. (We will cover the Site Survey Utility later in this chapter.) The Choose Configuration Tool window (shown in Figure 10-22) will appear.
3. Here you must choose whether you want your Cisco adapter to be controlled by the ADU software or remain under the control of a third-party tool, usually Windows Wireless Zero Configuration (WZC). The default is to use the ADU; select it for this example. Click Next to proceed. The installation will complete, and you will be prompted to reboot your system. After reboot, you will be ready to configure the ADU.

FIGURE 10-22

Choose Configuration Tool window

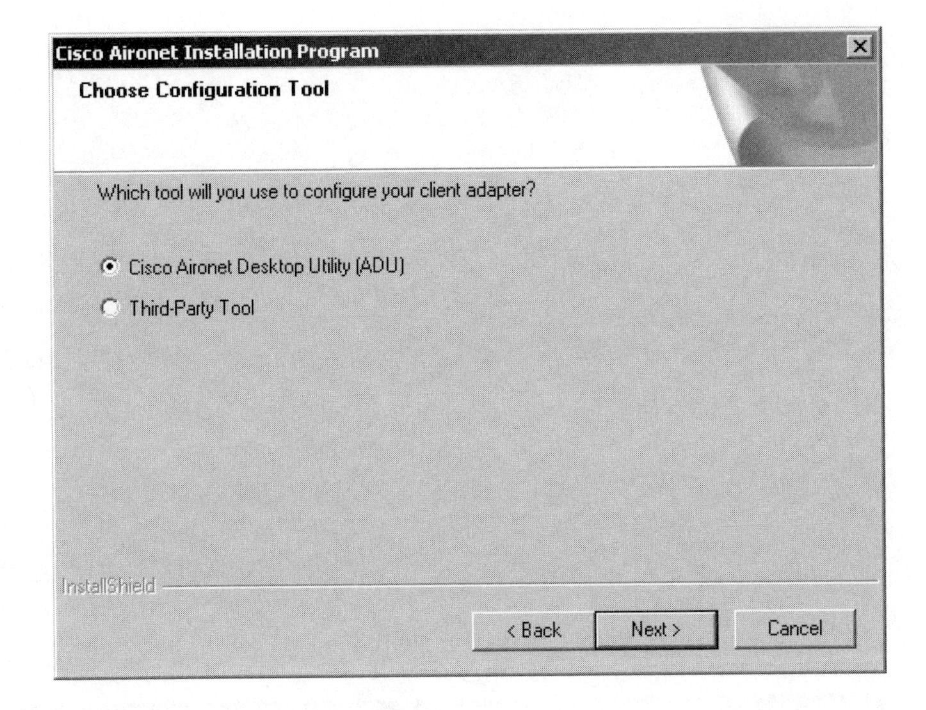

Configuration

When you first run the Aironet Desktop Utility (shown in Figure 10-23), you will have no wireless connection through your adapter and your adapter will be configured using a default user profile. You have the option of editing this profile or, as we will do in this example, creating a new profile to connect to the Enterprise SSID example we have been using throughout this chapter.

1. Click the Profile Management tab and the window shown in Figure 10-24 will appear.

FIGURE 10-23

Aironet Desktop Utility

Cisco Aironet Desktop Utility - Current Profile: Default
Action Options Help
Current Status
Profile Management
Diagnostics
CISCO
Profile Name: Default
Link Status: Not Associated
Network Type: Infrastructure
Wireless Mode: 2.4 GHz 54 Mbps
Current Channel: 6
Server Based Authentication: None
Data Encryption: None
IP Address: 0.0.0.0
Signal Strength:
No Link
Advanced

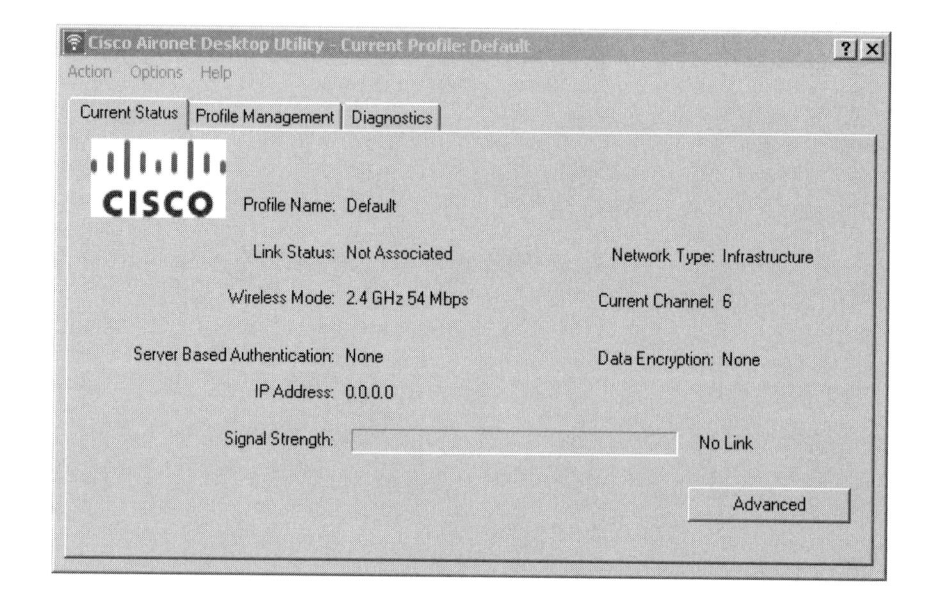

FIGURE 10-24

Profile Management tab

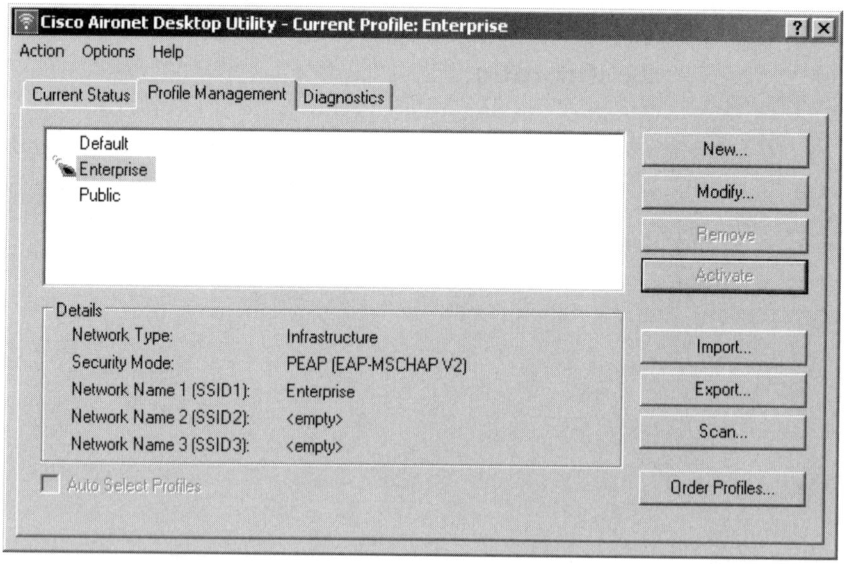

Cisco Aironet Desktop Utility - Current Profile: Enterprise
Action Options Help
Current Status
Profile Management
Diagnostics
Default
Enterprise
Public
New...
Modify...
Remove
Activate
Details
Network Type: Infrastructure
Security Mode: PEAP (EAP-MSCHAP V2)
Network Name 1 (SSID1): Enterprise
Network Name 2 (SSID2): <empty>
Network Name 3 (SSID3): <empty>
Import...
Export...
Scan...
Auto Select Profiles
Order Profiles...

2. Within the Profile Management tab, you will see a list of profiles. If this is your first time using ADU, chances are the only profile that will be shown is the default profile. In addition to having the option to activate these profiles for use, you can also click the Scan button to search for existing wireless networks that may not have a profile configured for them. Another feature of the ADU is the ability to prioritize already created profiles: click the Order Profiles button to see the Auto Profile Selection Management window (Figure 10-25).

   Click OK to go back to the Profile Management screen. From here you can create a new profile or modify an existing profile. In either case, the following steps for manually configuring a profile should be the same. Click either the New or Modify button, depending on your scenario. The Profile Management window (Figure 10-26) should appear.

3. Enter a profile name and the SSID (Enterprise) you wish to connect to for your profile. To configure wireless security settings for the profile, click the Security tab shown in Figure 10-27.

FIGURE 10-25

Auto Profile Selection Management window

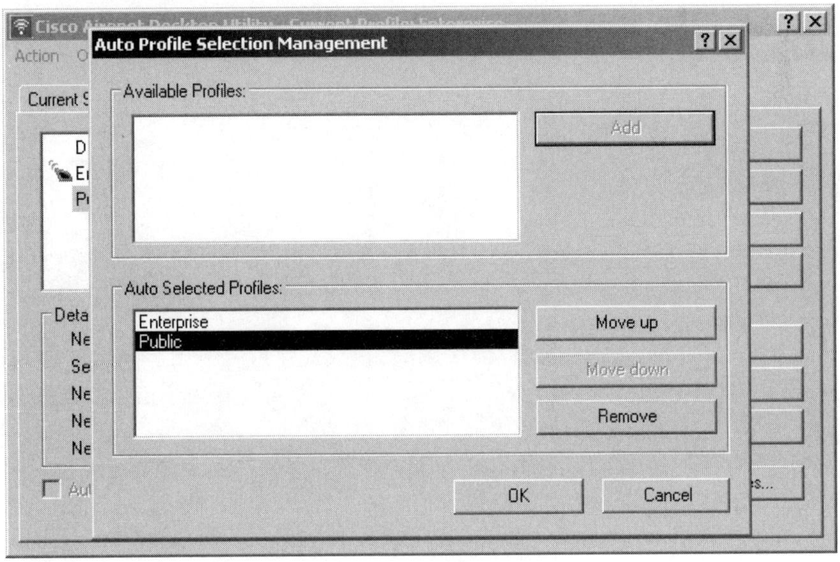

FIGURE 10-26

Add or modify a profile

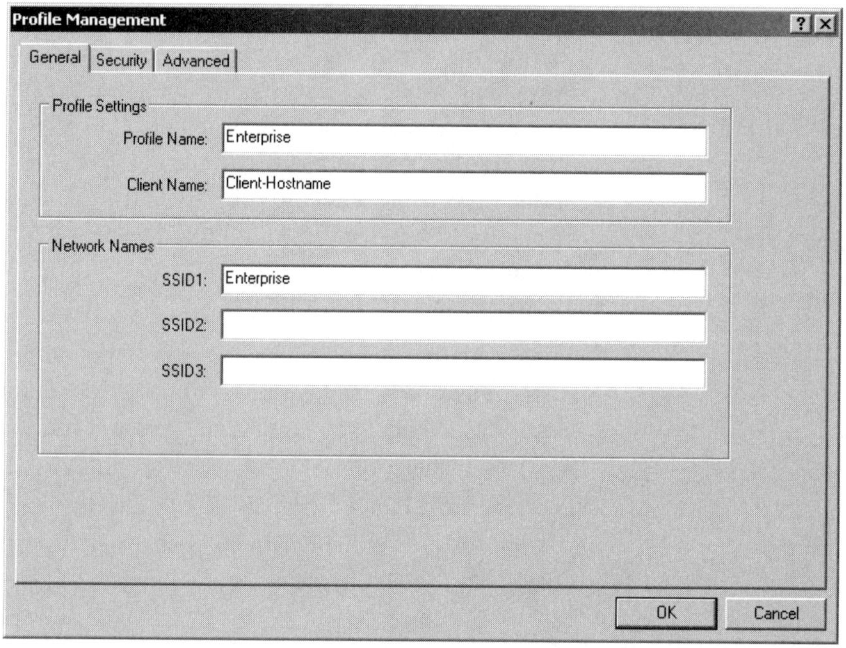

Profile Management

General | Security | Advanced

Profile Settings

Profile Name: Enterprise

Client Name: Client-Hostname

Network Names

SSID1: Enterprise

SSID2:

SSID3:

OK Cancel

FIGURE 10-27

Security

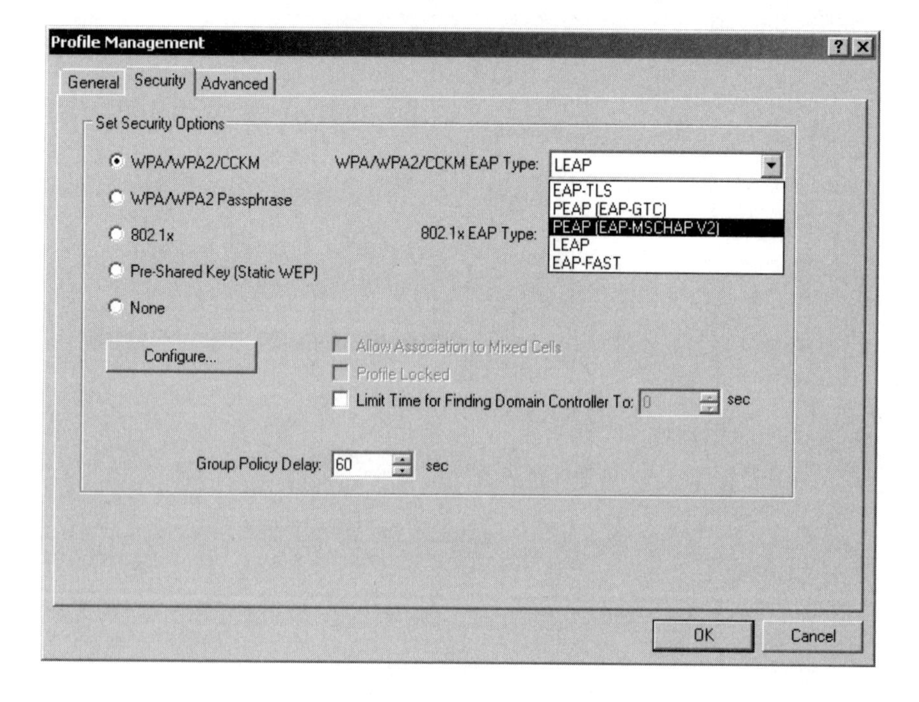

4. As with previous wireless client software, you can configure wireless encryption and authentication from this screen. The nomenclature on the ADU can be a bit confusing; each option is described here:
   - **WPA/WPA2/CCKM** WPA or WPA2 standard encryption paired with an 802.1X EAP type for authentication. This is the most secure option for an enterprise and takes advantage of the WPA four-way handshake discussed in Chapter 5.
   - **WPA/WPA2 Passphrase** WPA or WPA2 encryption with the Pre-Shared Key (PSK) option. If you select this option, you will be prompted to enter a pre-shared key string for encryption and authentication.
   - **802.1X** Rarely used, this option allows a client to use 802.1X authentication with no encryption or with WEP encryption. This is not a recommended choice and does not adhere to WPA or WPA2 standards.
   - **Pre-Shared Key (Static WEP)** For wireless networks that, for one reason or another, are still using the WEP encryption standard, the ADU gives clients the option of entering a static WEP key of the 64-bit or 128-bit variety.
   - **None** Open authentication with no encryption configured. This is typically used in guest access networks and is not recommended otherwise.

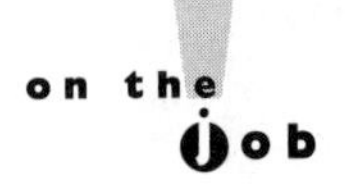

In modern wireless deployments, it makes little sense to deploy 802.1X without WPA or WPA2 encryption, as the technology is readily available on Cisco wireless and the majority of wireless clients.

5. For this Enterprise SSID , select the WPA/WPA2/CCKM radio button. From the drop-down box that appears, select the 802.1X EAP authentication type you want to use. For this example, that will be PEAP (EAP-MSCHAP V2). Then click the Configure button to configure user credentials for EAP authentication (Figure 10-28).
6. In the Configure PEAP window, you can configure server-side certificate information as well as the username and password that will be used to authenticate to the Enterprise SSID. In many cases, the Windows username

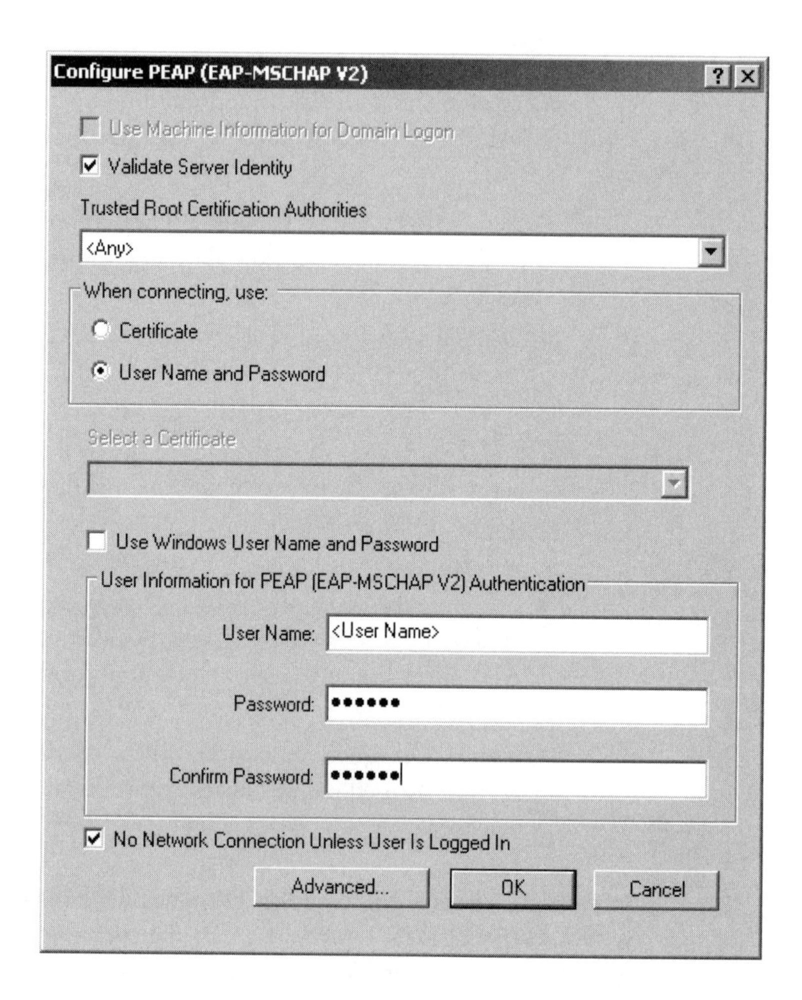

FIGURE 10-28 Configure EAP authentication

and password are used, but there are also scenarios where you can manually enter a valid username and password. To configure additional settings for this profile, click the Advanced tab (Figure 10-29) to see the Profile Management window.

FIGURE 10-29

Advanced tab

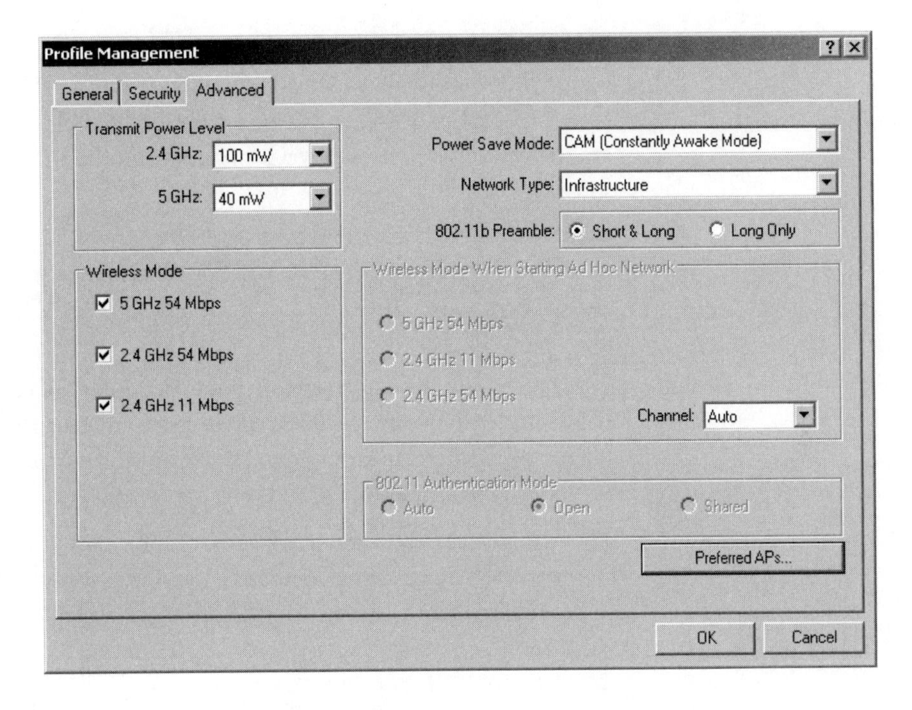

7. In this tab, you can configure radio settings for your client including the following:
   - **Transmit Power Level** Limits or raises the client's radio strength when communicating with an access point.
   - **Wireless Mode** Limits the client profile to operate only on certain 802.11 bands.
   - **Channel** Allows you to statically specify which RF channel you want the client to run on
   - **Advanced features** Includes Power Save Mode and the option to choose short or long 802.11b preambles for the client.
   - **Preferred APs** Enables you to choose preferred access points for association by manually entering their MAC addresses in an ordered list.

The Advanced tab is one of the sections where the Aironet Desktop Utility acts as client software while giving you features and options that are normally associated with advanced driver settings on a wireless NIC. These features are not seen in Wireless Zero Configuration or most other third-party client software.

8. When you have made all of your desired changes, click OK to save them and close the Profile Management Window. If you have successfully associated and authenticated, the Current Status tab for the ADU will look something like Figure 10-30.

   This tab gives basic information about the client association including the signal strength, acquired DHCP IP address, and security information. To see more advanced information about the wireless connection, click the Advanced button (Figure 10-31).

9. This window provides a great deal more information about the current client than Wireless Zero Configuration client software and can be used to provide troubleshooting information when needed.

FIGURE 10-30

Current Status tab

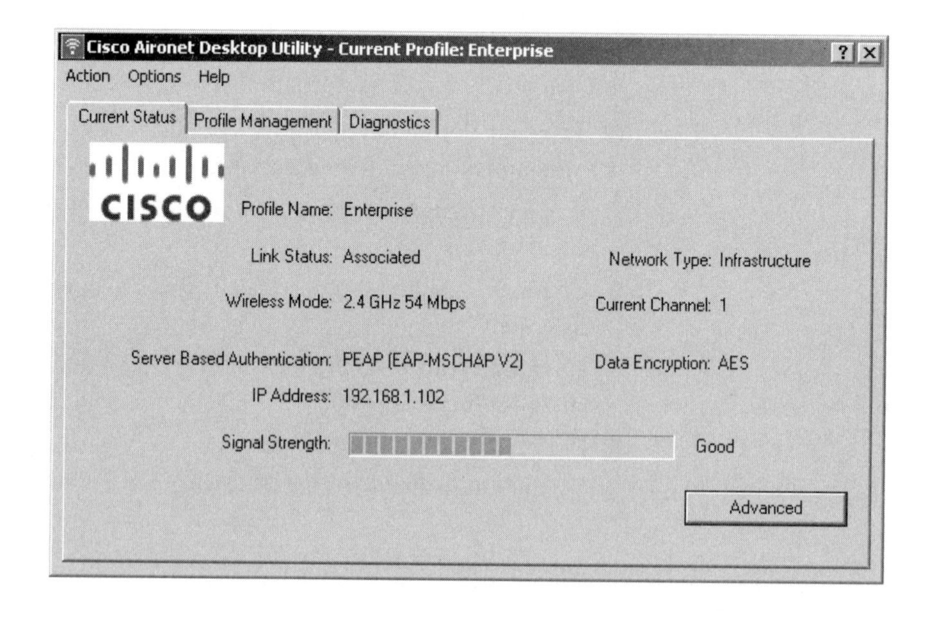

FIGURE 10-31

Advanced Status window

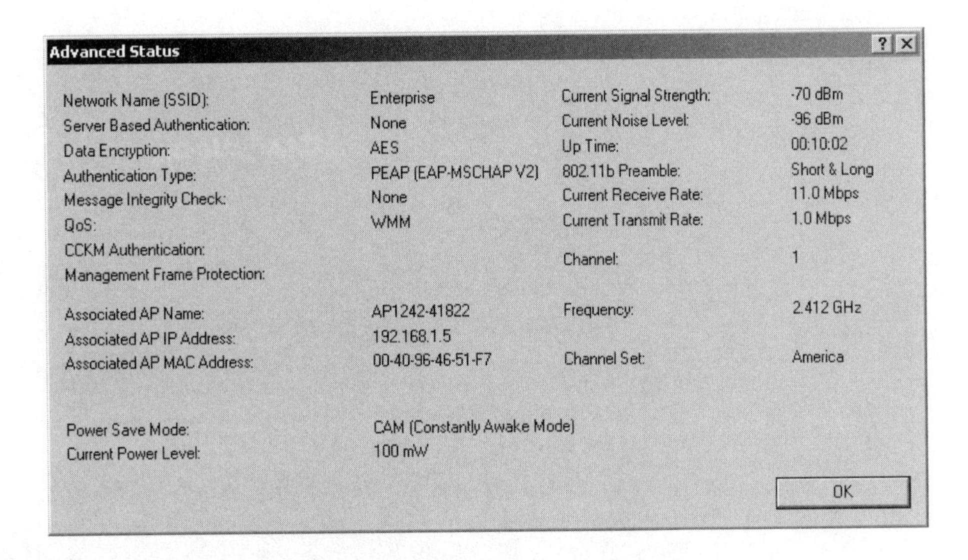

Aironet Site Survey Utility

A separate software utility that can be installed as a part of the Cisco Aironet Client Utility is the Cisco *Aironet Site Survey Utility (CSSU)*. The CSSU (Figure 10-32) provides basic site survey functionality by allowing a client to examine familiar RF properties for access points that it detects:

- Signal Strength (dBm)
- Noise Level (dBm)
- Signal-to-Noise Ratio (dB)
- Link Speed (mbps)
- AP Scan List of detected access points

Although far less feature-rich than a dedicated site survey software, the CSSU has enough customization and information to perform an on-the-fly survey without the cost or learning curve associated with a more advanced tool.

The ADU is a Cisco-specific application and will not work with wireless adapters from other vendors. The ADU is a package made up of a client adapter driver, a client supplicant, and a suite of tools designed to be used with Cisco CB21/PI21 adapters.

FIGURE 10-32

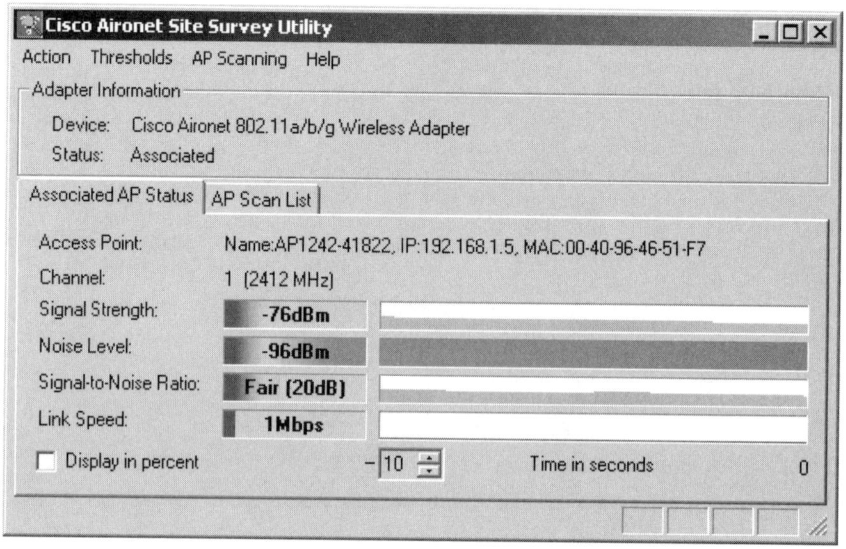

Cisco Aironet Site Survey Utility (CSSU)

Cisco Secure Services Client (CSSC)

The *Cisco Secure Services Client (CSSC)* was introduced in 2008 as a Windows 2000/XP/Vista client software supplicant for large enterprise wireless network deployments. Unlike the ACU, the CSSC supports almost all vendor wireless adapters and is designed as an authentication framework for both wired (802.3) and wireless (802.11a/b/g/n) network interface cards. Like other wireless supplicants, CSSC provides 802.1X user and device authentication and also has the ability to be centrally managed using the CSSC Management Utility as part of a larger IT-based solution. CSSC supports a very wide range of security and authentication types including all available EAP authentication methods and full support for all WPA and WPA2 standard encryption. In addition to these core features, CSSC also offers the following:

- **Broad user credentials support** CSSC allows users to authenticate using one-time password (OTP) tokens and a variety of smartcards, X.509 certificates, and regular passwords.
- **Automatic VPN and SoftToken integration** CSSC integrates with the Cisco IPSec VPN client and Secure Computing SoftToken to allow for quick access to VPNs using any of the CSSC supported adapters.

- **Client provisioning via an XML file** An administrator can easily push out the same CSSC client configuration to multiple clients using an XML configuration file via IT deployment tools (Microsoft Active Directory, SMS, Altiris) or by building an installation package that is pushed to user machines. This XML configuration is vendor neutral and can work regardless of what kind of hardware is on the client end.
- **Policy enforcement** The CSSC can be configured to
  - Filter unwanted SSIDs.
  - Require certain levels of wireless 802.1X security.
  - Mandate the shutdown of a wireless connection when a wired connection is present on a client.
  - Lock down client features and enforce specific preconfigurations.

CSSC, by default, has a nonexpiring license that allows for configuration of wired-only connections with a limited selection of EAP authentication methods. A 90-day trial license can be obtained to add wireless capabilities and support for all EAP methods. For full support, an unlimited wired and wireless license can also be purchased.

Installation

The installation of CSSC is a generic install, much like any other Windows software package. On a computer where you have administrator rights, you can run an executable package or MSI to complete installation. You will be prompted to reboot your system and CSSC will be active upon the next system start.

Configuration

Configuration of CSSC is a basic set of steps that is designed to be very intuitive. Follow the following steps to complete CSSC configuration:

1. When CSSC is installed, there should be a tray icon that you can right-click, allowing you to see the list of available wireless networks, similar to the menu in Figure 10-33.
2. From here, you can attempt to connect to one of the listed networks, turn the wireless radio on or off, connect to VPN, check connection status (shown in a later step), or access the CSSC config using the View Available Connections option. If you double-click the CSSC tray icon, you will get the same display shown in Figure 10-34.

FIGURE 10-33

CSSC tray menu

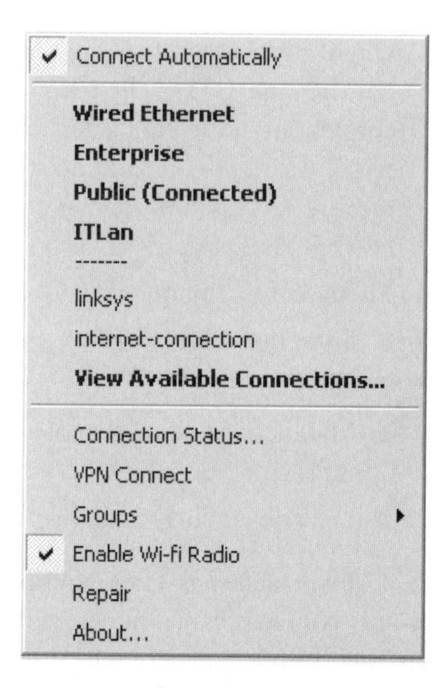

FIGURE 10-34

CSSC Configuration window

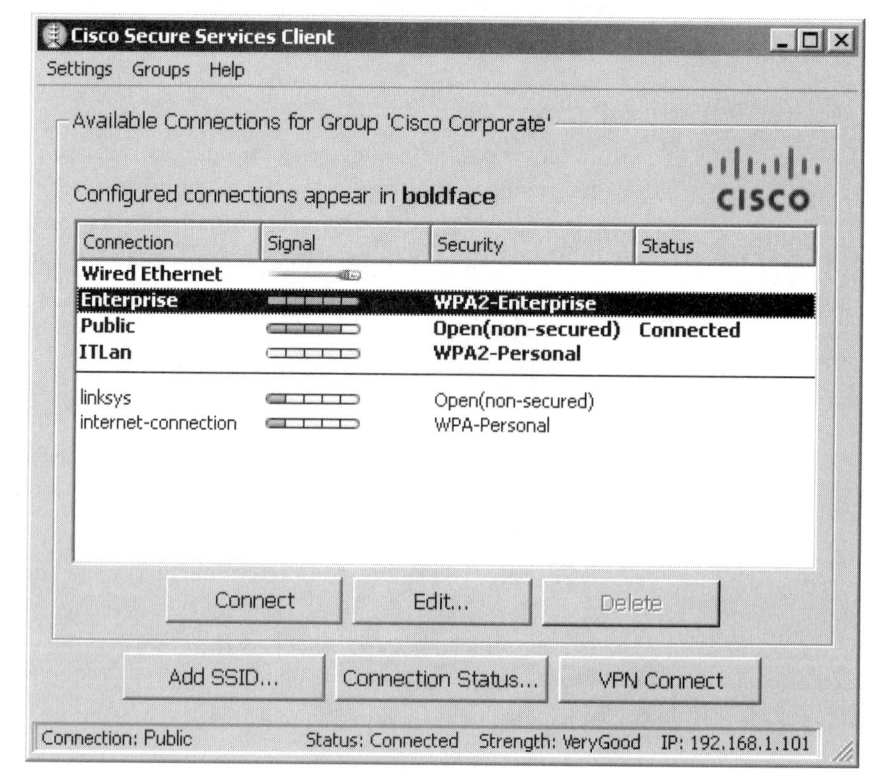

3. In this window you can add, edit, delete, or make changes to different configured connections (depending on what rights the administrator has given the client). To add or edit the Enterprise SSID, you can click either Add or Edit to reach the screen shown in Figure 10-35.
4. The CSSC has a very straightforward configuration format that allows you to configure the SSID for the connection as well as any security and 802.1X authentication information. You can also configure a specific Cisco VPN profile to be associated with the connection. If additional credentials are required, you will be prompted for them after you click OK in this window.
5. When you return to the main window, you can choose to connect to your configured connection and a status of Connected will appear after association, authentication, and acquiring of an IP address over your desired connection. If you so choose, you can also make an exclusive connection in the event that you have other connections available but want to lock in to a specific one. To check the status of the connection, click the Connection Status button. A window similar to Figure 10-36 will appear.

FIGURE 10-35

Enter Connection Info window

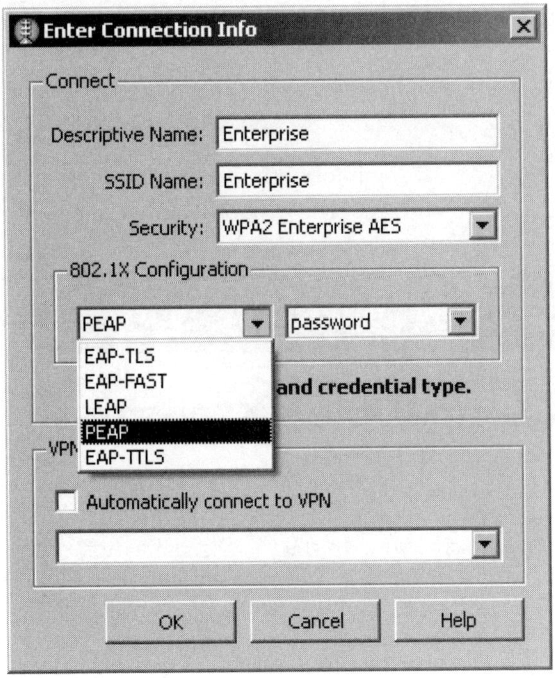

FIGURE 10-36

Connection Status window

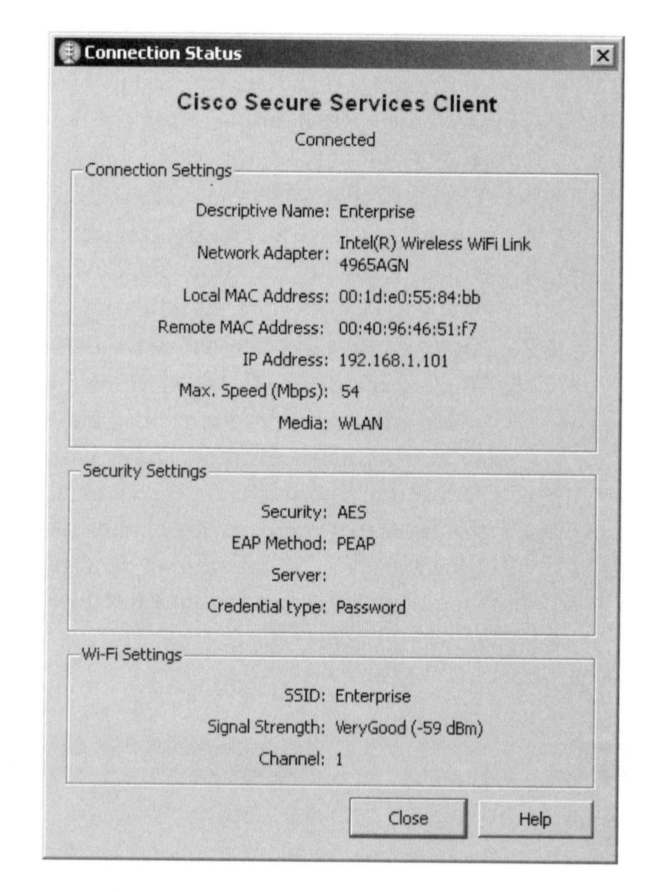

6. If the information shown is acceptable, you can close this window and begin utilizing the network connection you have chosen.

CERTIFICATION OBJECTIVE 10.03

Cisco Compatible eXtensions (CCX)

The *Cisco Compatible eXtensions (CCX)* program is a certification program for third-party wireless devices that verifies interoperability with a Cisco wireless LAN infrastructure. In the CCX certification process, Cisco has a set of common criteria for reaching CCX certification, which varies by version and will be discussed in more detail in this section.

CCX Certification Process

A vendor of WLAN client adapters or wireless-enabled devices engineers their product to reach these criteria before it is thoroughly tested in an independent lab. If the tests pass, then the device or adapter is CCX certified.

A CCX certified device is guaranteed to meet standards for interoperability with Cisco wireless devices as well as support for Cisco features around wireless security, roaming/mobility, and manageability.

Versions and Features

Cisco Compatible eXtension versions refer to different levels of specification. As wireless technologies and features have evolved, the CCX program has released versions of the certification to match. There are currently five versions, listed next. Each version builds upon previous versions, and all but a few features in older versions must be supported in new versions. The following is a list of what is featured in different CCX versions, taken from the Cisco CCX home page (www.cisco.com/web/partners/pr46/pr147/partners_pgm_concept_home.html):

- CCX Version 1 (V1)
  - Compliance with IEEE 802.11 and Wi-Fi
  - Support for the 802.1X authentication type: Cisco LEAP
  - Ability to interoperate with an access point that supports multiple Service Set Identifiers (SSIDs) tied to multiple VLANs, providing benefits such as flexible security schemes in a mixed client environment
- CCX Version 2 (V2)
  - Compliance with Wi-Fi Protected Access (WPA), including support for WPA Temporal Key Integrity Protocol (TKIP) encryption
  - Support for the 802.1X authentication type: Protected EAP (PEAP) with EAP-GTC

  - Fast, secure roaming through support of the 802.1X key management protocol: Cisco Centralized Key Management (CCKM)
  - Radio frequency (RF) scanning, with scanned data sent to the access point and ultimately to CiscoWorks Wireless LAN Solution Engine (WLSE) for analysis and performance of RF management functions such as intrusion detection, assisted site survey, and detection of interference sources
- CCX Version 3 (V3)
  - Compliance with WPA 2, including support for Advanced Encryption Standard (AES) encryption
  - Support for the 802.1X authentication type: EAP-FAST
  - Support for Wi-Fi Multimedia (WMM), a subset of the IEEE 802.11e quality of service (QoS) standard defined by the Wi-Fi Alliance
- CCX Version 4 (V4)
  - Support of Cisco Network Admission Control
  - Call admission control—addressing voice over IP (VoIP) stability, roaming, and other QoS-related issues
  - Support for a power-saving mechanism, U-APSD, in QoS environments
  - VoIP metrics for reporting to optimize WLAN VoIP performance
  - Enhanced roaming
  - Ability to function as an 802.11 location tag
- CCX Version 5 (V5)
  - Management Frame Protection (MFP) support
  - Support for diagnostic channel troubleshooting
  - Support for client reporting
  - Support for roaming/mobility diagnostics

INSIDE THE EXAM

Operating System Client Configuration

The majority of this chapter is based on understanding the configuration of different wireless client types. In the section on operating system clients, it is important to familiarize yourself with the windows and menus of the Windows, Apple, and Linux clients. A strong understanding of the wireless security concepts from Chapter 5 will contribute greatly to your understanding of what is needed on client configuration in this chapter. To best prepare for this piece of the exam, it is recommended that you physically access and work with the clients on the Windows, Apple, and Linux operating systems.

Cisco Client Configuration

It is important to familiarize yourself with the Cisco Aironet Desktop Utility and Secure Services Client windows and menus. Be sure to understand the features available on both products that cannot be found on traditional OS clients. Know that ADU is only supported on Cisco 802.11a/b/g CB21 and PI21 client cards and that the CSSC can support all wireless clients and operate as part of a centralized IT provisioning solution.

Cisco Compatible eXtensions (CCX)

You should know what the purpose and role of CCX is and be able to identify the differences between different CCX versions. Have a strong understanding of the CCX certification process and the benefits of wireless client devices that have been certified as a part of the CCX program.

CERTIFICATION SUMMARY

Operating System Client Configuration

The three major operating systems covered by this book are Windows, Mac, and Linux. Each of these operating systems has built-in wireless client software, which is also known as a supplicant. For Microsoft Windows, the operating system client is a limited package called Wireless Zero Configuration (WZC) that provides WPA and WPA 2 security along with the ability to automatically connect to available networks. For Apple, AirPort and AirPort Extreme provide the wireless client capabilities used in virtually all Mac computers. For Linux, the NetworkManager manages network connections, including wireless adapters using a simple applet GUI.

Cisco Client Configuration

Cisco, in the course of developing wireless products, has created a variety of wireless clients to support growing enterprise needs. The first of these is the Aironet Desktop Utility, which only supports the Cisco CB21 or PI21 802.11a/b/g wireless adapters. It is more powerful than traditional wireless client software and allows for more troubleshooting tools and information. ADU is also packaged with Cisco NIC drivers and the Site Survey utility. Cisco Secure Services Client (CSSC) is a more generalized wireless client that supports a great deal of customization and works with all wireless adapters. Unlike ACU and most other wireless clients, CSSC allows you to enforce security policies centrally and works with both wired and wireless connections.

Cisco Compatible eXtensions (CCX)

The Cisco Compatible eXtensions (CCX) program was created to establish standards for third-party wireless client vendors. By going through a certification process, wireless devices and adapters of all kinds can be certified to support a given range of Cisco wireless technologies. In the course of certification, the third-party client vendor undergoes a series of independent lab tests to verify interoperability with Cisco wireless devices before being certified as a part of the CCS program. At the time of this publication, there are five versions of CCX; each new version adds new requirements to its predecessor.

TWO-MINUTE DRILL

Operating System Client Configuration

- The Microsoft Windows wireless client is known as Wireless Zero Configuration (WZC).
- The Mac wireless client is known as AirPort or AirPort Extreme.
- The Linux wireless client is known as NetworkManager.

Cisco Client Configuration

- The Cisco Aironet Desktop Utility (ADU) is a Cisco client specifically designed for the Cisco CB21 and PI21 802.11a/b/g wireless adapters. It has a wide range of features and comes packaged with a site survey utility.
- The Cisco Secure Services Client (CSSC) is a full-featured client supplicant that can be heavily customized by IT administrators and has support for all wired and wireless network adapters.

Cisco Compatible eXtensions (CCX)

- CCX is Cisco's program for third-party wireless devices that verifies interoperability with a Cisco wireless LAN infrastructure.
- In the CCX certification process, Cisco has criteria for reaching CCX certification, which are fulfilled and then verified by independent lab tests.
- There are currently five versions of CCX with each subsequent version adding new requirements over its predecessor.

SELF TEST

The following self-test questions will help you measure your understanding of the material presented in this chapter. Read all the choices carefully, as there may be more than one correct answer. Choose all correct answers for each question.

Operating System Client Configuration

1. Which of the following things is not a supported security configuration on Windows Wireless Zero Configuration client?
   A. EAP-FAST with WPA2 AES encryption
   B. EAP-TLS with WPA2 AES encryption
   C. PEAP with WPA TKIP encryption
   D. WPA PSK with TKIP encryption
2. Which of these terms is not associated with the Windows wireless client?
   A. Wireless Zero Configuration
   B. PEAP with MSCHAP V2
   C. AirPort
   D. WPA
3. Which of these is not an operating system wireless client?
   A. WZC
   B. ProSet
   C. NetworkManager
   D. AirPort
4. NetworkManager has all but which of the following characteristics?
   A. Can run on multiple different Linux operating systems
   B. Supports WPA and WPA2 encryption
   C. Can support ad hoc wireless networks
   D. Configures only wireless network connections on a Linux platform

Cisco Client Configuration

5. Which of the following is not true of the Cisco Aironet Desktop Utility (ADU)?
   A. ADU, like CSSC, supports a variety of wireless clients.

B. ADU supports WPA2 with AES encryption.

C. ADU supports a wide variety of 802.1X EAP types including EAP-FAST and PEAP.

D. All of the above are true of the ADU.

6. When working with ADU, which of the following is not an advantage of ADU over other wireless clients?

A. You can configure the channel on the wireless client running ADU.

B. You can manually configure the transmit power setting of a wireless client running ADU.

C. The ADU is simpler to deploy through the use of XML configuration files.

D. The ADU allows you to build a list or preferred APs by MA C address.

7. Which of the following about CSSC is not true?

A. CSSC supports a large amount of 802.1X EAP types.

B. CSSC policies can be enforced by an IT administrator.

C. CSSC has a built-in site survey tool that allows you to perform RF diagnosis.

D. CSSC supports WPA and WPA2 encryption with Pre-Shared Keys (PSK).

8. CSSC is used on all but which of the following operating systems?

A. Windows XP

B. Windows Me

C. Windows 2000

D. Windows Vista

9. The CSSC is available in all but which of the following license configurations?

A. Wireless unlimited license with no wired support

B. Wired unlimited license with no wireless support

C. 90-day trial license with wired and wireless support

D. Wired and wireless unlimited license

Cisco Compatible eXtensions (CCX)

10. What is the minimum version of CCX compliance that requires support for 802.11i security with WPA2 encryption?

A. CCX v1

B. CCX v2

C. CCX v3

D. CCX v4

SELF TEST ANSWERS

Operating System Client Configuration

1. Which of the following things is not a supported security configuration on Windows Wireless Zero Configuration client?

A. EAP-FAST with WPA2 AES encryption

B. EAP-TLS with WPA2 AES encryption

C. PEAP with WPA TKIP encryption

D. WPA PSK with TKIP encryption

☑ **A.** Wireless Zero Configuration does not support the EAP-FAST 802.1X authentication type.

☒ **B, C,** and **D** are incorrect because they all configurable security types on WZC.

2. Which of these terms is not associated with the Windows wireless client?

A. Wireless Zero Configuration

B. PEAP with MSCHAP V2

C. AirPort

D. WPA

☑ **C.** AirPort is the wireless client for Mac and has nothing to do with the Windows wireless client.

☒ **B, C,** and **D** are all technologies/concepts associated with the Windows wireless client.

3. Which of these is not an operating system wireless client?

A. WZC

B. ProSet

C. NetworkManager

D. AirPort

☑ **B.** ProSet is actually an Intel-based wireless client but, unlike WZC (Wireless Zero Configuration) it is not one of the operating system wireless clients discussed in this chapter.

☒ **A, C,** and **D** are incorrect because they are all valid operating system clients.

4. NetworkManager has all but which of the following characteristics?

A. Can run on multiple different Linux operating systems

B. Supports WPA and WPA2 encryption

C. Can support ad hoc wireless networks

D. Configures only wireless network connections on a Linux platform

☑ **D.** NetworkManager for Linux is intended to configure both wired and wireless connections on a Linux platform.

☒ **A**, **B**, and C are incorrect because all of those statements are true of Linux's NetworkManager.

Cisco Client Configuration

5. Which of the following is not true of the Cisco Aironet Desktop Utility (ADU)?

A. ADU, like CSSC, supports a variety of wireless clients.

B. ADU supports WPA2 with AES encryption.

C. ADU supports a wide variety of 802.1X EAP types including EAP-FAST and PEAP.

D. All of the above are true of the ADU.

☑ **A.** ADU unfortunately only supports Cisco CB21 and P121 802.11a/b/g clients.

☒ **B**, **C**, and **D** are incorrect because they are all true statements about ADU.

6. When working with ADU, which of the following is not an advantage of ADU over other wireless clients?

A. You can configure the channel on the wireless client running ADU.

B. You can manually configure the transmit power setting of a wireless client running ADU.

C. The ADU is simpler to deploy through the use of XML configuration files.

D. The ADU allows you to build a list or preferred APs by MA C address.

☑ **C.** The ADU has never had the ability to be configured with an XML file. This is a feature that came about in CSSC.

☒ **A**, **B**, and **D** are incorrect because they are all advantages of ADU over traditional wireless client software.

7. Which of the following about CSSC is not true?

A. CSSC supports a large amount of 802.1X EAP types.

B. CSSC policies can be enforced by an IT administrator.

C. CSSC has a built-in site survey tool that allows you to perform RF diagnosis.

D. CSSC supports WPA and WPA2 encryption with Pre-Shared Keys (PSK).

☑ **C.** The CSSC, although it has RF capabilities, does not have a built-in site survey tool.

☒ **A**, **B**, and **D** are incorrect because they are all true statements about CSSC.

8. CSSC is used on all but which of the following operating systems?

A. Windows XP

B. Windows Me

C. Windows 2000

D. Windows Vista

☑ **B.** Windows Me is not an operating system supported by CSSC

☒ **A**, **C**, and **D** are incorrect because they are all operating systems that CSSC will run on.

9. The CSSC is available in all but which of the following license configurations?

A. Wireless unlimited license with no wired support

B. Wired unlimited license with no wireless support

C. 90-day trial license with wired and wireless support

D. Wired and wireless unlimited license

☑ **A.** The CSSC has support for wired connections but requires a license for wireless connections. There is no option for wireless only; wired is always included by default.

☒ **A**, **B**, and **C** are incorrect because they are all valid license configurations for CSSC.

Cisco Compatible eXtensions (CCX)

10. What is the minimum version of CCX compliance that requires support for 802.11i security with WPA2 encryption?

A. CCX v1

B. CCX v2

C. CCX v3

D. CCX v4

☑ **C.** CCX v3 is the first version of CCX to specify 802.11i standards compliance with WPA2 support.

☒ **A** and **B** are versions of CCX that were released before the 802.11i standard and therefore do not require them. **C** is incorrect because, although CCX v4 requires 802.11i and WPA2 compliance, it is not the earliest/minimum version to do so.

Part V

Maintaining and Troubleshooting a Cisco Wireless Network

CHAPTERS

11

Administering and Maintaining a Cisco Wireless Network

CERTIFICATION OBJECTIVES

In Chapter 8, we covered the concepts of configuring a Wireless LAN Controller (WLC) in a CUWN, autonomous and lightweight access points, as well as basic components in a WLC, including ports, interfaces, and WLANs (SSIDs). As mentioned earlier, the CCNA Wireless exam places focuses on the hands-on portion, which involves basic configuration of the elements needed to create a wireless network and familiarity with managing the wireless devices. Chapters 11 and 12 will build on top of concepts you're already familiar with and introduce the tasks of maintaining and troubleshooting a CUWN solution. There are a variety of terms and concepts involved in the day-to-day controller and access point maintenance and administration that will be an important part of the exam. It is necessary to fully comprehend these steps to successfully operate a CUWN. In this chapter, we will cover the different methods of accessing and maintaining a WLC, upgrading or downgrading a controller, upgrading or downgrading a lightweight access point, and maintaining the controller configurations.

CERTIFICATION OBJECTIVE 11.01

Cisco Wireless Network Administration and Access Methods

The Cisco Unified Wireless Network (CUWN) is designed to centrally provide 802.11 wireless services to clients and APs for large and complex networks. As mentioned briefly in previous chapters, a Wireless LAN Controller (WLC) can be managed via a CLI or GUI. Most administration and configuration tasks can be configured via a GUI, while a CLI can perform all of the administration, configuration, and monitoring tasks.

WLC CLI

Unlike other Cisco device CLIs, the WLC CLI does not run on top of Cisco IOS software. However, it does support commands such as the question mark (?) to make your life easier.

The WLC's CLI console access is a VT-100 terminal emulator that can be accessed through a PC, laptop, or workstation by using a null modem serial cable to connect the CLI console and WLC. To access the CLI console through the physical console port, perform the following steps:

1. Connect one end of a null modem serial cable to the controller's console port via a DB-9 interface.
2. Connect the other end of the serial cable to a PC via a COM port, and start a VT-100 terminal emulator, such as HyperTerminal, TeraTerm, or SecureCRT.
3. Configure the COM port using the following parameters:
   - 9600 baud rate
   - 8 data bits
   - No flow control
   - 1 stop bit
   - No parity
4. WLC refers to console port access as the serial port. Enter the CLI command for configuring the console (serial) port as shown next:

```
Cisco Controller) >config serial ?
baudrate        Configure the serial port baud rate.
timeout         Configure the serial port login inactivity timeout.
(Cisco Controller) >config serial baudrate ?

[1200/2400/4800/9600/19200/38400/57600/115200] Enter serial speed.
(Cisco Controller) >config serial timeout ?
[0-160]         Enter time in minutes.
(Cisco Controller) >
```

   Figure 11-1 illustrates configuration options for the console (serial port) configuration via WLC GUI.

Many new laptop computers are now removing the built-in serial port from the standard hardware configuration, such as the Lenovo Thinkpad series of laptops. For these laptop computers, a USB serial adaptor may be required, such as the CODI USB-to-serial adapter cable. Usually, USB serial adapters require a special driver from the vendor, and the driver is not typically included in the Windows Drivers Library.

FIGURE 11-1 Console (Serial) Port Configuration options via the GUI

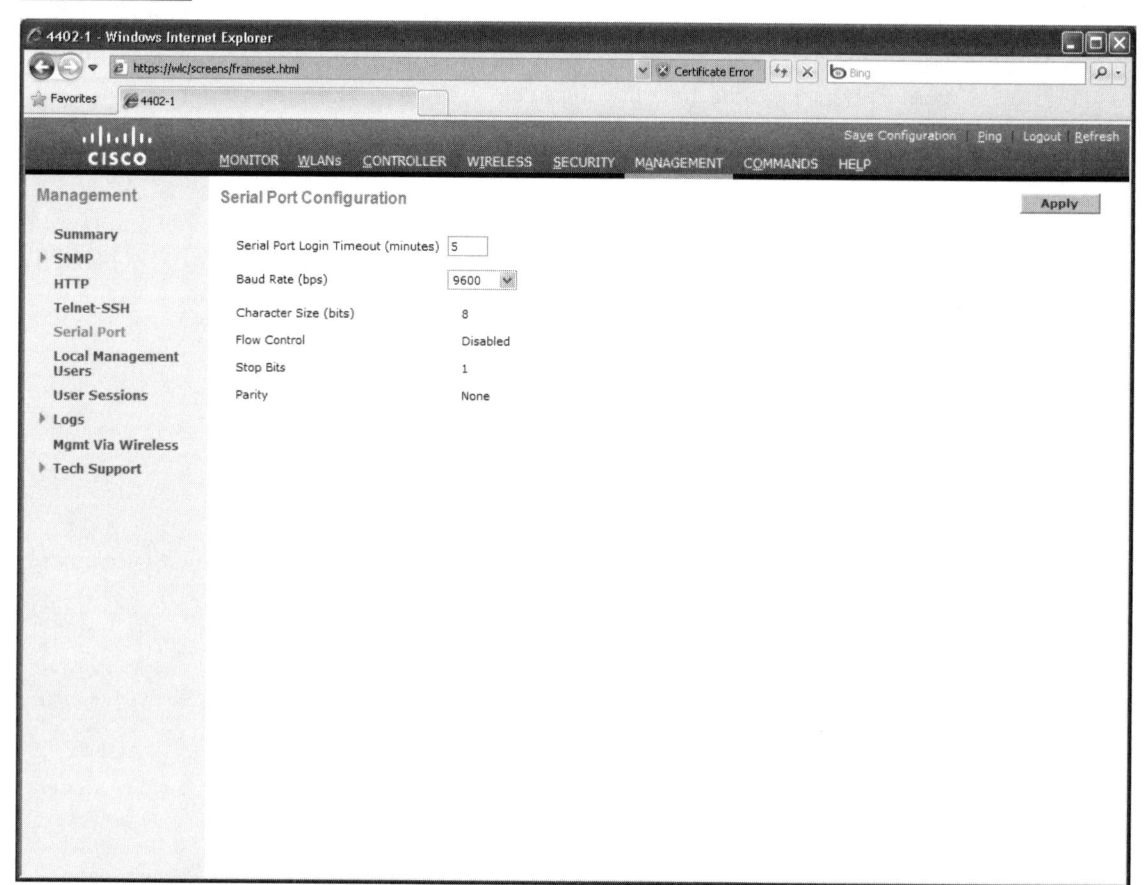

SSH is the default remote CLI access method for a WLC. SSH is enabled by default and can be enabled or disabled via the GUI or CLI. The SSH session should be initiated from the SSH client to the WLC's management interface.

Telnet is an optional access method for remote CLI access. Telnet is disabled by default and can be enabled or disabled via GUI or CLI. The Telnet session should be initiated from the Telnet client to the WLC's management interface.

The following CLI commands configure the different configuration options for SSH and Telnet access to WLC:

```
(Cisco Controller) >config network ssh ?
enable          Allow new ssh sessions.
disable         Disallow new ssh sessions.
host-key        Configure SSH access host key
(Cisco Controller) >config network telnet ?
[enable/disable] Enter mode.
```

Figure 11-2 illustrates the configuration options for the SSH/Telnet configuration options via the WLC GUI.

FIGURE 11-2 Telnet-SSH Configuration options via the GUI

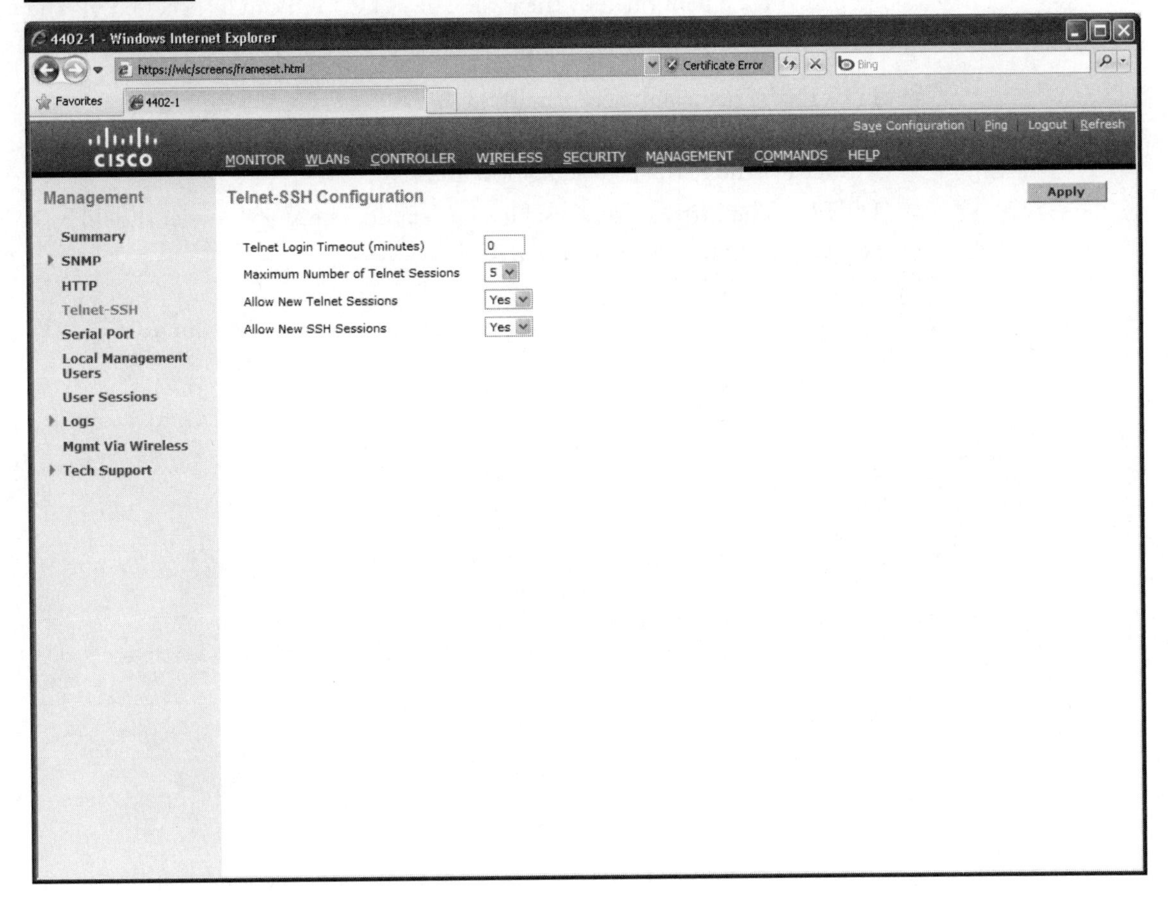

WLC GUI

The graphic user interface (GUI) is another way of accessing and administering a WLC. The Cisco WLC GUI requires the following Web browsers to be used: Microsoft Internet Explorer version 6 SP1 (or later) or Mozilla Firefox 2.0.0.11 (or later). Although other browsers such as Google Chrome and Apple Safari have been known to work with the WLC GUI, Cisco officially only supports Microsoft Internet Explorer and Mozilla Firefox. In addition, Opera and Netscape are two browsers that are specifically stated in WLC configuration documentation as not supported by Cisco. The WLC GUI supports most administrative tasks and is the preferred method of managing a WLC. There are some additional configurations and debug tasks that can only be performed by the CLI. The WLC GUI can be accessed via the following methods:

- **HTTPS** The default access method to the WLC GUI using TCP port 443. The Web browser should be pointed to the management interface IP address with either the HTTPS heading or port 443.
- **HTTP** The http access is disabled by default. The Web browser should be pointed to the management interface IP address.

```
(Cisco Controller) >config network secureweb ?
cipher-option  Configure cipher requirements for web admin and web auth.
enable         Enable the Secure Web (HTTPS) management interface.
disable        Disable the Secure Web (HTTPS) management interface.
To enable or disable secure web mode with increased security, use the following
commands:
(Cisco Controller) >config network secureweb cipher-option sslv2 ?
enable         Permits both SSLv2 and SSLv3 for both web administration and web
authentication.
disable        Permits only SSLv3 for both web administration and web authenti-
cation.
(Cisco Controller) >config network secureweb cipher-option ?
high           Configure whether or not 128-bit ciphers are required for web ad-
min and web auth.
sslv2          Enable or disable SSLv2 for both web administration
n and web authentication.
(Cisco Controller) >config certificate generate ?
webadmin       Generates a new web administration certificate
webauth        Generates a new web authentication certificate
```

The `config certificate generate webadmin` command will generate a self-signed certificate for the HTTPS access. Figure 11-3 illustrates the configuration options for the HTTP/HTTPS configuration options via WLC GUI.

FIGURE 11-3 HTTPS/HTTP Configuration options via the GUI

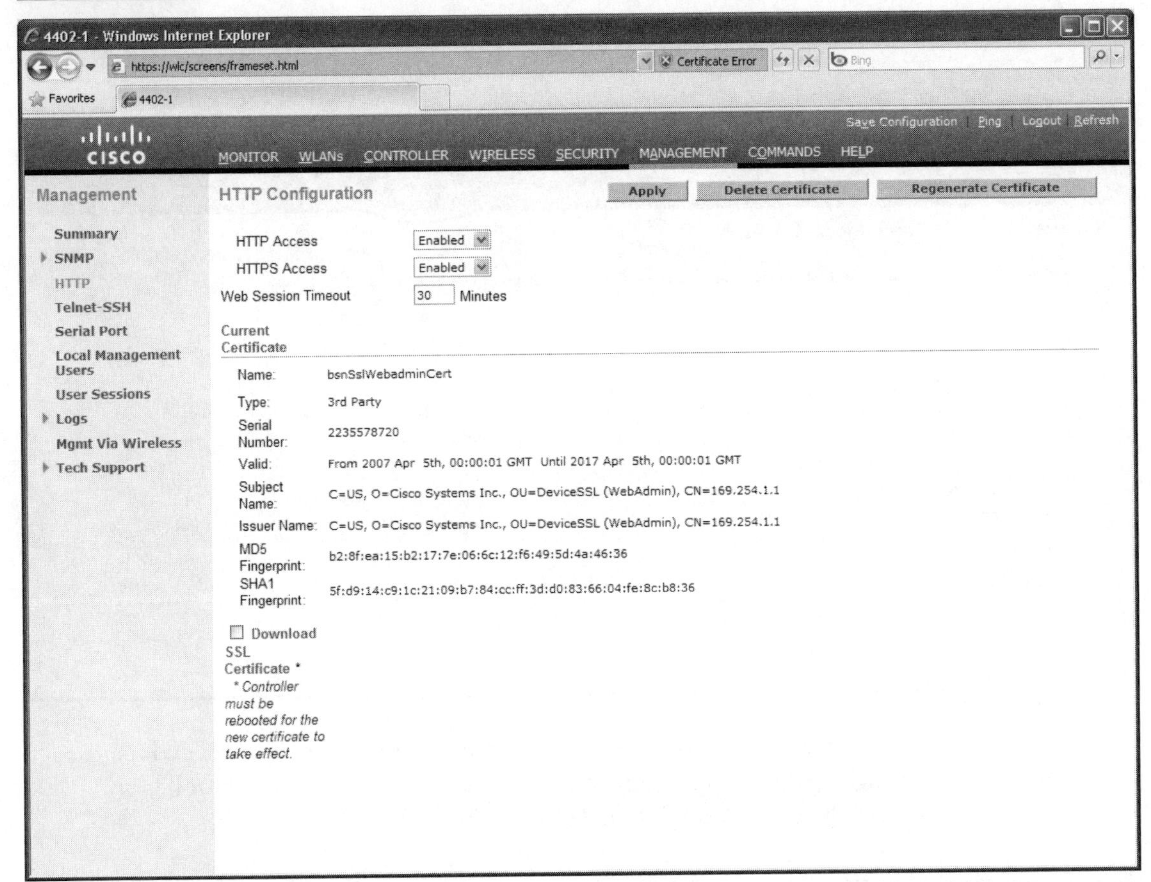

Wireless Connections to GUI and CLI

You can also administer and configure the WLC using a wireless client via the CLI or GUI. A wireless client can perform all management tasks except uploads from and downloads to a WLC. The wireless client access to WLC management interface is disabled by default. The following commands enable wireless connections to the GUI or CLI:

```
(Cisco Controller) >config network mgmt-via-wireless?
[enable/disable] Enter mode.
```

Figure 11-4 illustrates the configuration options for wireless client configuration options via the WLC GUI.

FIGURE 11-1 Wireless management configuration options via the GUI

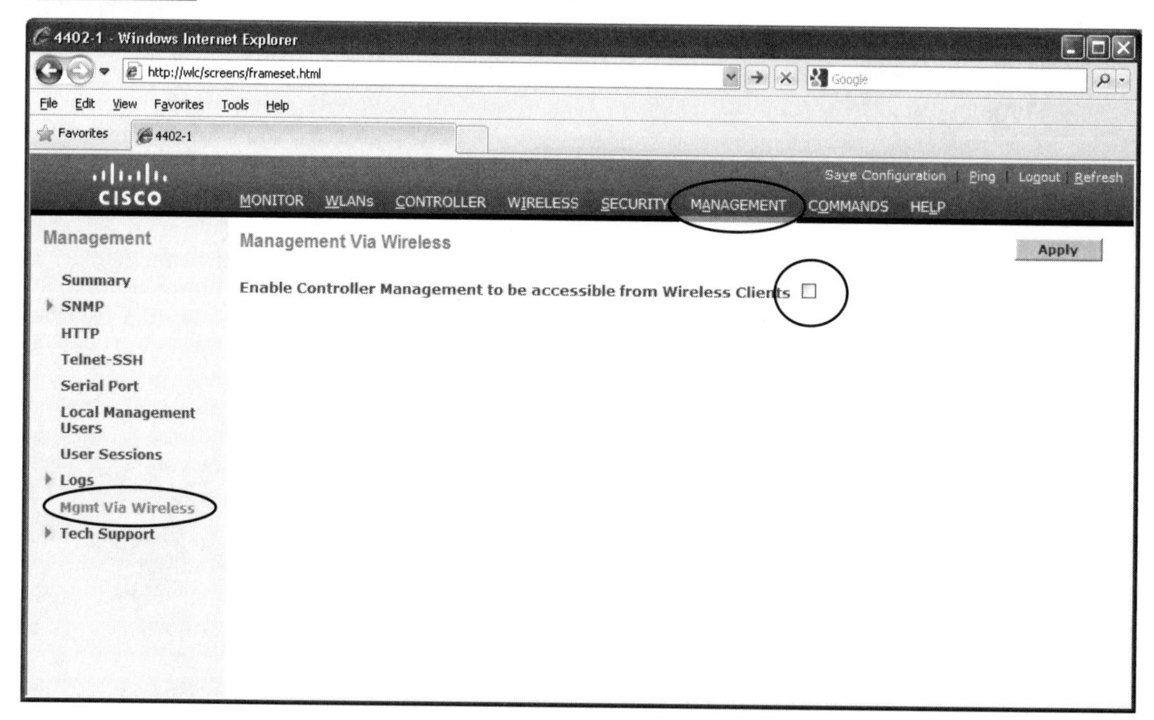

After management-via-wireless is enabled, a wireless client associated to a lightweight AP can then begin an SSH or Telnet session to the controller or browse to the controller GUI via HTTP or HTTPS.

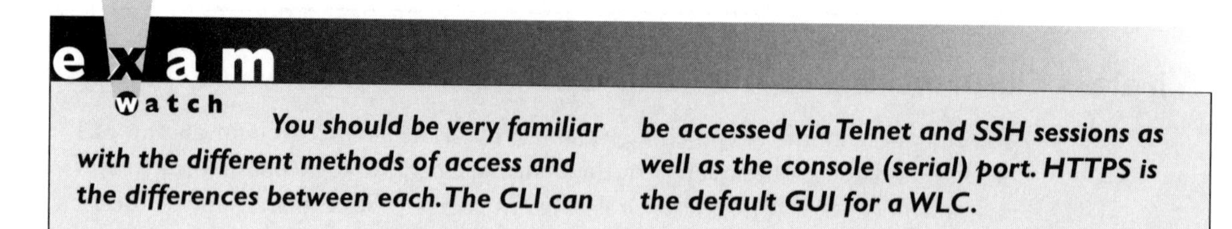

exam watch

You should be very familiar with the different methods of access and the differences between each. The CLI can be accessed via Telnet and SSH sessions as well as the console (serial) port. HTTPS is the default GUI for a WLC.

Managing the WLC through a Service-Port Interface

The 4400, 5500, and WiSM series of the WLC have a service-port interface. The WiSM will communicate with Supervisor 720 on a Catalyst 6500 through a service port. The service-port interface is designed for out-of-band communications and is mapped by the WLC to the service port. The service-port interface requires an IP address and allows CLI and GUI access to the WLC.

The service-port interface should be used only in the event of network failure when no other type of access is available for accessing the WLC. The service-port IP address should be on a different subnet from the management, AP manager, and any dynamic interfaces. The typical usage is to connect a laptop or workstation directly to the service port for access. This is the term out-of-band is referring to because the service port should not connect to a production network.

The service port can obtain an IP address using DHCP, or it can be assigned a static IP address, but a default gateway cannot be assigned to the service-port interface. Static routes can be defined through the controller for remote network access to the service port.

exam watch

The service port should only be used as an OOB access method and not usually used in normal operation. Only the 4400, 5500, and WiSM WLCs have services port. WiSM relies on the service port to communicate with Supervisor 720 on a Catalyst 6500.

CERTIFICATION OBJECTIVE 11.02

Cisco Wireless Network Maintenance Tasks

The basic WLC maintenance tasks include upgrading or downgrading the WLC image, uploading or downloading the WLC configuration files, and configuring WLC logging and traps for network management and monitoring stations. This section will provide an overview of these maintenance tasks.

Upgrade or Downgrade of WLC Images

The WLC images can be upgraded or downgraded using the WLC GUI or CLI. Different WLC platforms will load different image files, so it is crucial that you download the correct image files for the WLC based on the WLC model. You can find out about the WLC's model using the CLI or GUI. The following CLI output shows the WLC model; the WLC mode is in bold:

```
(Cisco Controller) >show inventory
Burned-in MAC Address............................ 00:0B:85:40:39:60
Crypto Accelerator 1............................. Absent
Crypto Accelerator 2............................. Absent
Power Supply 1................................... Absent
Power Supply 2................................... Present, OK
Maximum number of APs supported.................. 12
NAME: "Chassis"    , DESCR: "Cisco Wireless Controller"
PID: AIR-WLC4402-12-K9,  VID: V01,   SN: FLS0926009C
```

In the example, AIR-WLC4402-12-k9 is the model of the WLC. The GUI can also be used to display this information. Figure 11-5 shows the system information for WLC model via WLC GUI.

The correct WLC image can be downloaded from the Cisco.com software download center. You will need to have a valid Cisco.com user account to download software images for the controller and AP. 5500 series WLC also requires a proper license key to enable advanced features, such as OfficeExtend AP. After you log in with a valid username and password, from Cisco.com's main page, select Support

FIGURE 11-5 The wireless controller model information via GUI

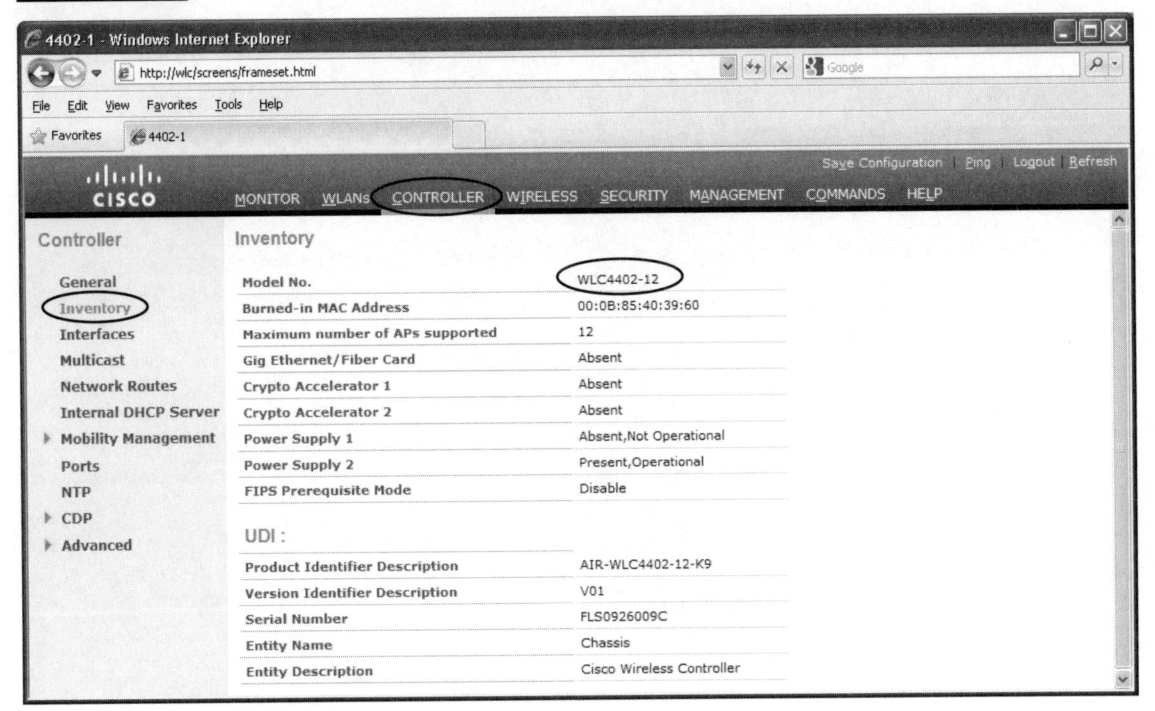

| Download Software | Wireless. You can also point your browser directly to www.cisco.com/go/software. Select the correct type of device and model using the software image tree structure. The WLC image file should carry a compressed archive .aes file extension, which includes the WLC Real-Time OS (RTOS), WLC boot loader, and firmware for lightweight AP. Figure 11-6 illustrates navigating through the software tree structure to find the device.

When the WLC image is upgraded or downgraded, APs that have joined the WLC are also automatically upgraded or downgraded. WLC release 5.1 and later allows WLC image to be upgraded using TFTP and FTP. Typically TFTP service supports file transfer for files that are less than 32MB. For WLC release 4.2 and later, the image size is greater than 32MB; thus, the TFTP server that supports files greater than 32MB is required if you choose to use TFTP to upgrade the WLC. TFTP

FIGURE 11-6 Select a wireless device through the software image tree on Cisco.com.

servers such as tftpd32 and 3CDaemon support TFTP transfer for files greater than 32MB. WCS also has an embedded TFTP server that supports TFTP transfer for files greater than 32MB. WLC allows image upgrade and downgrade using CLI and GUI. Here is a CLI output of an upgrade using WLC CLI:

```
(Cisco Controller) >transfer download datatype code
(Cisco Controller) >transfer download filename AIR-WLC4400-K9-5-2-193-0.aes
(Cisco Controller) >transfer download serverip 171.71.133.31
(Cisco Controller) >transfer download start
Mode........................................... TFTP
Data Type...................................... Code
TFTP Server IP................................. 171.71.133.31
TFTP Packet Timeout............................ 6
TFTP Max Retries............................... 10
TFTP Path...................................... ./
TFTP Filename.................................. AIR-WLC4400-K9-5-2-193-0.aes
This may take some time.
Are you sure you want to start? (y/N) y
TFTP Code transfer starting.
TFTP Code transfer starting.
TFTP receive complete... extracting components.
Writing new bootloader to flash.
Making backup copy of RTOS.
Writing new RTOS to flash.
Making backup copy of Code.
Writing new Code to flash.
TFTP File transfer operation completed successfully.
  Please restart the switch (reset system) for update to complete.
```

After the image has been successfully transferred via TFTP, you need to reset the system for WLC to load the new image. Here is a CLI output for resetting the WLC:

```
(Cisco Controller) >save config
Are you sure you want to save? (y/n) y
Configuration Saved!
(Cisco Controller) >reset system ?
(Cisco Controller) >reset system
Are you sure you would like to reset the system? (y/N) y
```

When the WLC is finished booting, you can issue a show sysinfo command and verify if the new image is now loaded. The following is output from CLI to verify an image upgrade:

```
(Cisco Controller) >show sysinfo
Manufacturer's Name............................ Cisco Systems Inc.
Product Name................................... Cisco Controller
Product Version................................ 5.2.193.0
RTOS Version................................... 5.2.193.0
Bootloader Version............................. 4.0.219.0
Emergency Image Version........................ 5.2.157.0
Build Type..................................... DATA + WPS
```

```
System Name...................................... 4402-1
System Location..................................
System Contact...................................
System ObjectID.................................. 1.3.6.1.4.1.14179.1.1.4.3
IP Address....................................... 10.10.100.131
System Up Time................................... 0 days 0 hrs 9 mins 17 secs
System Timezone Location.........................
Current Boot License Level.......................
Next Boot License Level..........................
Configured Country............................... US  - United States
Operating Environment............................ Commercial (0 to 40 C)
Internal Temp Alarm Limits....................... 0 to 65 C
```

The WLC GUI follows a similar process to upgrade the WLC image. From the WLC GUI, select Commands | Download File. Select File Type and Transfer Mode from the drop-down menu, and enter the TFTP IP address, directory, and filename, and click Download. Figure 11-7 illustrates the WLC GUI for this process.

FIGURE 11-7 Download a new image for upgrade or downgrade via WLC GUI.

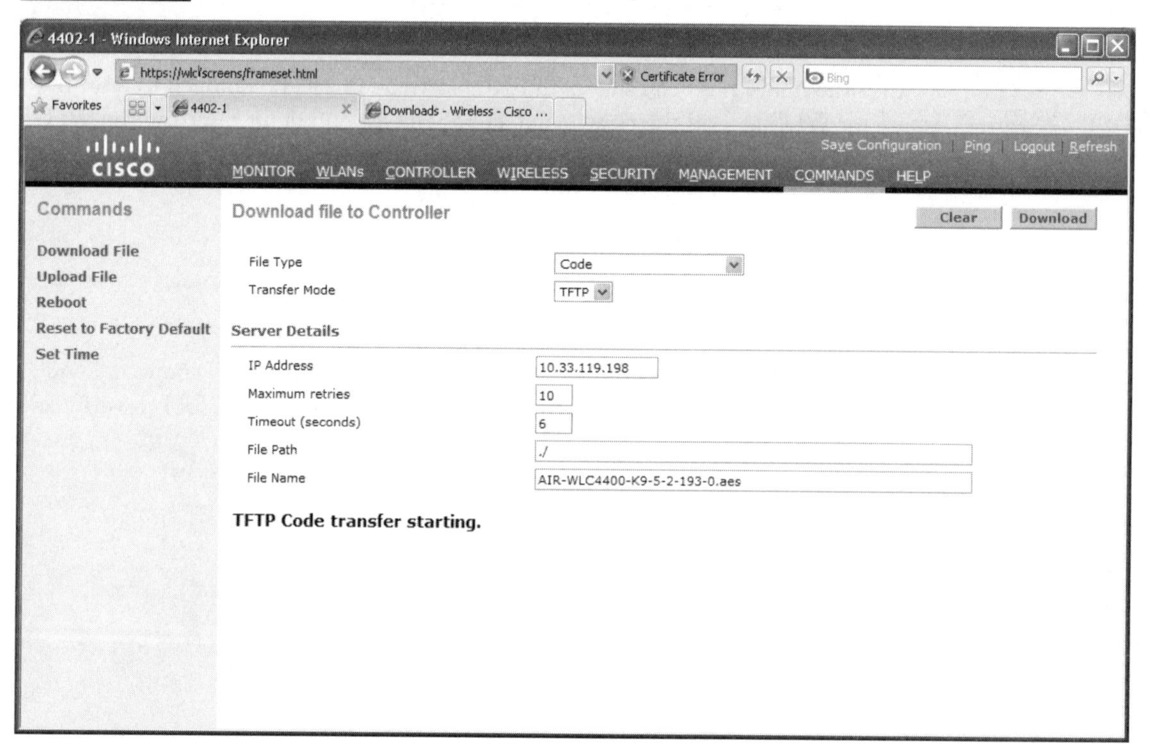

After the new image file is loaded to the WLC via TFTP, click Reboot. If there have been changes to the WLC configuration, WLC will prompt you to save the configuration; click Save and Reboot. After the WLC is rebooted, you can verify the code version running on WLC by going to the main WLC GUI, Monitor | Summary (see Figure 11-8).

FIGURE 11-8 Verify the WLC version via the WLC GUI.

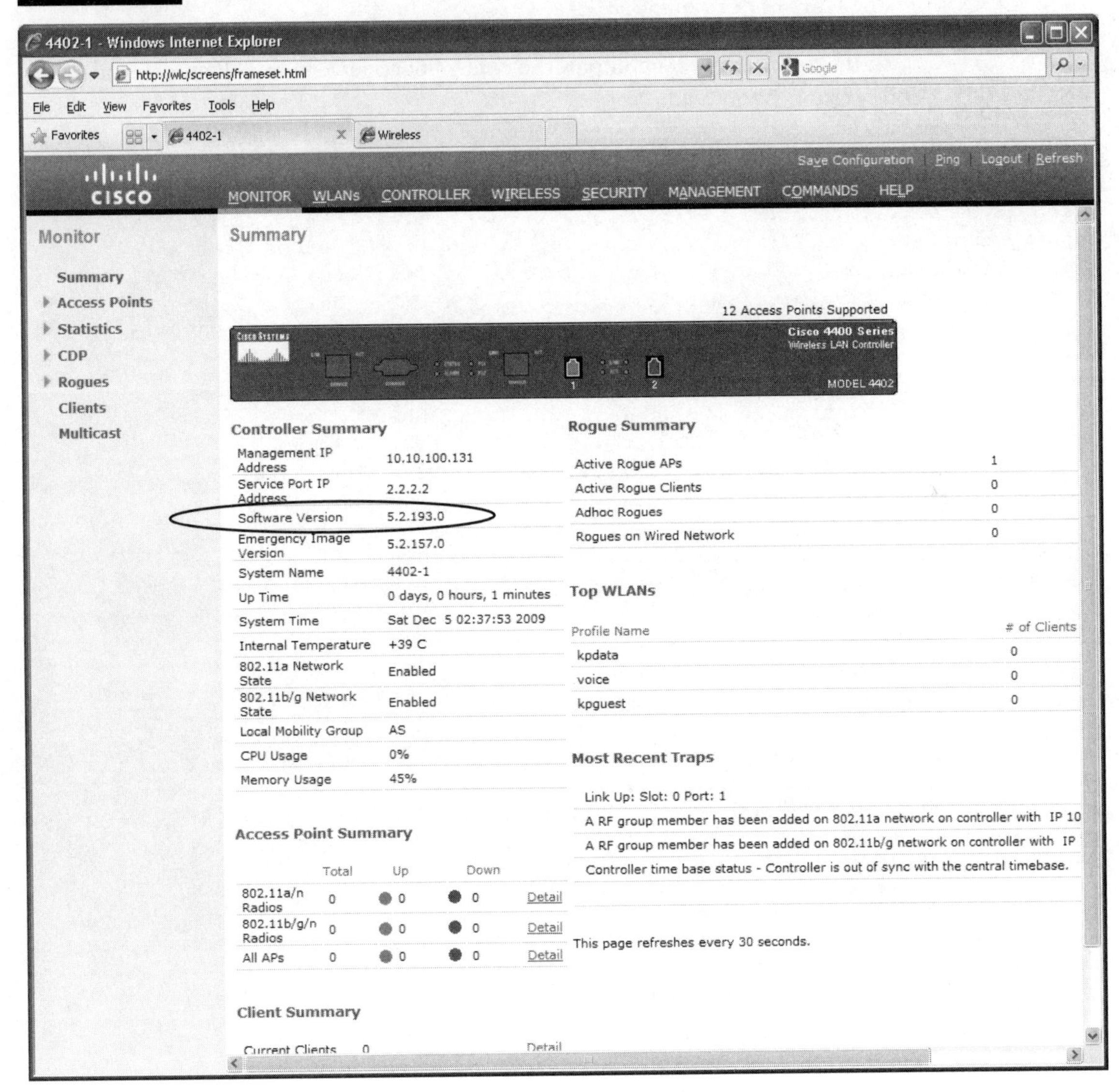

As mentioned earlier, the AP will download updated firmware from the WLC when it joins a WLC if the AP runs on a different version of code from WLC. It could be either an upgrade or a downgrade. WLC release 4.2 and later can support 10 simultaneous AP downloads, and each AP download takes about three minutes to complete. To verify the AP's model and code version, the following CLI command can be used for an AP named AP-1:

```
Show ap config general AP-1
```

In the following system output, the controller code version, AP model, and AP code version are in bold:

```
(Cisco Controller) >show ap config general AP-1
Cisco AP Identifier.............................. 89
Cisco AP Name.................................... AP-1
Country code..................................... US  - United States
Regulatory Domain allowed by Country............. 802.11bg:-A     802.11a:-A
AP Country code.................................. US  - United States
AP Regulatory Domain............................. 802.11bg:-A    802.11a:-A
Switch Port Number .............................. 29
MAC Address...................................... 00:17:94:cc:d9:34
IP Address Configuration......................... DHCP
IP Address....................................... 192.168.19.230
IP NetMask....................................... 255.255.255.0
Gateway IP Addr.................................. 192.168.19.1
CAPWAP Path MTU.................................. 1485
Telnet State..................................... Disabled
Ssh State........................................ Disabled
Cisco AP Location................................ default location
Cisco AP Group Name.............................. cdsacfdsa
Primary Cisco Switch Name........................ 4402-1
Primary Cisco Switch IP Address.................. 10.10.100.141
Secondary Cisco Switch Name......................
Secondary Cisco Switch IP Address................ Not Configured
Tertiary Cisco Switch Name.......................
Tertiary Cisco Switch IP Address................. Not Configured
Administrative State ............................ ADMIN_ENABLED
Operation State ................................. REGISTERED
```

```
Mirroring Mode .................................. Disabled
AP Mode ......................................... Local
Public Safety ................................... Disabled
AP SubMode ...................................... Not Configured
Remote AP Debug ................................. Disabled
Logging trap severity level ..................... informational
S/W  Version .................................... 5.2.178.0
Boot  Version ................................... 12.3.2.4
Mini IOS Version ................................ 3.0.51.0
Stats Reporting Period .......................... 180
LED State........................................ Enabled
PoE Pre-Standard Switch.......................... Disabled
PoE Power Injector MAC Addr...................... Disabled
Power Type/Mode.................................. Power injector / Normal mode
Number Of Slots.................................. 2
AP Model......................................... AIR-AP1251AG-A-K9
AP Image......................................... C1250-K9W8-M
IOS Version...................................... 12.4(18a)JA1
Reset Button..................................... Enabled
AP Serial Number................................. RFD10330727
AP Certificate Type.............................. Manufacture Installed
Management Frame Protection Validation........... Enabled (Global MFP Disabled)
AP User Mode..................................... AUTOMATIC
AP User Name..................................... Not Configured
AP Dot1x User Mode............................... Not Configured
AP Dot1x User Name............................... Not Configured
Cisco AP system logging host..................... 255.255.255.255
AP Up Time....................................... 42 days, 18 h 37 m 23 s
AP LWAPP Up Time................................. 42 days, 18 h 36 m 15 s
Join Date and Time............................... Mon Jun 15 21:16:04 2009
Join Taken Time.................................. 0 days, 00 h 01 m 07 s
Ethernet Port Duplex............................. Auto
Ethernet Port Speed.............................. Auto
AP Link Latency.................................. Disabled
(Cisco Controller) >
```

Figure 11-9 illustrates the AP code version via the WLC GUI, and Figure 11-10 illustrates the AP model information via the WLC GUI.

FIGURE 11-9 AP code version information via the WLC GUI

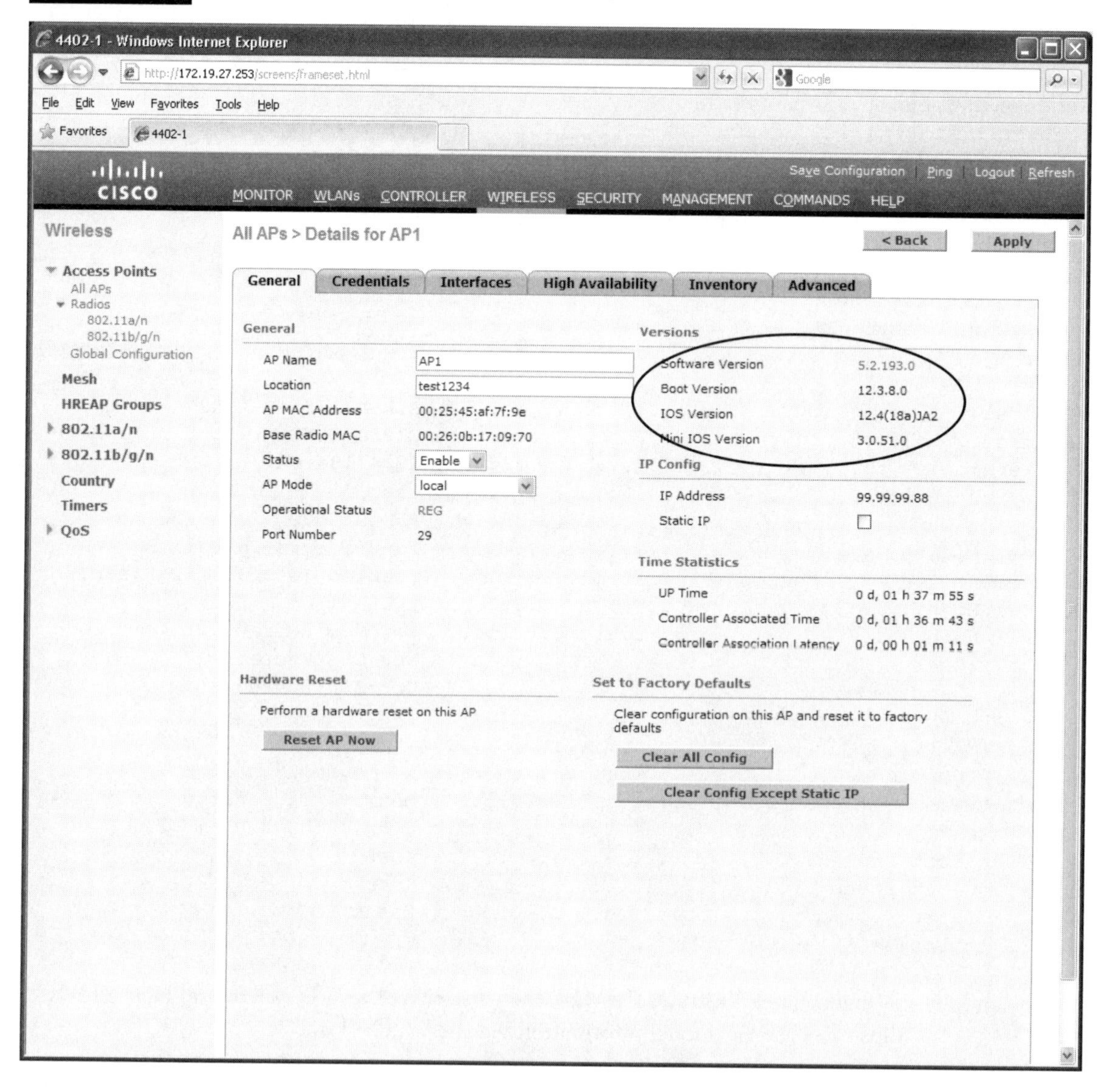

FIGURE 11-10 AP model information via the WLC GUI

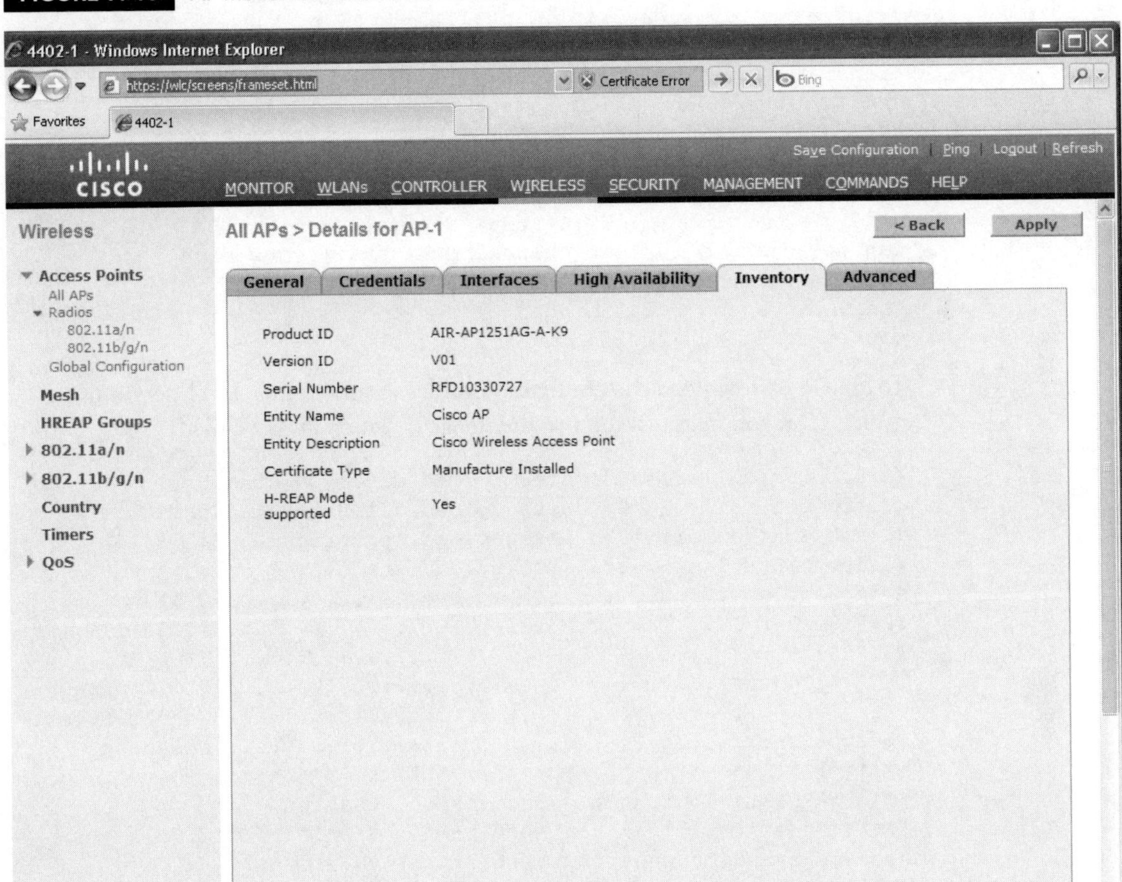

Upload or Download of WLC Configurations

WLC configurations are stored in configuration files using XML. The WCS can periodically download WLC XML configuration files and store them in a configuration depository. The following is a sample excerpt of WLC XML configuration file:

```
transfer download path ./
transfer download serverip 10.33.119.198
transfer download filename temp.xml
```

```
transfer upload path ./
transfer upload serverip 10.33.119.198
transfer upload filename temp.xml
transfer upload datatype config
config custom-web weblogo disable
config custom-web webtitle "TEST GUEST WEB REDIRECT"
config custom-web redirecturl http://www.cisco.com
config custom-web webmessage "TEST GUEST WEB REDIRECT"
config advanced 802.11b channel dca interval 8
config advanced 802.11b channel dca sensitivity high
config advanced 802.11b channel add 1
config advanced 802.11b channel add 6
config advanced 802.11b channel add 11
```

To upload or download the configuration files to and from a TFTP server using the WLC CLI, commands similar to the image download will be used:

```
(Cisco Controller) >transfer upload datatype config
(Cisco Controller) >transfer upload filename WLC_Config
(Cisco Controller) >transfer upload serverip 10.33.119.198
(Cisco Controller) >transfer upload start
Mode........................................... TFTP
TFTP Server IP................................. 10.33.119.198
TFTP Path...................................... ./
TFTP Filename.................................. WLC_Config
Data Type...................................... Config File
Encryption..................................... Disabled
**************************************************
***  WARNING: Config File Encryption Disabled  ***
**************************************************
Are you sure you want to start? (y/N) y
TFTP Config transfer starting.
File transfer operation completed successfully.
```

The WLC GUI follows a similar process as upgrading the WLC image for uploading configuration files to a TFTP server. From WLC GUI, select Commands | Upload File. Select File Type as Configuration and Transfer Mode from the drop-down menu, and enter the TFTP IP address, directory, and file name, and click Upload. Figure 11-11 illustrates the WLC GUI for this process.

FIGURE 11-11 Upload a configuration file to the TFTP server via WLC GUI.

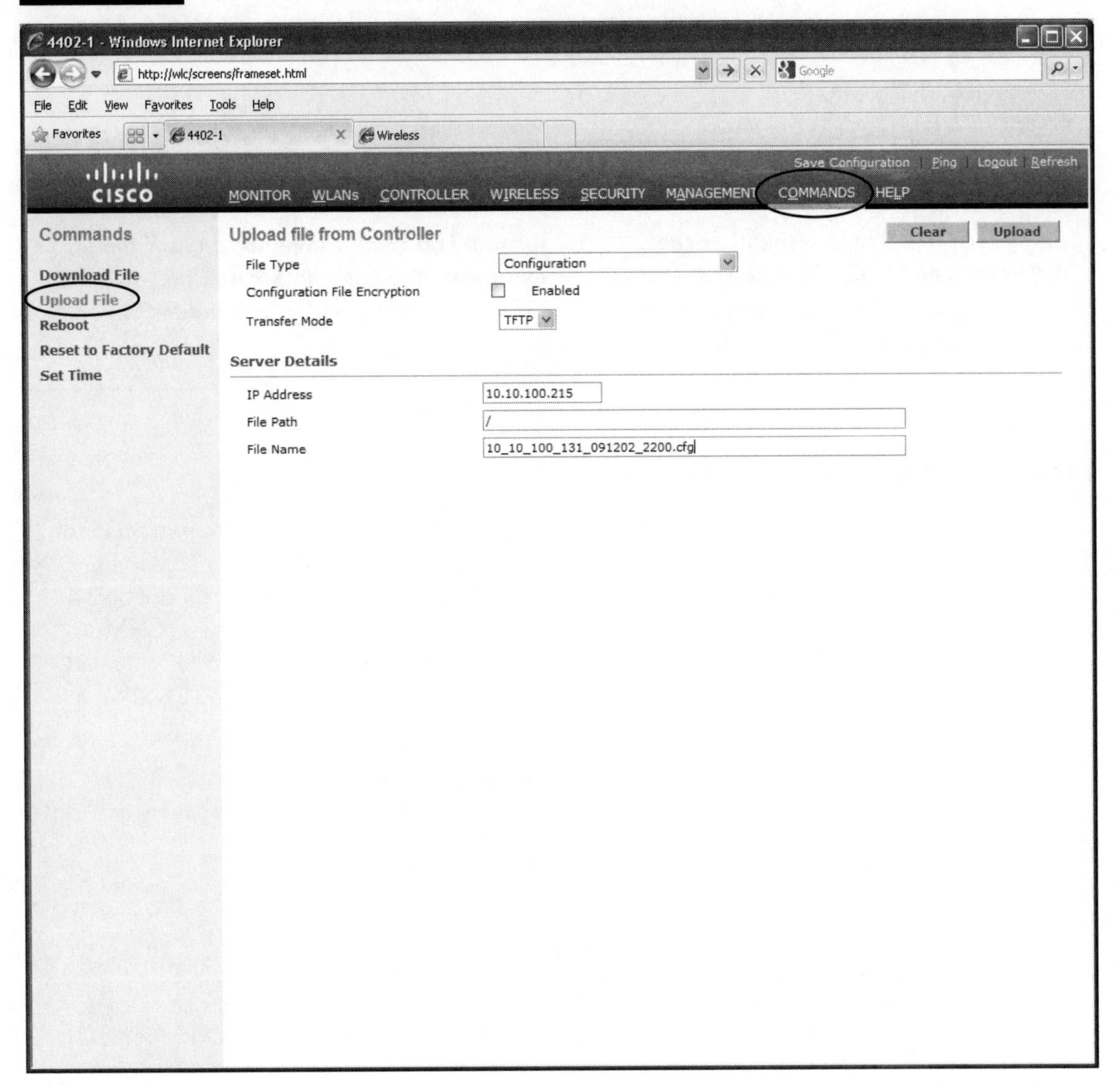

A similar process exists for downloading configuration files from the TFTP server to WLC. The WLC will prompt you to reboot the WLC after the download process is completed.

Try to remember the different types of files associated with WLC images and configurations. The configuration files are based on XML format. The WLC image has an .aes file extension and is a compressed file with the OS, AP firmware, and a bootloader image.

Monitoring WLC

WLC allows the administrator to configure the syslog and SNMP traps to help the network administrator monitor the health of the WLC. The WLC needs to have valid SNMP credentials and/or community strings defined in order for the SNMP traps to work correctly. In WLC release 5.2, you can configure up to six SNMP trap receivers. You can also specify different types of SNMP traps to be generated. The follow CLI commands configure the WLC to a syslog server and SNMP traps collector:

```
(Cisco Controller) >config syslog ?
[<ip-address>/disable] Set syslog destination, or disable syslog.
(Cisco Controller) >config snmp trapreceiver ?
create          Add a new SNMP trap receiver.
delete          Delete an SNMP trap receiver.
mode            Enable or disable an SNMP trap receiver.
(Cisco Controller) >config snmp trapreceiver create 10.10.100.11 ?
<IP addr>       Enter SNMP community IP addr.
(Cisco Controller) >config snmp trapreceiver create ?
<name>          Enter SNMP community name up to 32 characters.
```

```
(Cisco Controller) >config snmp trapreceiver create private ?
<IP addr>      Enter SNMP community IP addr.
```

Figures 11-12 and 11-13 illustrate the WLC GUI for configuring the syslog and SNMP traps collector to receive syslog and trap messages.

FIGURE 11-12 Configure the syslog server via the WLC GUI.

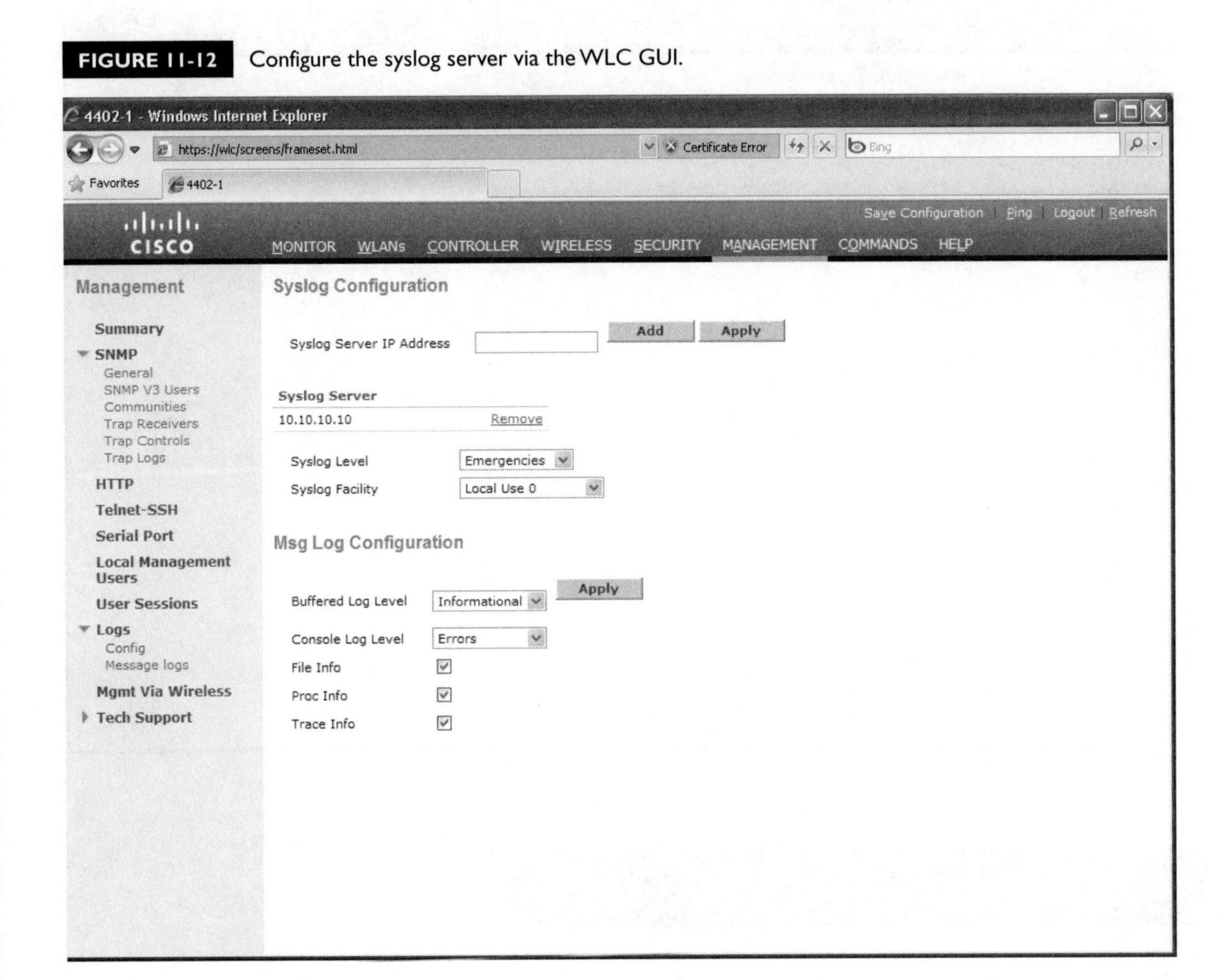

FIGURE 11-13 Configure SNMP trap receivers via the WLC GUI.

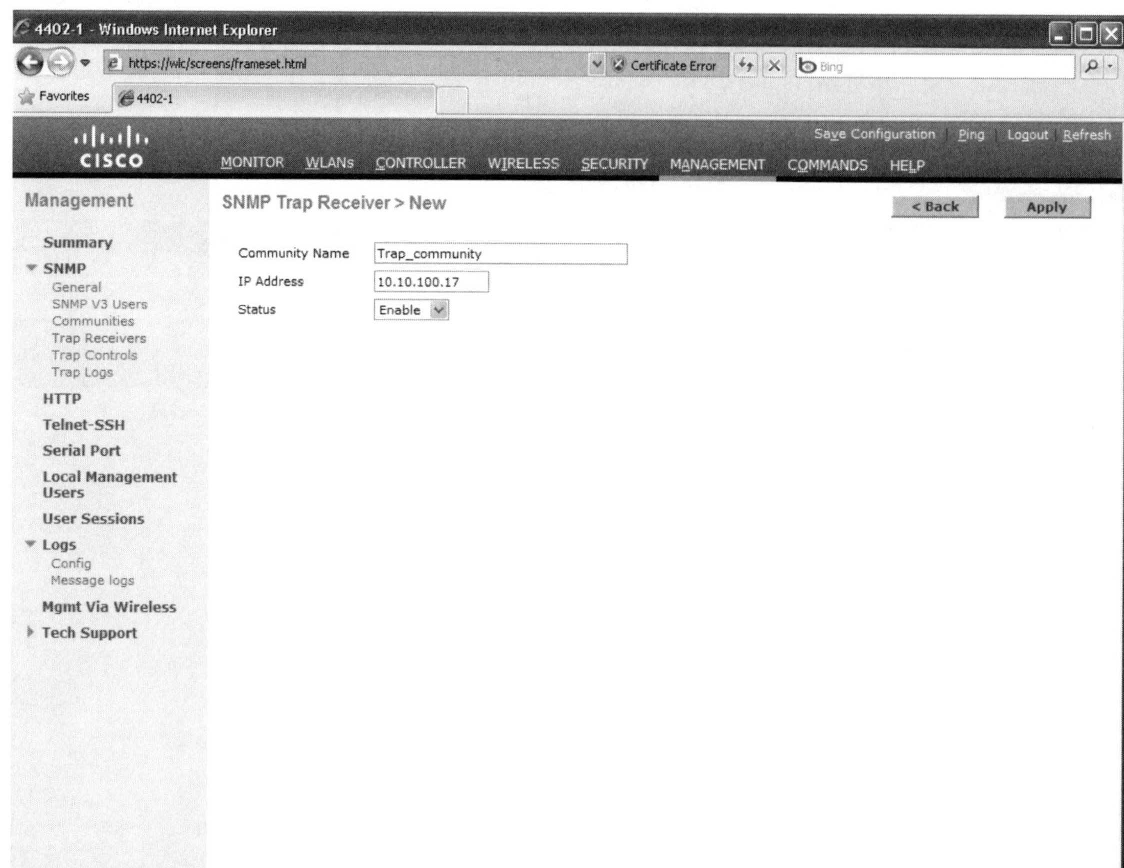

CERTIFICATION OBJECTIVE 11.03

Cisco Wireless Network Deployment Considerations

There are two tasks that are important to a wireless network deployment: RF planning and RF site survey.

RF Planning

RF planning usually helps identify possible AP placements and coverage areas based on predefined RF characteristics in the early stage of a wireless network deployment, and it is usually performed prior to a RF site survey. The administrator can create a map by importing a floor plan, adding walls, windows, doors, and cabinets, and adjusting the attenuation characteristics of different types of obstructions to create an RF prediction. The predicted RF coverage is usually displayed in the form of a heat map to provide a visual aid to the RF engineer to develop a deployment plan for the wireless network. Planning tools, such as the Motorola LAN Planner and AirMagnet Planner, usually have built-in simulators, so the engineer can try different prediction scenarios by changing the type of AP, AP placements, and antenna, as well as RF attenuation characteristics of objects, such as walls and doors.

WCS planning mode provides tools and information to help the administrator determine the number of APs and AP placements and RF coverage in a WLAN. The WCS map editor allows the administrator to create floor and building maps and enter RF characteristics of different building materials. The planning tool will then calculate a projected RF heat map and place hypothetical APs on the map for a given coverage area. The WCS also provides a tool to validate if the current and projected APs can sufficiently support advanced applications, such as voice over WLAN and context-aware location-based services.

RF Site Survey

After RF planning, an RF site survey is conducted to provide actual empirical data and to validate results from the RF planning exercise. A good RF site survey is considered one of the most important tasks in a wireless network deployment. Site survey and RF design techniques can vary among RF engineers depending on experience and training, resulting in wide ranging solutions for the same physical area. Although these RF designs may lack uniformity, each of these different designs may be sufficient and acceptable for a given physical area.

The CUWN is based on IEEE 802.11-based standards and technologies. To conduct a site survey for wireless LAN, a basic understanding of LAN technologies and an advanced understanding of wireless LAN theory, technology, and products are desired. The following basic knowledge and skill sets are recommended for the engineers conducting the site survey:

- IEEE 802.11a, b, and g standards (802.11n is optional, however, it is becoming increasingly important in new WLAN.)
- Basic spread spectrum RF technology

- Antenna theory
- Site survey skills
- Basic configurations of Cisco access points, controllers, and other devices
- Basic LAN technology including TCP/IP
- Awareness of local regulations regarding wireless links and antenna installation

The following site survey tools are typically employed by field engineers:

- **RF Site Surveyor** Performs RF data collection activities; procurement and use of an RF site surveyor is required. Many surveyor tools run on Microsoft-based Windows platform, which allows the capture of live RF data emitted by the test AP during a site survey. Vendors providing surveyor tools include AirMagnet, Ekahau, Network Chemistry, BVS, Motorola LAN Planner, among others. Client adapter vendors usually offer simple site survey tools that can be used to provide basic site survey information. Cisco has bundled such a tool in its client utility suite, Cisco Aironet Desktop Utility (ADU).
- **RF Spectrum Analyzer** Inspects and analyzes the RF spectrum; procurement and use of an RF spectrum analyzer is required. Cisco Spectrum Expert for 2.4 GHz and 5 GHz is an easy to use Microsoft Windows–based tool that allows inspection of the RF spectrum in the physical area that requires WLAN coverage. Other vendors providing similar tools include Agilent, BVS Systems, among others.

The 802.11 standards–based wireless networks operate in the unlicensed frequency bands. In general, there are no restrictions on the types of devices that operate in these bands provided that they all conform to a common set of rules. As a result, the 2.4 GHz and 5 GHz bands are available for anyone to use. Interference from these and other sources adds to the *noise floor*, which is a mixture of all the background RF energy found in the environment surrounding the system in use. RF signals from the infrastructure AP must be higher than the noise floor in order to be detectible as a valid, useful signal by a receiver.

The spectrum analyzer will show the channel utilization. With Cisco Spectrum Expert, the duty cycle plot displays the percentage of the time that the overall RF power—due to both 802.11 devices and interferences—is 20 dB or more above the noise floor (SNR). Duty cycle is one metric that should be monitored. WLAN operations are normally impacted to a minimal extent beginning at just around 20 percent, with the impact increasing as the RF interference duty cycle increases. Consider accounting for the effects on RF propagation of the following common wireless interferences and inhibitors frequently found at the intended deployed areas:

- Metal pipes
- Load bearing metal beams
- Fluorescent lighting
- Microwave ovens
- Bluetooth devices
- Large (reflective) metal cabinets
- Large or wall mounted shelving that might restrict RF
- Restrooms with metal plumbing
- Other RF systems, such as a wireless PA system

The site survey results should provide sufficient information to conduct an RF baseline analysis to better understand RF propagation with the selected antenna and a given unique environment. Criteria to consider are acceptable antenna types and orientation, AP placement, target signal coverage, maximum or minimal AP power settings, channel settings, and general distance between access points to maintain as low of a noise floor as possible. This analysis is especially important in multifloor building environments where wireless LAN is deployed in a three-dimensional plane. Identifying between floor RF propagation and attenuation characteristics may impact overall locations of access points on each floor. If attenuation values are sufficient between floors, APs could potentially be located in the same general area on adjacent floors (with different frequency configurations), which could speed up the site survey process by providing a template for floors that may be physically similar. While determining if a template can be created, it is important to remember that the best practice guideline as denoted in the Cisco VoWLAN guideline is to maintain 19dBm of separation between co-channel access points to avoid co-channel interference (CCI).

Upon completion of an RF site survey, the following tasks should be performed:

- Review site survey documentation.
- Review and evaluate site survey results.
- Develop WLAN design plans and documentation.
- Identify resources and skill sets for WLAN deployment activities.
- Estimate resource(s) requirements and prepare deployment plan budget.
- Review and develop deployment project plan and schedule.
- Execute deployment plans.
- Initiate the post deployment assessment process.

The RF site survey is arguably the most important step in a wireless network deployment. Many RF problems can be effectively prevented if a thorough RF site survey is conducted in the design and deployment stage. RF site surveys not only provide insightful information to help engineers select APs, antenna types, and AP placements, but they also profile different sources of interferences in the environment that may impact the wireless network. RF site surveys can be performed without any prior RF planning and analysis.

INSIDE THE EXAM

Cisco Wireless Network Administration and Access Methods

The goal of this exam is make sure you have a basic understanding of the WLC architecture and the different methods of accessing and administering a WLC. A WLC offers both a CLI and a GUI so the administrator can perform all necessary administrative tasks. The CLI interface can be accessed through a console (serial) port interface and through Telnet or SSH remote access. The GUI interface can be accessed through HTTP and HTTPS as long as the WLC maintains connectivity to an IP-based LAN network. The 4400, 5500, and WiSM-based WLC also comes with a service port, which can be used for out-of-band (OOB) access to the WLC GUI interface. The service port should not be used if regular network connectivity is active.

The WLC GUI is officially supported using Microsoft Internet Explorer 6 and Mozilla Firefox 2 and later. The Opera and Netscape Web browsers are not supported by Cisco. The WLC GUI supports most administrative tasks and is the preferred method of managing a WLC. There are some additional configuration and debug tasks that can only be performed by the CLI.

INSIDE THE EXAM

Cisco Wireless Network Maintenance Tasks

For the CCNA Wireless exam, you should have familiarity with basic maintenance tasks, including upgrading or downgrading a WLC image through TFTP or FTP, uploading or downloading a WLC configuration file through TFTP or FTP, and configuring syslog and SNMP traps. All of these maintenance tasks can be performed using the CLI and GUI. Typically, network devices use TFTP as a way of transferring device images. The WLC image for release 4.2 and later is greater than 32MB, which is not supported by some TFTP servers. You should be familiar with some of the TFTP servers that are capable of supporting image files greater than 32MB. In addition, WCS has a built-in TFTP server that will support transfer of image files greater than 32MB. The upgrade and downgrade tasks can be performed using both CLI and GUI.

Uploading and downloading configuration files is another important maintenance task regularly performed by the administrator. Syslog and SNMP traps are also important network management tools used by many administrators to manage and monitor the health of their networks.

Cisco Wireless Network Deployment Considerations

Two important tasks for a successful wireless network deployment discussed in this chapter are RF planning and RF site survey. The RF planning usually involves building network maps and computing projection of RF coverage in the form of an RF heat map based on hypothetical AP placements, types, and antennas. The WCS has built-in planning tools and is capable of modeling different types of RF projection for different applications, including voice over WLAN and context-aware location-based services. The RF planning result may provide useful RF propagation reference information for an RF site survey and make the whole design and deployment process more efficient.

RF site survey is arguably the most important step in the whole design and deployment process. RF site survey basically validates the RF environment and provides empirical data for the design engineer to select and place APs and antennas for a wireless network.

CERTIFICATION SUMMARY

The day-to-day operations and management of a wireless network involve managing WLC and AP devices in a network, maintaining device configurations, and performing upgrade or downgrade device images. Cisco provides multiple methods to manage a WLC, including a CLI and GUI. This chapter provides the basic steps of accessing a WLC's CLI and GUI through Telnet, SSH, and a console (serial) port for a terminal emulation session access and HTTP and HTTPS for GUI access.The WLC CLI and GUI support nearly all maintenance tasks, and through use of TFTP and FTP, you can easily upgrade or downgrade the controller image and upload and download the controller configuration files. The CLI and GUI can also be used for WLC to send syslog and SNMP trap messages to network management stations for controller health monitoring and event logging purposes. RF planning and RF site survey are complementary steps in a wireless network design and deployment process. RF planning can provide valuable information to the design and deployment in selecting the appropriate types of APs and antennas. RF site survey will further validate the projection of RF planning results and provide empirical data to the RF engineer.

TWO-MINUTE DRILL

Cisco Wireless Network Administration and Access Methods

- ❑ A Cisco wireless controller can be managed through a CLI and GUI.
- ❑ Administrators can use WLC's console (serial) port to access the controller CLI directly.
- ❑ For remote CLI access, Telnet and SSH can be used.
- ❑ For WLC GUI access, HTTP and HTTPS can be used. HTTPS is the default method of accessing WLC GUI.
- ❑ Some models of WLC come with a service port, which should only be used as an out-of-band access method to the WLC's GUI.

Cisco Wireless Network Maintenance Tasks

- ❑ Both the CLI and GUI can be used to perform image upgrade and downgrade tasks on a WLC.
- ❑ WLCs allow use of TFTP and FTP to transfer image files.
- ❑ If TFTP is used for transfer of a WLC image file, the TFTP server needs to support files greater than 32MB.
- ❑ The WLC allows use of TFTP and FTP to transfer WLC configuration files.
- ❑ Both the CLI and GUI can be used to configure the syslog server and SNMP trap receivers on a WLC.

Cisco Wireless Network Deployment Considerations

- ❑ RF planning provides projected data to help a network engineer select types of AP, antennas, and AP placement.
- ❑ WCS can be used as an RF planning tool and provide projected heat maps to illustrate RF coverage based on selected APs and antennas.
- ❑ RF site survey is the most important step in the wireless network design and deployment process as it provides empirical data on RF propagation in the intended coverage area.
- ❑ RF planning and RF site survey are complementary tasks in the design and deployment process.

SELF TEST

The following self-test questions will help you measure your understanding of the material presented in this chapter. Read all the choices carefully, as there may be more than one correct answer. Choose all correct answers for each question.

Cisco Wireless Network Administration and Access Methods

1. What is not a valid method to access a Wireless LAN Controller remotely?

A. Telnet to the WLC through a service-port IP address.

B. Telnet to the WLC through a management interface IP address.

C. Point a Web browser to the WLC's management interface IP address using HTTP.

D. SSH to the WLC through a management interface IP address.

Cisco Wireless Network Maintenance Tasks

2. What WLC CLI commands do you use to verify WLC's model and code version? (Choose all that apply.)

A. Show version

B. Show inventory

C. Show interface

D. Show sysinfo

3. How is controller configuration file being managed in the WLC 5.1 release?

A. XML

B. Clear text

C. AES encryption

D. MD5 hash function

4. After upgrading a controller image, how does the administrator know if the WLC is running the new image?

A. Save the configuration, reboot the controller, and use the CLI command show sysinfo to verify the new image is running on the controller.

B. The controller will automatically run the new image after an upgrade.

C. The controller will run the new image after you set the controller back to the factory default.

D. The controller will only run the new image at the next power outage.

5. What protocols can be used to transfer image file and configuration files to a wireless controller in version 5.1? (Choose all that apply.)

A. SCP

B. RDP

C. TFTP

D. FTP

6. What file extension is used for the WLC image file?

A. .bin

B. .exe

C. .dat

D. .aes

7. What happens when a lightweight AP is running a different version of code from the WLC it joins?

A. The AP will download firmware from the WLC and reboot.

B. If the AP is running a newer code, the WLC will download the new code from the AP.

C. Nothing happens; the AP can run independently from the WLC regardless of the code level it runs, as long as the WLC runs on LWAPP.

D. The AP will continue to reboot itself until a WLC is running the same version of code as the AP.

8. How many SNMP trap receivers can be configured for a controller running release 5.2?

A. 20

B. 10

C. 6

D. 2

9. How should a WLC service port be used?

A. It can be configured to be on the same subnet as the management interface.

B. It can be configured to be on the same subnet as APs.

C. It can be configured to be on the same subnet as DHCP and NTP servers.

D. It can be configured to be on a segregated network that is not routable from the network.

Cisco Wireless Network Deployment Considerations

10. When should an RF site survey be performed?

A. An RF site survey cannot be performed without RF planning.

B. An RF site survey is usually performed after RF planning.

C. An RF site survey is not important in a design and deployment process and can be ignored.

D. An RF site survey is very costly; thus, it should only be performed in a large RF deployment.

SELF TEST ANSWERS

Cisco Wireless Network Administration and Access Methods

1. What is not a valid method to access a Wireless LAN Controller remotely?

A. Telnet to the WLC through a service-port IP address.

B. Telnet to the WLC through a management interface IP address.

C. Point a Web browser to the WLC's management interface IP address using HTTP.

D. SSH to the WLC through a management interface IP address.

☑ **A.** The service port should not be used as a remote access method to a WLC

☒ **B, C,** and **D** are incorrect. Telnetting to the WLC through management interface IP address, pointing a Web browser to WLC's management interface IP address using HTTP, and using SSH to WLC through the management interface IP address are all valid methods to access the WLC remotely.

Cisco Wireless Network Maintenance Tasks

2. What WLC CLI commands do you use to verify WLC's model and code version? (Choose all that apply.)

A. Show version

B. Show inventory

C. Show interface

D. Show sysinfo

☑ **B, D.** The show inventory command will display the WLC model and the show sysinfo command will display the current code version running on the controller.

☒ **A** and **D** are incorrect. "The show version" is a valid command for Cisco IOS CLI but not a valid command for the WLC CLI. Show interface does not provide the controller model and code version information.

3. How is controller configuration file being managed in the WLC 5.1 release?
   A. XML
   B. Clear text
   C. AES encryption
   D. MD5 hash function

☑ **A.** The controller's configuration file is managed in XML format.

☒ **B, C,** and **D** are incorrect. The controller configuration file is never managed in clear text and is not encrypted using AES or hashed using MD5.

4. After upgrading a controller image, how does the administrator know if the WLC is running the new image?
   A. Save the configuration, reboot the controller, and use the CLI command show sysinfo to verify the new image is running on the controller.
   B. The controller will automatically run the new image after an upgrade.
   C. The controller will run the new image after you set the controller back to the factory default.
   D. The controller will only run the new image at the next power outage.

☑ **A.** After loading a new image to the WLC, a reboot is required for the controller to run on the new image. The CLI command `show sysinfo` will display the current code version running on a WLC.

☒ **B, C,** and **D** are incorrect. The controller will not automatically load the new image without a reboot. A manual reboot will trigger the WLC to run the newly loaded image.

5. What protocols can be used to transfer image file and configuration files to a wireless controller in version 5.1? (Choose all that apply.)
   A. SCP
   B. RDP
   C. TFTP
   D. FTP

☑ **C, D.** TFTP and FTP can be used to transfer image and configuration files to a wireless controller in version 5.1.

☒ **A** and **B** are incorrect. SCP and RDP cannot be used to transfer image and configuration files to a wireless controller in version 5.1.

6. What file extension is used for the WLC image file?
   A. .bin
   B. .exe
   C. .dat
   D. .aes

☑ **D**. The correct file extension for the WLC image file is .aes.

☒ **A**, **B**, and **C** are incorrect. The file extensions .bin, .exe, and .dat are not correct for a WLC image file.

7. What happens when a lightweight AP is running a different version of code from the WLC it joins?

A. The AP will download firmware from the WLC and reboot.

B. If the AP is running a newer code, the WLC will download the new code from the AP.

C. Nothing happens; the AP can run independently from the WLC regardless of the code level it runs, as long as the WLC runs on LWAPP.

D. The AP will continue to reboot itself until a WLC is running the same version of code as the AP.

☑ **A**. The AP will download firmware from the WLC and reboot. The AP will always attempt to run the same version of firmware as the WLC.

☒ **B**, **C**, and **D** are incorrect. The WLC will never download an image from an AP. APs cannot run on a different version of code from the WLC. Once an AP joins a WLC it will check the WLC code version. If different, AP will download the firmware from WLC, then reboot.

8. How many SNMP trap receivers can be configured for a controller running release 5.2?

A. 20

B. 10

C. 6

D. 2

☑ **C**. The WLC running 5.2 release can configure up to six SNMP trap receivers.

☒ **A**, **B**, are **D** incorrect. The SNMP trap can only support up to 6 SNMP trap receivers in WLC running 5.2 release. A and B both have too many trap receivers for WLC, and D has too few trap receivers.

9. How should a WLC service port be used?

A. It can be configured to be on the same subnet as the management interface.

B. It can be configured to be on the same subnet as APs.

C. It can be configured to be on the same subnet as DHCP and NTP servers.

D. It can be configured to be on a segregated network that is not routable from the network.

☑ **D**. The service port should only be configured to be used in out-of-band scenarios, and the IP address should be on a nonroutable subnet.

☒ **A**, **B**, and **C** are incorrect. The service port should have no network connectivity to the rest of the LAN network.

Cisco Wireless Network Deployment Considerations

10. When should an RF site survey be performed?

A. An RF site survey cannot be performed without RF planning.

B. An RF site survey is usually performed after RF planning.

C. An RF site survey is not important in a design and deployment process and can be ignored.

D. An RF site survey is very costly; thus, it should only be performed in a large RF deployment.

☑ **B**. The RF site survey is usually performed after RF planning, which usually runs a simulation based on projected RF propagation data. The information provided by RF planning makes RF site survey more efficient.

☒ **A**, **C**, and **D** are incorrect. An RF site survey does not require RF planning to be completed ahead of time. An RF site survey is very important to the wireless network design and deployment. Cisco recommends an RF site survey for all wireless network deployments.

12

Cisco Wireless Network Troubleshooting Tasks

CERTIFICATION OBJECTIVES

12.01	Wireless LAN Troubleshooting Methods	✓	Two-Minute Drill
12.02	Use of Wireless LAN Controller Show, Debug, and Logging Commands	Q&A	Self Test
12.03	Use of the WCS Client Troubleshooting Tool		

In Chapter 11, we covered the basic day-to-day tasks of maintenance and monitoring. In this chapter, we will discuss basic wireless network troubleshooting methods, tools, and tips. We will talk about troubleshooting wireless related problems in a systematic fashion and using the WLC CLI, GUI, and WCS. An 802.11-based wireless LAN network is primarily a Layer 1 and Layer 2 technology, and the network assumes all upper layer networks and applications work correctly. From a network engineer's perspective, most problems occur at Layer 1 through Layer 3. For example, is the medium clear for sending traffic? Does the client obtain an IP address? Does the client have the correct default gateway configured? When one of these parameters is not set correctly, the wireless network will not work for a client device. Thus, it is highly recommended that you possess basic networking skills and understand the OSI reference model and TCP/IP. We will focus our discussion on Layers 1 through 3 as IP is synonymous with computer networking today.

CERTIFICATION OBJECTIVE 12.01

Wireless LAN Troubleshooting Methods

We believe the best practice in troubleshooting a wireless LAN network is to take a systematic approach to problem isolation and resolution. This applies not only to troubleshooting a wireless LAN network, but also to any other types of computer networks. An unsystematic approach to troubleshooting may result in wasting valuable time and resources and can sometimes worsen the problem at hand. The systematic approach consists of defining specific symptoms, identifying all potential problems that could be causing the symptoms, and then systematically eliminating each potential cause (from most likely to least likely) until the problem is resolved. Figure 12-1 is a simple troubleshooting flow that follows a systematic troubleshooting methodology.

Although each person may take a different approach to troubleshooting, this book provides an overview of common troubleshooting methods widely adopted by IT professionals. The wireless LAN troubleshooting process should consist of the following steps:

1. **Identify general symptoms of the problem** To properly troubleshoot a problem, one must identify the general symptoms first and then ascertain what kinds of problems (issues) could result from these symptoms. For example, a wireless client (adapter) may experience a problem obtaining an IP address (a symptom). Possible problems may include an incorrectly configured client device, a bad wireless connection, or a mis-configured AP, WLC, or network services (such as the DHCP server).

FIGURE 12-1

Wireless LAN troubleshooting methodology

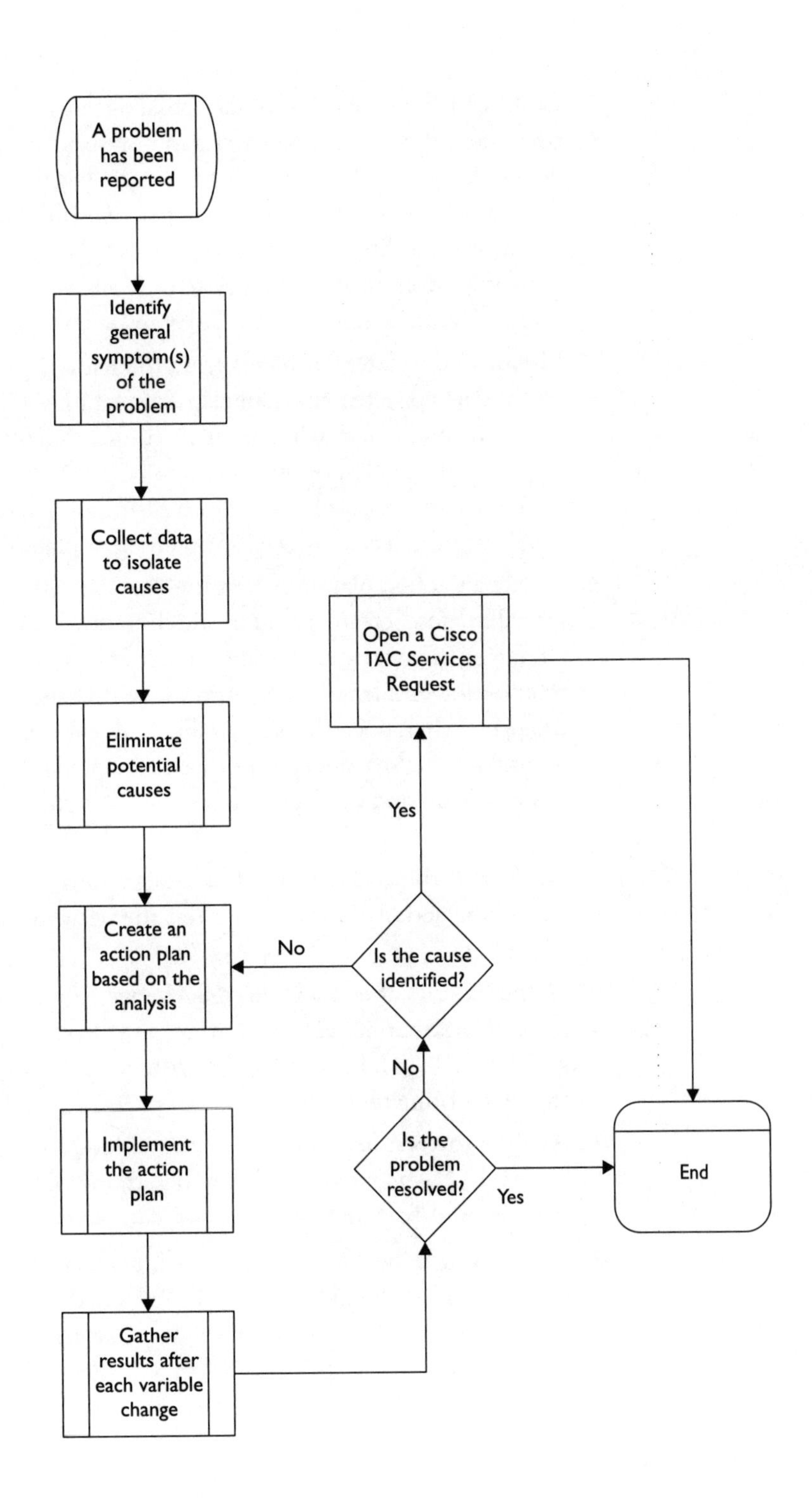

2. **Collect sufficient data to isolate possible causes** Once the symptoms have been identified, collect more data to isolate possible causes. There is no clear definition of what "sufficient data" means, but the data collected should help determine causes of the symptoms. Ask questions of affected users, network administrators, IT managers, and other key people. Collect data from sources such as network management systems (such as WCS), protocol analyzer traces, or output from AP/WLC diagnostic commands.
3. **Eliminate unrelated problems from the list of possible causes** Depending on the data collected, for example, one might be able to eliminate a lightweight AP as the problem in wireless client connectivity symptoms; therefore allowing all troubleshooting efforts to be directed toward other areas, such as client software problems. At every opportunity, the number of potential problems should be narrowed so an efficient plan of action can be created.
4. **Create an action plan based on the analysis** Beginning with the most probable cause, create a plan in which network variables are manipulated. Changing one variable at a time enables one to reproduce a given symptom to a specific problem. It's always possible that the symptom or problem will disappear when one or more variable is altered simultaneously; nonetheless, identifying the specific change that eliminated the symptom becomes far more difficult and will not help resolve the same issue if it occurs in the future.
5. **Implement the action plan** Implement the action plan and perform each step in the action plan carefully. Test the network and check the result to see whether the symptom disappears.
6. **Gather results after each variable change** Generally, one should use the same data gathering method that was used in step 2 (i.e., working with the key stakeholders), in conjunction with utilizing available resources and troubleshooting tools.
7. **Analyze the results** Sometimes the action plan does not resolve the issue in one shot. Analyze the results to determine whether the problem has been resolved. If it has, then the process is complete.
8. **If the problem has not been resolved, change another variable and repeat the process** If the problem is not resolved, create an action plan based on the next most likely cause from the list. Return to step 4, change one variable at a time, and repeat the process.

If all the common causes and actions have been checked or tested and the problem still cannot be resolved, contact the Cisco Technical Assistance Center (TAC) and open a service request (SR) for further assistance on troubleshooting the issue.

exam watch

No specific troubleshooting methodology should be used, as different people may have different but equally effective methodologies for troubleshooting a problem. You should understand where most problems occur in a networked environment and be aware of basic troubleshooting techniques and tools.

CERTIFICATION OBJECTIVE 12.02

Use of Wireless LAN Controller Show, Debug, and Logging Commands

As with all Cisco networking devices, the Cisco Wireless Controller contains a variety of methods for performing troubleshooting. Some common events on a controller to perform troubleshooting on include:

- **An access point cannot associate to the controller** When this occurs, an access point may, for a variety of reasons, not be able to associate to a controller to download configuration and/or software. This is commonly seen in the early stages of access point deployment.
- **Unexpected mobility/roaming behavior** In the case of either client roaming or establishing mobility relationships with other controllers, being able to monitor and debug mobility events and status is essential to keeping a wireless network healthy.
- **Client association/authentication failure** This will be the most common ongoing concern in any wireless network. For a variety of reasons, a wireless client or user may have issues connecting to the wireless network. There are tools on both the wireless LAN controller and WCS to troubleshoot these occurrences.

In the following sections we will look at different classes of commands that can be run on the wireless LAN controller's command line interface (CLI) and viewed on the graphical user interface (GUI).

Wireless LAN Controller Show Commands

The most commonly used commands on the wireless LAN controller's CLI for first-level troubleshooting are *show* commands. Show commands provide information about currently configured properties and real-time status of various activities that occur on the controller. This can be done with no risk of making unwanted changes or modifications to current controller configuration. Show commands are only current for the moment that they are run and, with a dynamic network, may present misleading information. We will cover the most commonly utilized show commands in this section.

- **show running-config, show run-config** This command is identical to the one in Cisco IOS routers and switches. By issuing this command, you can view the entire configuration of the controller and verify the configuration of every access point, interface, WLAN, and external server on the controller in detail. A capture of this show command can be very large, but it will tell a support representative a great deal about what is being done on the wireless network. In newer versions of wireless LAN controller software, the command **show run-config** has replaced **show running-config**.
- **show interface summary** Issuing this command is similar to clicking the Controller | Interfaces on the controller GUI. In one screen you can immediately see which interfaces are configured on the controller, including the interface name, VLAN ID, IP address, and type (static or dynamic). To see more information about individual interfaces, you can issue the **show interface detailed <interface-name>** command.

- **show wlan summary** Issuing this command is similar to clicking the WLAN menu item on the controller GUI. In this screen you can see key information about every configured WLAN on the controller, including the WLAN ID, WLAN profile name/SSID, status (enabled, disabled), and associated interface name. To see more information about individual WLANs, you can issue the **show wlan <wlan-ID>** command.
- **show ap summary** Issuing this command is similar to clicking the Wireless menu item on the controller GUI. In this display, you can get basic information about all of the access points associated to the controller, including AP Name, AP Model, Ethernet MAC, and location.
- **show ap join stats summary all** While checking AP associations to the controller during a wireless deployment, this command can provide timely and valuable information about the status of access points that are joining or trying to join the controller. As seen in Figure 12-2, you can very quickly determine if a controller is seeing an AP attempting to join and also monitor whether it successfully completes the join process.

FIGURE 12-2 Sample AP join stats summary

```
(WiSM-slot2-1) >show ap join stats summary all

Number of APs............................................ 11


00:0b:85:28:bb:90........................................ Not joined
00:0f:f7:be:e7:40........................................ Not joined
00:17:df:35:96:a0........................................ Joined
00:17:df:a4:4d:d0........................................ Joined
00:17:df:a4:ac:d0........................................ Not joined
00:17:df:a6:e2:a0........................................ Joined
00:17:df:a7:66:80........................................ Joined
00:22:90:92:9e:a0........................................ Not joined
00:22:90:95:d3:20........................................ Not joined
00:22:90:95:d3:70........................................ Not joined
```

- **show mobility summary** Unlike previously mentioned commands, this one will efficiently present information that is not easily found on the GUI of the controller. The mobility summary will present all of the controller settings for mobility-related intervals and ports. It will also quickly tell you about other controllers configured for the same mobility group and whether or not they have successfully joined that group. As seen in Figure 12-3, you can quickly determine which controllers in the mobility group may be misconfigured and minimize client roaming issues by fixing those controllers.
- **show client summary** The client summary is similar to clicking Monitor | Clients. In this display, you can get basic information about all clients seen by the controller, including the associated/detected AP name, status (associated, probing, disassociated), and protocol (802.11b, 802.11g, 802.11a, or 802.11n) that the client is associated with. Figure 12-4 shows a sample client summary.

FIGURE 12-3 Sample show mobility summary

```
(WiSM-slot2-1) >show mobility summary

Symmetric Mobility Tunneling (current) ......... Enabled
Symmetric Mobility Tunneling (after reboot) ..... Enabled
Mobility Protocol Port.......................... 16666
Mobility Security Mode.......................... Disabled
Default Mobility Domain......................... AS
Multicast Mode ................................. Disabled
Mobility Domain ID for 802.11r.................. 0x7d65
Mobility Keepalive Interval..................... 10
Mobility Keepalive Count........................ 3
Mobility Group Members Configured............... 6
Mobility Control Message DSCP Value............. 0

Controllers configured in the Mobility Group
 MAC Address        IP Address       Group Name                        Multicast IP     Status
 00:0b:85:40:39:60  10.10.100.131    AS                                0.0.0.0          Up
 00:0b:85:43:51:40  10.10.100.133    AS                                0.0.0.0          Control and Data Path Down
 00:0b:85:44:5f:80  10.10.100.135    AS                                0.0.0.0          Control and Data Path Down
 00:11:92:ff:7c:00  10.10.100.141    AS                                0.0.0.0          Up
 00:11:92:ff:7c:20  10.10.100.143    AS                                0.0.0.0          Up
 00:24:97:cc:72:60  10.10.100.151    AS                                0.0.0.0          Control and Data Path Down
```

FIGURE 12-4 Sample show client summary

```
Number of Clients................................ 11

MAC Address       AP Name           Status        WLAN/Guest-Lan Auth Protocol Port Wired
----------------- ----------------- ------------- -------------- ---- -------- ---- -----

00:02:8a:a2:2e:60 AP1242-AP3        Probing       N/A            No   802.11b  29   No
00:1e:e5:dd:f6:bf AP1242-AP3        Probing       N/A            No   802.11b  29   No
00:1f:3a:68:83:f7 AP1242-AP3        Probing       N/A            No   802.11a  29   No
00:21:5d:e8:ba:6e AP1242-AP3        Probing       N/A            No   802.11a  29   No
00:40:96:a0:b4:f4 AP1242-AP3        Probing       N/A            No   802.11a  29   No
00:40:96:a7:f9:e7 AP1242-AP3        Probing       N/A            No   802.11a  29   No
00:40:96:ac:1c:6f AP1242-AP3        Probing       N/A            No   802.11b  29   No
00:40:96:ad:51:0c AP1242-AP3        Probing       N/A            No   802.11a  29   No
00:40:96:b4:c8:39 Location_1        Probing       N/A            No   802.11a  29   No
00:40:96:b6:9d:0c AP1242-AP3        Probing       N/A            No   802.11b  29   No
00:40:96:b6:9d:15 AP1242-AP3        Probing       N/A            No   802.11b  29   No
```

- **show client detail <mac-address>** One of the best methods for diagnosing a client that is having trouble authenticating or associating to the wireless network is to run this command. Similar to clicking an individual client in the Monitor | Clients menu item, this command gives an array of details about individual clients and can tell you if the client is successfully associating or authenticating to the wireless network. Figure 12-5 shows a sample client detail readout. You can immediately determine the current client state, the client's RADIUS username if 802.1X EAP authentication is used, the wireless LAN ID (SSID, if any) that the client is associated to, the CCX version that the client supports, and important RF information such as the client's signal strength as seen by multiple access points. In this particular example, the client is in a probing state and is not associated to any access point at all.
- **show tech-support** This command, along with the **show running-config** command, will produce a great deal of output and information about the controller. Although this command is often used by the Cisco Technical Assistance Center (TAC) to help troubleshoot a controller during a support case, administrators can make use of this command's output as well.

FIGURE 12-5

Sample show client detail802.1X

```
(WiSM-slot2-1) >show client detail 00:40:96:b6:9d:15
Client MAC Address............................... 00:40:96:b6:9d:15
Client Username ................................. N/A
AP MAC Address................................... 00:17:df:35:96:a0
Client State..................................... Probing
Client NAC OOB State............................. Invalid
Wireless LAN Id.................................. N/A
BSSID............................................ 00:17:df:35:96:9f
Connected For ................................... 8144086 secs
Channel.......................................... 1
IP Address....................................... Unknown
Association Id................................... 0
Authentication Algorithm......................... Open System
Reason Code...................................... 0
Status Code...................................... 0
Session Timeout.................................. 0
Client CCX version............................... No CCX support
Mirroring........................................ Disabled
QoS Level........................................ Silver
Diff Serv Code Point (DSCP)...................... disabled
802.1P Priority Tag.............................. disabled
WMM Support...................................... Disabled
Supported Rates..................................
Mobility State................................... None
Mobility Move Count.............................. 0
Security Policy Completed........................ No
Policy Manager State............................. START
Policy Manager Rule Created...................... Yes
NPU Fast Fast Notified........................... No
Policy Type...................................... N/A
Encryption Cipher................................ None
Management Frame Protection...................... No
EAP Type......................................... Unknown
Interface........................................ management
VLAN............................................. 0
Quarantine VLAN.................................. 0
Access VLAN...................................... 0
Client Capabilities:
      CF Pollable................................ Not implemented
      CF Poll Request............................ Not implemented
      Short Preamble............................. Not implemented
      PBCC....................................... Not implemented
      Channel Agility............................ Not implemented
      Listen Interval............................ 0
      Fast BSS Transition........................ Not implemented
Fast BSS Transition Details:
Client Statistics:
      Number of Bytes Received................... 0
      Number of Bytes Sent....................... 0
      Number of Packets Received................. 0
      Number of Packets Sent..................... 0
      Number of EAP Id Request Msg Timeouts...... 0
      Number of EAP Request Msg Timeouts......... 0
      Number of EAP Key Msg Timeouts............. 0
      Number of Policy Errors.................... 0
      Radio Signal Strength Indicator............ Unavailable
      Signal to Noise Ratio...................... Unavailable
Nearby AP Statistics:
      AP1242-AP3(slot 0) .......................
antenna0: 2 seconds ago -59 dBm.................. antenna1: 644375 seconds ago
-128 dBm
      AP1242-AP3(slot 1) .......................
antenna0: 1 seconds ago -56 dBm.................. antenna1: 644375 seconds ago
-128 dBm
      Location_1(slot 1) .......................
antenna0: 1 seconds ago -82 dBm.................. antenna1: 1 seconds ago -83
dBm
```

Wireless LAN Controller Debug Commands

Although the show commands that we discussed will give valuable data about the processes that occur on a controller, there are often scenarios where more timely information is needed in order to resolve an issue. For more advanced troubleshooting of the controller and viewing real-time interactions that occur on the controller and with other elements of the wireless network, there are many debug commands that are available. The most commonly utilized ones are the following:

- **debug lwapp/capwap events enable/disable** This debug command can either be "lwapp" or "capwap," depending on the controller code version that you are running. In all controller software versions 5.2 and up this has been changed to "capwap" due to the adoption of a standards-based tunneling protocol for wireless traffic. When running this debug, you can view all the LWAPP/CAPWAP-related events and errors that occur on a controller. This debug command is useful for monitoring access points when they attempt to join/associate to a controller.
- **debug ap enable/disable <AP name>** The output of this debug command will allow you to view debugging messages from the LWAPP AP that you select. This debug can provide you with the access point's "point of view" in activities involving client association and joining a wireless LAN controller. The **debug ap** command variant of this line will also allow you to run certain CLI commands on the AP.
- **debug client <mac address> enable/disable** This debug command is very useful for viewing client interactions; because many controllers have multiple clients on them, the debugs from them can be extremely verbose. By specifying a single client MAC address, you can directly examine a single client without being overwhelmed with extraneous debug information. This command enables multiple debug commands, which aid in viewing authentication and association processes.
- **debug dot11 all enable/disable** This command allows you to see a variety of security and mobility related messages that occur on the wireless network. Like the **debug client** command, this particular debug contains information from multiple debug commands and aids in troubleshooting rogue AP activity, load-balancing, mobility states of clients, and 802.11 MAC management traffic.
- **debug aaa events enable/disable** This command allows you to see interactions that occur between the controller and RADIUS or EAP server. It can be useful for troubleshooting client authentication by verifying proper connectivity between the authenticator (wireless controller) and authentication server.

- **debug dot1x events enable/disable** This debug allows you to see 802.1X-related events that can be helpful for troubleshooting failed client authentications. In these debugs you can discover authentication successes and failures. The debug output often offers valuable clues as to why failures in authentication of an 802.1X wireless client occur.
- **debug dhcp message enable/disable** This debug allows you to view any dhcp-related event on the controller. This is most useful when troubleshooting a client that cannot acquire an IP address via DHCP after association and authentication has completed. When viewing this debug, you can monitor the process that a client and DHCP server go through when the client device attempts to obtain an IP address after association and authentication to the wireless network has completed.

Note that in all of the debug types, there are two variations of the command (enable or disable) that turn the debugging process on or off. It is important to remember that debugging, when enabled, can be very processor intensive; it takes priority over all other processing that occurs on the controller. To quickly end all debug processes, the **debug disable-all** command is very important to be aware of. You also have the option of allowing the CLI session from which you ran the debug to time out. Once the session times out, all associated debugs will be disabled as well.

Wireless LAN Controller Logging Commands

When working with the wireless LAN controller, you have the ability to leverage the GUI to examine various logs with error messages and notifications. These logs provide a history of events and errors that occur over time and can be useful for historical troubleshooting of a controller. In addition to this, controller logs provide clues about events that may have occurred in the past. We will examine two logs that are typically kept on a controller, the syslog and the SNMP Log.

- **show logging** This command displays the syslog of the controller, which can also be easily viewed using the GUI by navigating to Management menu on the wireless LAN controller GUI and selecting Logs | Message Logs. The messages seen in this particular log are typically sent to an external syslog server and will notify administrators of system-related errors or notifications on the controller. Figures 12-6 and 12-7 show the respective CLI and GUI outputs of the show logging command.

FIGURE 12-6 Sample show logging, CLI output

```
(WiSM-slot2-1) show logging

Logging to buffer :
- Logging of system messages to buffer :
 - Logging filter level.......................... informational
 - Number of system messages logged............. 2763492
 - Number of system messages dropped............ 2502
- Logging of debug messages to buffer .......... Disabled
 - Number of debug messages logged.............. 0
 - Number of debug messages dropped............. 81677
Logging to console :
- Logging of system messages to console :
 - Logging filter level.......................... errors
 - Number of system messages logged............. 0
 - Number of system messages dropped............ 2765994
- Logging of debug messages to console ......... Enabled
 - Number of debug messages logged.............. 81677
 - Number of debug messages dropped............. 0
Logging to syslog :
- Syslog facility................................ local0
- Logging of system messages to syslog :
 - Logging filter level.......................... emergencies
 - Number of system messages logged............. 0
--More-- or (q)uit
 - Number of system messages dropped............ 2765994
- Logging of debug messages to syslog .......... Disabled
 - Number of debug messages logged.............. 0
 - Number of debug messages dropped............. 81677
- Number of remote syslog hosts.................. 1
  - Host 0....................................... 10.10.10.10
  - Host 1....................................... Not Configured
  - Host 2....................................... Not Configured
Logging of traceback............................. Enabled
- Traceback logging level........................ emergencies
Logging of process information................... Enabled
Logging of source file informational............. Enabled
Timestamping of messages.........................
- Timestamping of system messages................ Enabled
 - Timestamp format.............................. Date and Time
- Timestamping of debug messages................. Enabled
 - Timestamp format.............................. Date and Time

 Logging buffer (2763492 logged, 84179 dropped)

*Aug 23 11:35:07.065: %ETHOIP-3-PKT_RECV_ERROR: ethoip.c:337 ethoipSocketTask: ethoipRecvPkt returned error
*Aug 23 11:35:07.065: %ETHOIP-3-INVALID_PKT_RECVD: ethoip.c:110 Ethernet over IP pkt too short; pkt size=66, expected
min=82
*Aug 23 11:34:57.065: %ETHOIP-3-PKT_RECV_ERROR: ethoip.c:337 ethoipSocketTask: ethoipRecvPkt returned error
*Aug 23 11:34:57.064: %ETHOIP-3-INVALID_PKT_RECVD: ethoip.c:110 Ethernet over IP pkt too short; pkt size=66, expected
min=82
Previous message occurred 2 times.
*Aug 23 11:34:51.383: %CAPWAP-3-REASSEM_SPACE: capwap_ac_reassembly.c:656 Unable to store capwap fragment from
00:15:2c:7b:4e:40.
*Aug 23 11:34:47.064: %ETHOIP-3-PKT_RECV_ERROR: ethoip.c:337 ethoipSocketTask: ethoipRecvPkt returned error
*Aug 23 11:34:47.064: %ETHOIP-3-INVALID_PKT_RECVD: ethoip.c:110 Ethernet over IP pkt too short; pkt size=66, expected
min=82
*Aug 23 11:34:37.569: %AAA-5-AAA_AUTH_ADMIN_USER: aaa.c:1206 Authentication succeeded for admin user 'admin'
*Aug 23 11:34:37.064: %ETHOIP-3-PKT_RECV_ERROR: ethoip.c:337 ethoipSocketTask: ethoipRecvPkt returned error
*Aug 23 11:34:37.064: %ETHOIP-3-INVALID_PKT_RECVD: ethoip.c:110 Ethernet over IP pkt too short; pkt size=66, expected
min=82
*Aug 23 11:34:31.866: %WCP-3-KEEPALIVE_LOST: wcp_controller_server.c:1761 Lost keepalives from Cat 6K Supervisor.
*Aug 23 11:34:27.065: %ETHOIP-3-PKT_RECV_ERROR: ethoip.c:337 ethoipSocketTask: ethoipRecvPkt returned error
*Aug 23 11:34:27.064: %ETHOIP-3-INVALID_PKT_RECVD: ethoip.c:110 Ethernet over IP pkt too short; pkt size=66, expected
min=82
*Aug 23 11:34:17.063: %ETHOIP-3-PKT_RECV_ERROR: ethoip.c:337 ethoipSocketTask: ethoipRecvPkt returned error
*Aug 23 11:34:17.063: %ETHOIP-3-INVALID_PKT_RECVD: ethoip.c:110 Ethernet over IP pkt too short; pkt size=66, expected
min=82
*Aug 23 11:34:07.063: %ETHOIP-3-PKT_RECV_ERROR: ethoip.c:337 ethoipSocketTask: ethoipRecvPkt returned error
*Aug 23 11:34:07.063: %ETHOIP-3-INVALID_PKT_RECVD: ethoip.c:110 Ethernet over IP pkt too short; pkt size=66, expected
min=82
*Aug 23 11:33:57.062: %ETHOIP-3-PKT_RECV_ERROR: ethoip.c:337 ethoipSocketTask: ethoipRecvPkt returned error
*Aug 23 11:33:57.062: %ETHOIP-3-INVALID_PKT_RECVD: ethoip.c:110 Ethernet over IP pkt too short; pkt size=66, expected
min=82
*Aug 23 11:33:47.082: %ETHOIP-3-PKT_RECV_ERROR: ethoip.c:337 ethoipSocketTask: ethoipRecvPkt returned error
*Aug 23 11:33:47.082: %ETHOIP-3-INVALID_PKT_RECVD: ethoip.c:110 Ethernet over IP pkt too short; pkt size=66, expected
min=82
```

FIGURE 12-7 Sample message logs, GUI window

- **show traplog** This command displays the SNMP log, which shows errors and messages associated with access point, rogue access point, and client association/authentication transactions on the controller. The messages in these logs are sent to the Wireless Control System (WCS) to generate the faults and alarms you typically see in the monitoring menus. In addition to viewing the logs on the command line interface, you can see these logs on the GUI by navigating to Monitor | View All. Figures 12-8 and 12-9 show the respective CLI and GUI outputs of the **show traplog** command.

FIGURE 12-8 Sample show traplog, CLI output

```
(WiSM-slot2-1) show traplog

Number of Traps Since Last Reset ............ 1230636
Number of Traps Since Log Last Displayed .... 1

Log System Time              Trap
--- ------------------------ -------------------------------------------------
  0 Sun Aug 23 11:44:31 2009 Lost heartbeat with the Supervisor
  1 Sun Aug 23 11:43:43 2009 Rogue AP : 00:16:9c:91:55:f1 detected on Base Rad
                             io MAC : 00:17:df:35:96:a0  Interface no:0(802.11
                             b/g) with RSSI: -74 and SNR: 15 and Classificatio
                             n: unclassified
  2 Sun Aug 23 11:43:43 2009 Rogue AP : 00:19:a9:0c:e9:b0 detected on Base Rad
                             io MAC : 00:17:df:35:96:a0  Interface no:0(802.11
                             b/g) with RSSI: -53 and SNR: 18 and Classificatio
                             n: unclassified
  3 Sun Aug 23 11:43:43 2009 Rogue AP : 00:1d:46:fe:b7:71 detected on Base Rad
                             io MAC : 00:17:df:35:96:a0  Interface no:0(802.11
                             b/g) with RSSI: -68 and SNR: 29 and Classificatio
                             n: unclassified

Would you like to display more entries? (y/n) y


Log System Time              Trap
--- ------------------------ -------------------------------------------------
  4 Sun Aug 23 11:43:43 2009 Rogue AP : 00:22:90:38:f9:30 detected on Base Rad
                             io MAC : 00:17:df:35:96:a0  Interface no:0(802.11
                             b/g) with RSSI: -73 and SNR: 11 and Classificatio
                             n: unclassified
  5 Sun Aug 23 11:43:43 2009 Rogue AP : 00:1b:8f:89:de:80 detected on Base Rad
                             io MAC : 00:17:df:35:96:a0  Interface no:1(802.11
                             a) with RSSI: -90 and SNR: 10 and Classification:
                             unclassified
  6 Sun Aug 23 11:43:43 2009 Rogue AP : 00:1c:57:41:3e:21 detected on Base Rad
                             io MAC : 00:17:df:35:96:a0  Interface no:1(802.11
                             a) with RSSI: -79 and SNR: 16 and Classification:
                             unclassified
  7 Sun Aug 23 11:43:43 2009 Rogue AP : 00:1d:e6:24:67:0f detected on Base Rad
                             io MAC : 00:17:df:35:96:a0  Interface no:1(802.11
                             a) with RSSI: -81 and SNR: 18 and Classification:
                             unclassified
```

FIGURE 12-9 Sample trap logs, GUI window

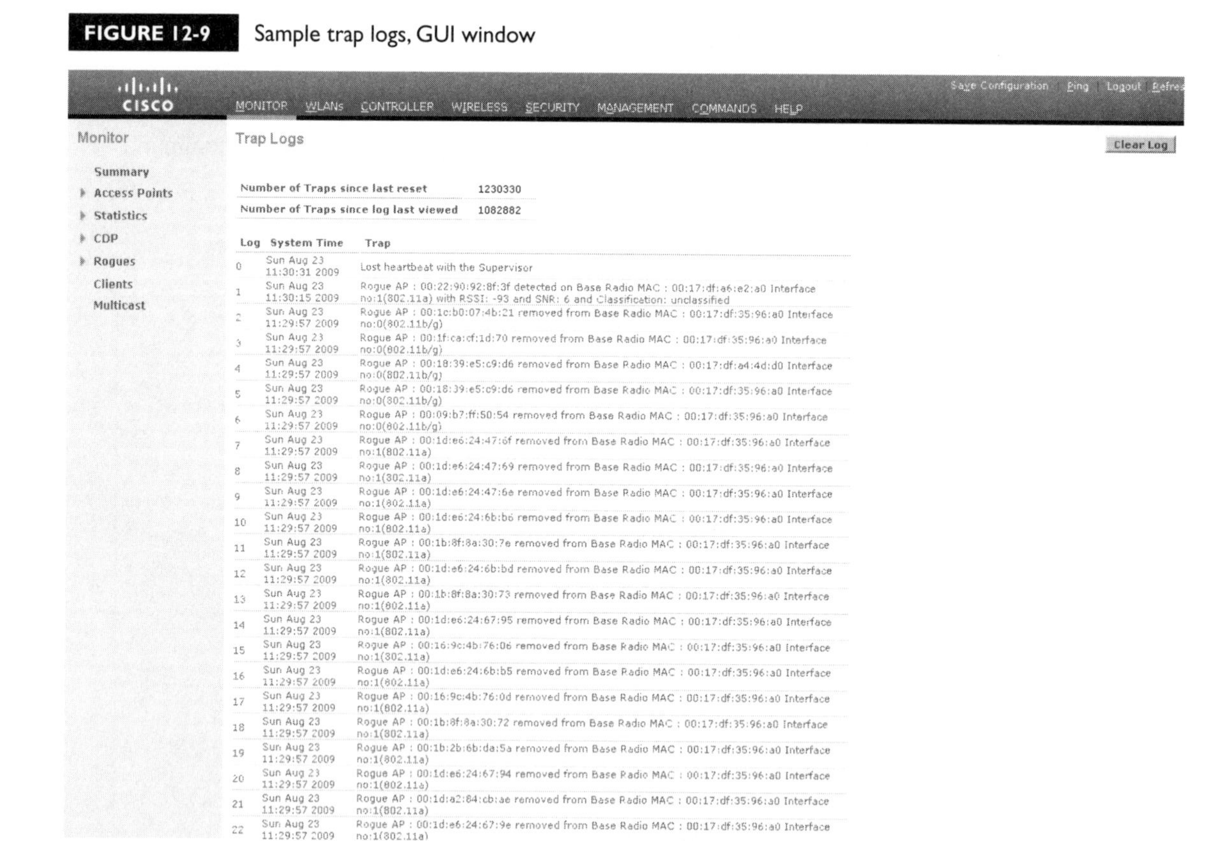

CERTIFICATION OBJECTIVE 12.03

Use of the WCS Client Troubleshooting Tool

In Chapter 9, an overview of many of the Wireless Control System (WCS) features was presented. In that chapter, we discussed various monitoring and alarm tools, including the client monitoring window. One of the most useful features for system administrators is the WCS Client Troubleshooting tool. The Client Troubleshooting tool provides an easy to use graphical interface for tracking clients that may have difficulties associating to the wireless network. In many ways, the Client Troubleshooting tool provides much of the information that can be found in show, debug, and logging commands in one simple turnkey tool that is standard with the Cisco Wireless Control System. On the WCS, select the Monitor | Clients menu item so you can access the Client Troubleshooting section in the lower right hand side from the main Clients Summary window (Figure 12-10).

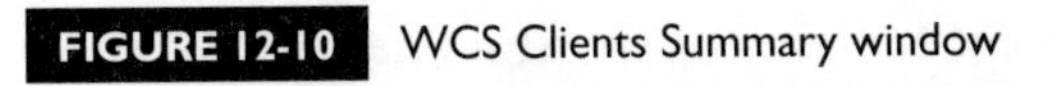

FIGURE 12-10 WCS Clients Summary window

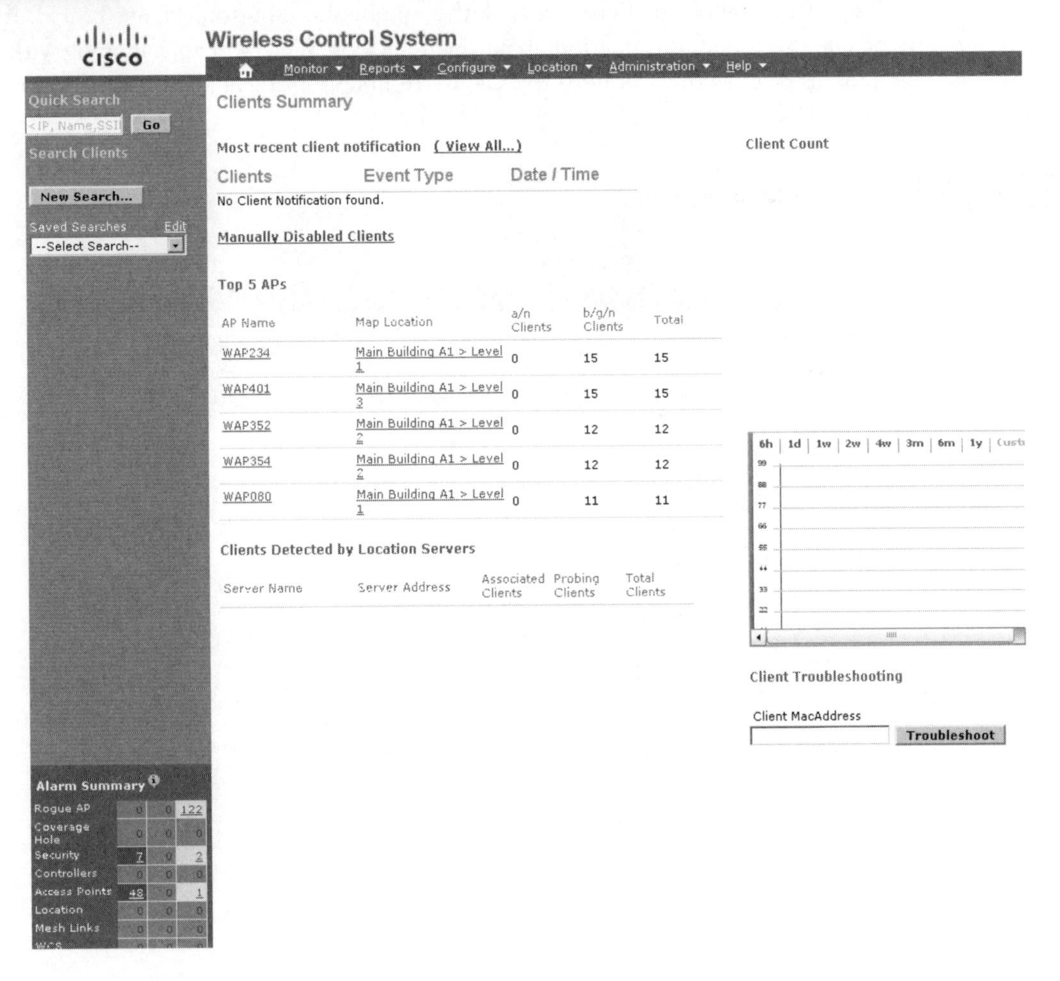

FIGURE 12-11 Client Troubleshooting prompt

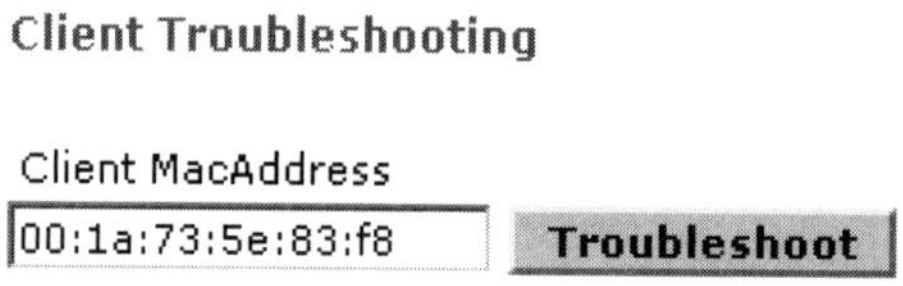

Client Troubleshooting

Client MacAddress

00:1a:73:5e:83:f8 Troubleshoot

In the course of determining a client you want to monitor or troubleshoot, it is important that you get that client's MAC address off its wireless adapter/client software. Once you have found this information, you can enter it directly into the Client Troubleshooting field and click the Troubleshoot button (Figure 12-11). If the client is found, information about the device on WCS will appear along with a pop-up window that contains the Client Troubleshooting application, shown in Figure 12-12.

FIGURE 12-12 Client Troubleshooting tool

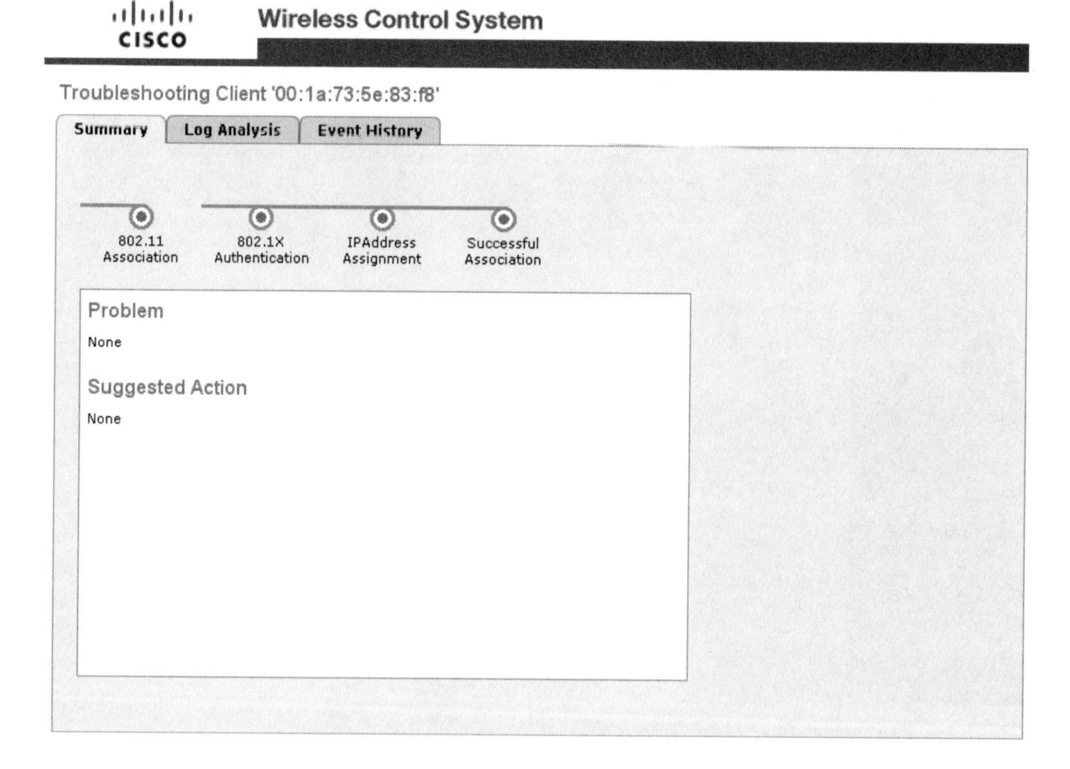

After this window pops up, the client is tested for its ability to successfully associate and authenticate to the network with its given credentials. As this battery of tests is performed, the main window will run through three "status lights" that will show where (if anywhere) the full client association process has failed. The three major areas where a client can fail are as follows:

- **802.11 Association** This piece of the client association process is where the client finds the proper SSID of the access point and is able to achieve the signal strength that it requires to successfully begin the authentication process. Most association failures occur as a result of incorrectly entered SSIDs, security access restrictions on an SSID, or lack of proper RF coverage for the client. Further manual troubleshooting of this process can be accomplished with the **debug dot11 all** or **debug client** commands discussed in the "Wireless LAN Controller Debug Commands" section of this chapter.
- **802.1X Authentication** This piece of the client association process is where the client undergoes 802.1X authentication and the WPA four-way handshake to establish encryption (if using a specified EAP-type). When this process fails, it usually points to a RADIUS configuration issue or an issue with incorrectly entered login credentials on the client end. Further investigation can be completed by running the **debug dot1x events** command.
- **IP Address Assignment** The final piece of the client association process simply checks to see whether or not an associated and authenticated client is able to acquire an IP address from the DHCP server. If this fails, it usually means that there is a statically configured IP address on the client, a lack of connectivity to a DHCP server or an incorrectly configured DHCP server on the dynamic interface within the controller. The **debug dhcp message** command can be used to further diagnose the problem.

After a client has undergone the association tests, you can also view log messages and events related to the client on the Client Troubleshooting tool tabs. The Log Analysis tab is shown in Figure 12-13.

As you can see, there are several log messages that are shown in this tab and can be filtered by the type of message/error. Viewing these log messages in the event of a client association failure can provide valuable clues and error details to allow

FIGURE 12-13 Client Troubleshooting tool Log Analysis tab

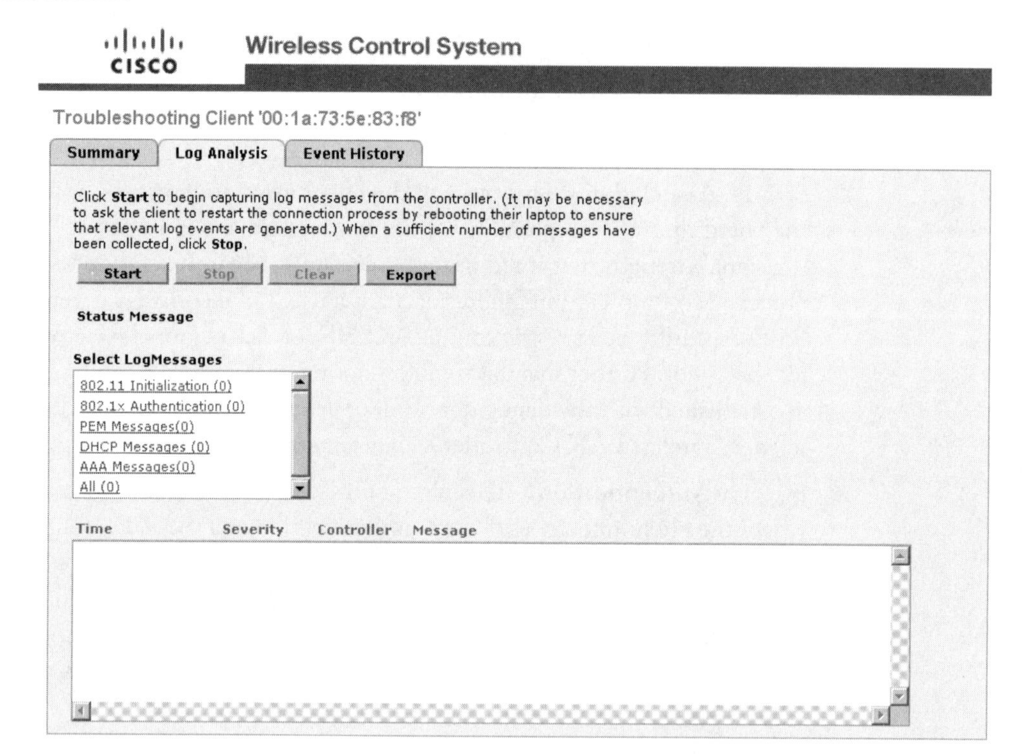

an administrator to diagnose an issue without having to touch the client itself. Historical information about client and access point events can be seen in the Event History tab shown in Figure 12-14.

FIGURE 12-14 Client Troubleshooting tool Event History tab

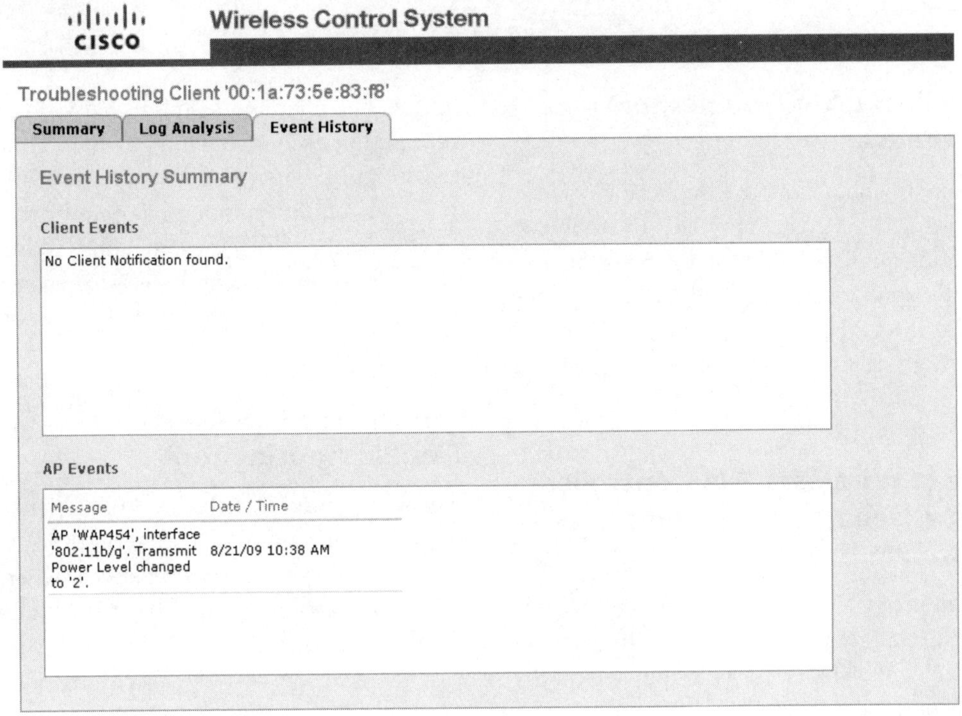

In this tab, you can view historical events associated with the access point and client that are not necessarily related to this particular test. By viewing the event histories of both the client and access point, you can gain further clues as to the possible cause for a client association failure.

INSIDE THE EXAM

Wireless LAN Troubleshooting Methods

Be sure to understand the importance of following a systematic approach to troubleshooting network issues for a wireless deployment. Familiarize yourself with the eight steps of wireless LAN troubleshooting. In some situations, it is necessary to contact the Cisco Technical Assistance Center (TAC).

Use of Wireless LAN Controller Show, Debug, and Logging Commands

Be familiar with the role of show, debug, and logging commands in troubleshooting a wireless network. The best way to understand the value of these commands is to spend a substantial amount of time working with them on the controller and reading what kind of information each command provides. Be able to distinguish between the use, format, and timeliness of information presented by the show, debug, and logging commands. Also make sure you understand the value each of these commands can provide in the troubleshooting process.

Use of the WCS Client Troubleshooting Tool

Familiarize yourself with the WCS Client Troubleshooting tool's location on the WCS and the various windows that appear when you run the tool. Be aware that much of the tool's information can be used in conjunction with show, debug, and logging commands to provide holistic information in the process of troubleshooting a client.

CERTIFICATION SUMMARY

The troubleshooting process is a systematic approach to diagnosing and solving network-related issues. A wireless network is no different, and it is important to always follow a troubleshooting strategy that involves isolating and solving problems by identifying and tracking down the causes of certain network symptoms and eliminating the causes of these problems. The eight steps include identifying the symptoms of the problem, collecting sufficient data to isolate possible causes, eliminating unrelated problems from the possible causes, creating an action plan, implementing the action plan, gathering the results, evaluating the results, and changing the variables and repeating the process until you reach resolution. Following a systematic approach will yield better troubleshooting results in less time and is used by all experienced network engineers.

When performing troubleshooting tasks on the controller, it is important to note some of the major issues that you are likely to troubleshoot in a wireless LAN deployment. These include access point to controller communication issues, mobility/roaming issues, and client association and authentication issues. Show, debug, and logging commands can help with all of these issues, plus any others you may run into when deploying a wireless network. Show commands are commonly used and can provide information about currently configured properties and the status of various activities that occur on the controller. Debug commands allow for more timely and detailed information about transactions and activities that occur on a controller. Logging commands show a history of messages and provide information about events that may have occurred in the past in the form of syslogs or SNMP trap logs.

The Wireless Control System (WCS) Client Troubleshooting tool is a useful tool that combines many of the features of show, debug, and logging commands to provide a single easy-to-use GUI method of troubleshooting wireless LAN clients. There are various windows that can provide valuable information toward repairing wireless client issues with a minimum of time while eliminating the need for IT administrators to touch individual clients.

TWO-MINUTE DRILL

Wireless LAN Troubleshooting Methods

- The systematic approach to troubleshooting involves the following eight steps:
  1. Identify the symptoms of the problem.
  2. Collect sufficient data to isolate possible causes.
  3. Eliminate unrelated problems from the possible causes.
  4. Create an action plan.
  5. Implement the action plan.
  6. Gather results of the actions performed.
  7. Evaluate the gathered results.
  8. Change variables and repeat the process until you reach resolution.

Use of Wireless LAN Controller Show, Debug, and Logging Commands

- Access point to controller communication issues, mobility/roaming issues, and client association and authentication issues are three common items that troubleshooting on the controller is used to diagnose.
- Show commands provide information about current configurations and status of various activities that occur on the controller.
- Debug commands provide a real-time and detailed display of transactions and activities that occur on a controller during various processes.
- Logging commands show a history of events and errors that have occurred on the controller; these commands come in the form of syslogs and SNMP trap logs.

Use of the WCS Client Troubleshooting Tool

- The Wireless Control System (WCS) Client Troubleshooting tool features many elements of show, debug, and logging commands to provide a GUI method of troubleshooting wireless LAN clients.
- The multiple tabs within the Client Troubleshooting tool allow for different levels of wireless client diagnosis and troubleshooting.

SELF TEST

The following self test questions will help you measure your understanding of the material presented in this chapter. Read all the choices carefully, as there may be more than one correct answer. Choose all correct answers for each question.

Wireless LAN Troubleshooting Methods

1. In a troubleshooting situation, should you change multiple variables to accelerate the process?
   - A. No, changing multiple variables is not the most efficient way to find the root cause.
   - B. Yes, changing multiple variables may have compounded effects to make things work better.
   - C. No, the problem should be contained at Layer 2, so only two variables should be changed at a time.
   - D. Yes, changing multiple variables is a quicker method for eliminating false positives.
2. When the client is configured with the wrong default gateway IP address, what layer of OSI model is not configured correctly?
   - A. 1
   - B. 2
   - C. 3
   - D. 4

Use of Wireless LAN Controller Show, Debug, and Logging Commands

3. Which of these is not a troubleshooting CLI show command that will help you monitor an AP to controller join process?
   - A. show AP summary
   - B. show AP join stats summary all
   - C. show client summary
   - D. show tech-support
4. Which of the following show commands is best suited to troubleshooting a client mobility/roaming process between different controllers?
   - A. show running-config
   - B. show mobility summary
   - C. show client summary
   - D. show interface summary

5. When working with a Cisco TAC engineer, which of the following show commands will give the engineer the most comprehensive information about the wireless controller and wireless LAN you are troubleshooting?
   A. show tech-support
   B. show run-config
   C. show mobility summary
   D. show wlan summary
6. Which of the following debug commands would you be most likely to use when troubleshooting an AP to controller join process in conjunction with necessary show commands?
   A. debug dot11 all enable
   B. debug aaa events enable
   C. debug dot1x events enable
   D. debug lwapp events enable
7. Which of the following debug commands would you be least likely to use when diagnosing a client issue that is directly related to authentication?
   A. debug dot1x events enable
   B. debug dhcp message enable
   C. debug aaa events enable
   D. debug client <mac address> enable
8. Which of the following commands would not be useful in troubleshooting the full client association process (including authentication and IP addressing)?
   A. show lwapp events enable
   B. show client summary
   C. debug client <mac address> enable
   D. debug dot1x events enable
9. Which of the following is not a characteristic of logging commands seen on the wireless LAN controller?
   A. They can be viewed from both the controller CLI and GUI.
   B. The logs are utilized by various external sources.
   C. Logging commands can be turned on or off depending on when they are needed.
   D. They provide a history of events and errors that occur on the controllers.

Use of the WCS Client Troubleshooting Tool

10. Which of the following processes is not verified by the WCS Client Troubleshooting tool?

A. Association

B. RADIUS authentication

C. DHCP address assignment

D. RF signal strength

SELF TEST ANSWERS

Wireless LAN Troubleshooting Methods

1. In a troubleshooting situation, should you change multiple variables to accelerate the process?

A. No, changing multiple variables is not the most efficient way to find the root cause.

B. Yes, changing multiple variables may have compounded effects to make things work better.

C. No, the problem should be contained at Layer 2, so only two variables should be changed at a time.

D. Yes, changing multiple variables is a quicker method for eliminating false positives.

☑ A. When performing troubleshooting it is recommended that you only change one variable at a time to determine the root cause of a problem.

☒ **B, C,** and **D** are incorrect because they all recommend the change of more than one variable at a time.

2. When the client is configured with the wrong default gateway IP address, what layer of OSI model is not configured correctly?

A. 1

B. 2

C. 3

D. 4

☑ C. When you are working with IP-related issues such as a default gateway address, it is a Layer 3 troubleshooting issue.

☒ **A, B,** and **D** are incorrect because these layers of the OSI model are not directly related to the IP default gateway address.

Use of Wireless LAN Controller Show, Debug, and Logging Commands

3. Which of these is not a troubleshooting CLI show command that will help you monitor an AP to controller join process?

A. show AP summary

B. show AP join stats summary all

C. show client summary

D. show tech-support

☑ **C.** The show client summary command provides no information that can assist in the troubleshooting of AP to controller join process.

☒ **A, B,** and **D** are incorrect because all of the show commands specified provide information that can be used to troubleshoot the AP to controller join process.

4. Which of the following show commands is best suited to troubleshooting a client mobility/roaming process between different controllers?

A. show running-config

B. show mobility summary

C. show client summary

D. show interface summary

☑ **B.** The show mobility summary command can allow you to quickly examine the mobility states of controllers to troubleshoot issues with client roaming/mobility between different controllers.

☒ **A** and **D** are incorrect because they give no information about client roaming. **C** is incorrect because, although it gives information about clients on the controller, it gives few clues about roaming/mobility behavior.

5. When working with a Cisco TAC engineer, which of the following show commands will give the engineer the most comprehensive information about the wireless controller and wireless LAN you are troubleshooting?

A. show tech-support

B. show run-config

C. show mobility summary

D. show wlan summary

☑ **A.** The show tech-support command provides nearly all diagnostic and configuration information about a wireless LAN to a TAC engineer and is the most comprehensive show command available on the controller.

☒ **B, C,** and **D** are incorrect because, although they provide information that is valuable to a TAC engineer, they do not provide nearly the depth of information that the show tech-support command does.

6. Which of the following debug commands would you be most likely to use when troubleshooting an AP to controller join process in conjunction with necessary show commands?

A. debug dot11 all enable

B. debug aaa events enable

C. debug dot1x events enable

D. debug lwapp events enable

☑ **D.** The debug LWAPP (or CAPWAP) events enable command allows you to view LWAPP and CAPWAP transactions associated with the AP join process.

☒ **A, B,** and **C** are incorrect because the outputs of those debugs provide little to no information about the AP to controller join process.

7. Which of the following debug commands would you be least likely to use when diagnosing a client issue that is directly related to authentication?

A. debug dot1x events enable

B. debug dhcp message enable

C. debug aaa events enable

D. debug client <mac address> enable

☑ **B.** The debug dhcp message enable command is only applicable after authentication has successfully completed and plays no part in troubleshooting a wireless authentication process.

☒ **A, C,** and **D** are incorrect because they are all debugs that can be used to troubleshoot the wireless client authentication processes.

8. Which of the following commands would not be useful in troubleshooting the full client association process (including authentication and IP addressing)?

A. show lwapp events enable

B. show client summary

C. debug client <mac address> enable

D. debug dot1x events enable

☑ **A.** The debug lwapp events enable command does not provide information about the client association process

☒ **B, C** and **D** are incorrect because they, along with the debug dhcp message enable command, provide information directly related to the client association process.

9. Which of the following is not a characteristic of logging commands seen on the wireless LAN controller?
   A. They can be viewed from both the controller CLI and GUI.
   B. The logs are utilized by various external sources.
   C. Logging commands can be turned on or off depending on when they are needed.
   D. They provide a history of events and errors that occur on the controllers.

☑ **C.** Controller trap logs and SNMP logs are not an item that an administrator would turn on or off.

☒ **A, B,** and **D** are incorrect because they are all characteristics of logging on a wireless LAN controller.

Use of the WCS Client Troubleshooting Tool

10. Which of the following processes is not verified by the WCS Client Troubleshooting tool?
    A. Association
    B. RADIUS authentication
    C. DHCP address assignment
    D. RF signal strength

☑ **D.** Although approximate RF signal strength can be seen in the Client Details window, it is not a function of the WCS Client Troubleshooting tool.

☒ **A, B,** and **C** all represent tasks and processes that are verified by the WCS Client Troubleshooting tool.

Part VI

Appendixes

APPENDIXES

A

Cisco Wireless LAN Controller

This appendix will provide a brief overview of the Wireless LAN Controllers offered by Cisco for the Cisco Unified Wireless Network solution. The intent is to help you select the right controller appropriate for your networks and design scenarios. We will introduce the branch controllers first, which include both the Wireless LAN Controller module for the 2800 and 3800 Series Integrated Services Router (ISR) and the standalone Wireless LAN Controller appliance, the Cisco 2100 Series controller. We will then discuss the Cisco 4400 Series controller and the integrated Wireless Services Module (WiSM) for the Catalyst 6500/7600 Series switches. We will close the appendix with a brief discussion on the next generation controller platform, the Cisco 5500 Series controller. We will also briefly discuss the upgrade path for each controller and how the controller of choice may impact existing wireless LAN design and deployment.

Cisco 2100 Series Wireless LAN Controller

The Cisco 2100 Series Wireless LAN Controller is an entry level controller appliance designed for the branch office environment. The Cisco 2100 Series controllers come with a few flavors and support up to 6, 12, or 25 access points, making it a cost-effective solution for the small wireless LAN environment. The controller comes with eight 10/100 Ethernet ports; two can provide 802.3af power over Ethernet (PoE) directly to Cisco lightweight access points. The Cisco 2100 includes the models shown in Table A-1.

Up to seven Cisco lightweight access points can connect directly through the controller's physical 10/100 Ethernet ports, while one should be reserved for the controller to connect to the LAN. For Cisco 2112 and Cisco 2125 controllers, lightweight access points not physically connected to the controller will use normal LWAPP/CAPWAP discovery and join processes to connect to the controller. Figure A-1 illustrates the front view of a Cisco 2100 Series controller.

TABLE A-1 Cisco 2100 Series Wireless LAN Controller Models and Number of Access Points Supported

Cisco 2100 Series Wireless LAN Controller	Number of Lightweight Access Points Supported
2106	6
2112	12
2125	25

FIGURE A-1 Cisco 2100 Series Wireless LAN Controller, front view (Courtesy of Cisco Systems, Inc. Unauthorized use is not permitted.)

Figure A-2 illustrates the back view of a Cisco 2100 Series controller. Ports 7 and 8 are PoE ports, and the controller can be locked by a regular laptop lock (also known as a Kensington lock). The power port and console port are also attached to the controller from the back of the controller. The 2100 Series controller supports most of the same features as a 4400 Series controller. However, it does not support anchor controller functionality in a guest access scenario. Thus, for a centralized guest access design, a 4400 Series or equivalent controller should be used as the guest access anchor controller.

FIGURE A-2 Cisco 2100 Series Wireless LAN Controller, back view (Courtesy of Cisco Systems, Inc. Unauthorized use is not permitted.)

Cisco Wireless LAN Controller Module

The Cisco Wireless LAN Controller Module (WLCM) is another entry level controller designed for the branch office environment to be used in conjunction with the Cisco 2800/3800 Series Integrated Services Router. Cisco WLCM, like the Cisco 2100 Series controller, comes with a few flavors and supports up to 6, 8, 12, or 25 access points, making it a cost-effective solution for the small wireless LAN environment. The WLCM does not come with any usable physical port. The 10/100 Ethernet port on the front panel of WLCM is not a supported port. The Cisco WLCM includes the models shown in Table A-2.

Because there is no external interface available on a WLCM, all communications, including lightweight access points, client traffic, and console connection, need to use the internal logical ports and interfaces through the ISR. The lightweight access points will need to connect either to a standalone PoE switch or through a switch module on ISR to connect to the controller. Lightweight access points will use normal LWAPP/CAPWAP discovery and join processes to connect to the controller. Figure A-3 illustrates the front view of a Cisco Wireless LAN Controller Module (WLCM).

TABLE A-2 Cisco Wireless LAN Controller Modules and Number of Access Points Supported

Cisco 2100 Series Wireless LAN Controller	Number of lightweight Access Points Supported
NME-AIR-WLC6-K9	6
NME-AIR-WLC8-K9	8
NME-AIR-WLC12-K9	12
NME-AIR-WLC25-K9	25

FIGURE A-3 Cisco Wireless LAN Controller Module (Courtesy of Cisco Systems, Inc. Unauthorized use is not permitted.)

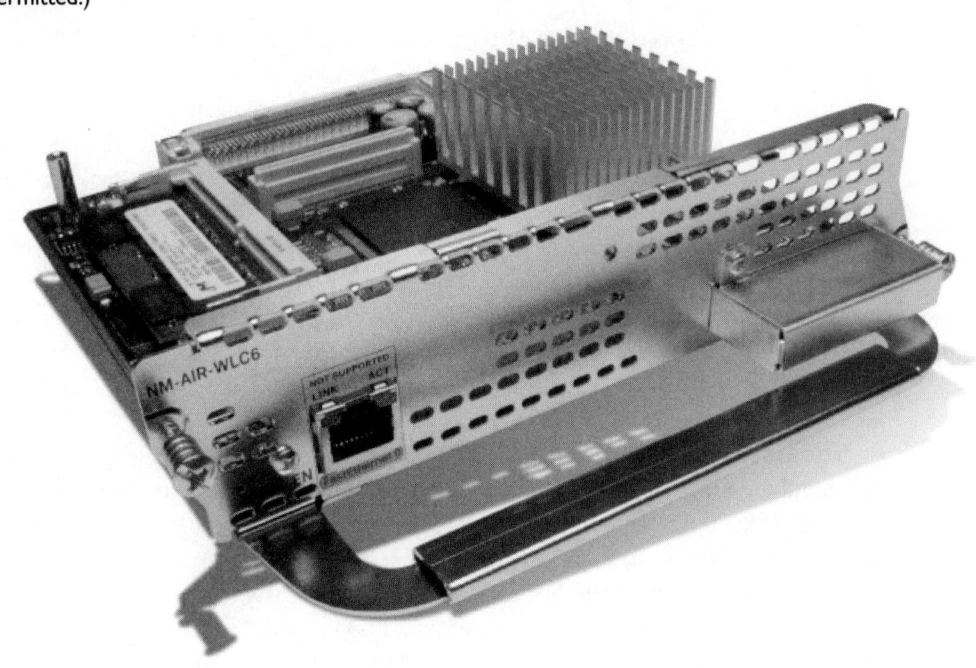

Like the 2100 Series controller, the WLCM supports most of the same features as a 4400 Series controller. However, it does not support anchor controller functionality in a guest access scenario. Thus, for a centralized guest access design, a 4400 Series or equivalent controller should be used as the guest access anchor controller. Also, it is worth mentioning that, although WLCM is supported by the 2800 and 3800 Series ISR, it is only supported on ISR platforms with Network Module slots. Smaller ISR platforms, such as ISR 2801, do not have a Network Module slot; hence, they do not support WLCM. You should always verify compatibility of ISR before making the purchase.

Cisco 4400 Series Wireless LAN Controller

The Cisco 4400 Series Wireless LAN Controller is a full featured Cisco Wireless LAN Controller designed to support branch, campus, and remote guest access functionalities. Cisco 4400 Series controllers are dedicated 1-RU appliances for wireless LAN control, provisioning, and services. The 4400 Series controllers come with two flavors and support up to 12, 25, 50, and 100 access points, making it ideal for medium to large wireless LAN environments. The Cisco 4402 Series controller comes with two gigabit Ethernet-based small form-factor pluggable (SFP) ports, a 10/100 Ethernet-based service port, a console port, and two power supply slots. The controller needs only one power supply to operate, and the second power supply is hot-swappable for redundancy purposes. In the front panel of the 4402 controller, there is an RJ-45 based utility port currently not in use. The Cisco 4402 includes the models shown in Table A-3.

The Cisco 4402 wireless controller hardware is not field upgradeable, so when the number of access points reaches the maximum number allowed by the 4402 controller, the upgrade path is either to replace the existing 4402 controller with another controller with more capacity or to add another controller to accommodate increased access point loads. For example, if you currently have one AIR-WLC4402-25-K9 and 24 access points on your network and you plan to add two more access points, you will need to either replace the AIR-WLC4402-25-K9 with an AIR-

TABLE A-3 Cisco Wireless 4400 Series Wireless LAN Controller and Number of Access Points Supported

Cisco 4402 Series Wireless LAN Controller	Number of Lightweight Access Points Supported
AIR-WLC4402-12-K9	12
AIR-WLC4402-25-K9	25
AIR-WLC4402-50-K9	50

WLC4402-50-K9, or purchase another controller to support the two additional access points. Your existing AIR-WLC4402-25-K9 cannot be upgraded to support more access points beyond 25. Figure A-4 illustrates the front view of a Cisco 4402 Series Wireless LAN Controller.

There is only one flavor of Cisco 4404 wireless controller, which supports up to 100 Cisco lightweight access points. The main differences between a 4402 and a 4404 controller are as follows:

- The 4404 controller has four SFP ports, with an aggregate uplink of up to 4 Gbps; the 4402 controller has two SFP ports, with an aggregate uplink of up to 2 Gbps..
- The 4404 controller supports up to 100 Cisco lightweight access points; the 4402 controller has three flavors and supports 12, 25, and 50 Cisco lightweight access points respectively.
- The 4404 controller supports up to 5000 clients, where 4402 only supports up to 2500 clients.

Figure A-5 illustrates the front view of a Cisco 4402 Series Wireless LAN Controller.

FIGURE A-4 Cisco 4402 Series Wireless LAN Controller (Courtesy of Cisco Systems, Inc. Unauthorized use is not permitted.)

FIGURE A-5 Cisco 4404 Series Wireless LAN Controller (Courtesy of Cisco Systems, Inc. Unauthorized use is not permitted.)

The Cisco 4400 Series Wireless LAN controller manages all of the Cisco access points, including the discontinued Aironet 1000 Series access points, Aironet 1200 Series access points, Aironet 1130 and 1240 Series access points, Aironet 1140 and 1250 Series access, and Aironet 1500 and 1520 Series outdoor mesh access points. You should always verify software compatibility for support of any lightweight access points.

Cisco Wireless Services Module

The Cisco Wireless Services Module (WiSM) is a featured Cisco wireless LAN solution designed to support branch, campus, and remote guest access functionalities with the Cisco Catalyst 6500 Series switches/Cisco 7600 Series routers. The Cisco WiSM supports up to 300 access points and occupies an open slot in the Catalyst 6500 switch or Cisco 7600 router. The WiSM operates as two distinct logical controllers, equivalent to a 4404 Series controller and supports up to 150 access points on each logical controller. The WiSM can support up to 10,000 clients and is ideal for large wireless LAN environment that require high reliability, scalability, and security. WiSM allows the wireless LAN to take advantage of rich services offered by the Catalyst 6500/7600 through other services modules, such as FWSM, ACE, and NAM, and it eliminates the complexity of wireless networks, helping to ensure smooth performance, enhanced security, and maximized network availability. The WiSM controller comes with two RJ-45 based console ports and no SFP or any other physical ports on the front panel. The WiSM communicates with the Catalyst 6500/7600 via a total of 8 logical gigabit Ethernet interfaces. There are also two logical services ports connecting the Supervisor engine of the Catalyst 6500/7600 to each logical controller. The controller does not require a separate power supply to operate. Figure A-6 illustrates the front view of a Cisco Wireless Services Module.

FIGURE A-6 Cisco Wireless Services Module (Courtesy of Cisco Systems, Inc. Unauthorized use is not permitted.)

The WiSM offers tremendous flexibility for deploying anywhere in the network and complex network designs. Similar to 4400 Series controllers, the WiSM manages all of the Cisco access points, including the discontinued Aironet 1000 Series access points, Aironet 1200 Series access points, Aironet 1130 and 1240 Series access points, Aironet 1140 and 1250 Series access, and Aironet 1500 and 1520 Series outdoor mesh access points.

Cisco Catalyst 3750 Series Integrated Wireless LAN Controller

The Cisco Catalyst 3750 Series Integrated Wireless LAN Controller is a featured Cisco wireless LAN solution added to the stackable and highly resilient Cisco Catalyst 3750G Series switches. The 3750 Series Integrated Controller is designed to support branch, campus, and remote guest access functionalities. Cisco 3750 Series controllers are 2-RU appliances, each supporting up to 24 devices via 802.3af compliant PoE ports and two SFP uplink ports. The internal logical ports will switch and route traffic for wireless LAN control, provisioning, and services. The

3750 Series Integrated Controllers come with two flavors and support up to 25 and 50 Cisco lightweight access points, making it ideal for medium-sized business or an enterprise branch LAN environment. The 3750 Series Integrated Controller operates as one distinct logical controller, equivalent to a 4402 Series controller, and supports up to 50 access points on the logical controller and up to 200 access points in a 3750 stack.

The 3750 Series Integrated Controller does not come with a separate console port or SFP ports for the controller, and it does not require a separate power supply to operate. Figure A-7 shows the front view of a Cisco 3750 Series Integrated Controller.

Similar to 4400 Series controllers, the 3750 Series Integrated Wireless Controller manages all of the Cisco access points, including the discontinued Aironet 1000 Series access points, Aironet 1200 Series access points, Aironet 1130 and 1240 Series access points, Aironet 1140 and 1250 Series access, and Aironet 1500 and 1520 Series outdoor mesh access points.

FIGURE A-7 Cisco 3750 Series Integrated Wireless LAN Controller (Courtesy of Cisco Systems, Inc. Unauthorized use is not permitted.)

Cisco 5500 Series Wireless LAN Controller

The Cisco 5500 Series Wireless LAN Controller is the next generation Cisco Unified Wireless Network platform for scalable and flexible wireless LAN deployment. The 5500 Series controller is a 1-RU controller appliance with the highest oversubscription ratio of all Cisco Unified Wireless Network controllers. The Cisco 5500 Series Wireless LAN Controller is designed to support large and complex campus and remote guest access functionalities. The Cisco 5500 Series Wireless LAN Controller has 8 SFP uplink ports and the capacity to support up to 250 Cisco lightweight access points. The 5500 Series supports a higher density of clients and delivers more efficient roaming, with at least nine times the throughput of existing 802.11a/g networks. One important fact about the 5500 Series controller is that it does not support legacy Cisco Aironet 1000 Series lightweight access points, and it only operates in CAPWAP mode. Aironet lightweight access points that you wish to connect to the 5500 Series controller will be upgraded to CAPWAP before establishing actual connection with the controller.

A new licensing scheme was introduced when Cisco introduced the 5500 Series controller. Two types of licenses are now available:

- Wireless base license
- Wireless plus license

The wireless base license determines the number of access points a 5500 Series controller can support. Each controller is capable of supporting up to 250 lightweight access points, but control of support for a different number of access points is enabled by software license key. Customers can purchase different wireless base license keys to increase the number of access points supported on the 5500 Series controller as the wireless LAN grows. Table A-4 illustrates the license key scheme for a different number of access point support.

TABLE A-4 Cisco Wireless 5500 Series Wireless LAN Controller and Number of Access Points Supported

Cisco 5508 Series Wireless LAN Controller	Number of Lightweight Access Points Supported
AIR-CT5508-12-K9	12
AIR-CT5508-25-K9	25
AIR-CT5508-50-K9	50
AIR-CT5508-100-K9	100
AIR-CT5508-250-K9	250

Upgrade licenses are also available for purchase as your network grows beyond the current access point support level.

The second type of software license for the 5500 Series controller is the wireless plus license, which allows the controller to support advanced services, such as mesh networking and OfficeExtend. Figure A-8 illustrates the front view of a Cisco 5500 Series Wireless LAN Controller.

Similar to the Cisco 4400 Series controller, this controller appliance is capable of supporting up to two power supplies, and the power supplies are hot swappable. Figure A-9 illustrates the back view of a Cisco 5500 Series Wireless LAN Controller.

FIGURE A-8 Cisco 5500 Series Wireless LAN Controller, front view (Courtesy of Cisco Systems, Inc. Unauthorized use is not permitted.)

FIGURE A-9 Cisco 5500 Series Wireless LAN Controller, back view (Courtesy of Cisco Systems, Inc. Unauthorized use is not permitted.)

Discontinued Wireless LAN Controllers

After Cisco acquired Airespace, three models of Cisco Wireless LAN Controllers were discontinued:

- Cisco 2000 Series Wireless LAN Controller
- Cisco 4000 Series Wireless LAN Controller
- Cisco 4100 Series Wireless LAN Controller

We will briefly discuss the three models and upgrade path for people owning these controllers.

Cisco 2000 Series Wireless LAN Controller

The Cisco 2000 Series Wireless LAN Controller was designed for branch and small office wireless access. The controller can support up to 6 lightweight access points, but it does not support 802.3af PoE for access points. The 2000 Series controller has four 10/100 Ethernet ports and one RS-232 serial console port. Cisco announced end-of-life of the 2000 Series wireless controller in April 2007 and will stop support of this platform in April 2012. The replacement product for the 2000 Series controller is the Cisco 2100 Series Wireless LAN Controller. The last supported controller software release for the 2000 Series wireless controller is controller release version 4.2. The Cisco 2000 Series wireless controller information can be viewed at www.cisco.com/en/US/products/ps6308/index.html.

The detailed Cisco 2000 Series controller end-of-life announcement can be viewed at www.cisco.com/en/US/prod/collateral/wireless/ps6302/ps8322/ps6308/prod_end-of-life_notice0900aecd805d22b0.html.

Cisco 4000 Series Wireless LAN Controller

The Cisco 4000 Series Wireless LAN Controller is a rebranded wireless controller from Airespace. Cisco discontinued this model of controller shortly after it acquired Airespace. The 4000 Series wireless controller supports up to 24 lightweight access points with 802.3af PoE. Cisco announced end-of-life of the 4000 Series wireless controller in March 2005 and will stop support of this platform in September 2010. The replacement product for the 4000 Series controller is the Cisco 4400 Series Wireless LAN Controller. The last supported controller image for the 4000 Series wireless controller is controller release 3.2.215. The detailed Cisco 4000 Series controller end-of-life announcement can be viewed at www.cisco.com/en/US/prod/collateral/wireless/ps6302/ps8322/ps6307/prod_end-of-life_notice0900aecd803ee6ef.html.

Cisco 4100 Series Wireless LAN Controller

The Cisco 4100 Series Wireless LAN Controller is designed for medium sized facility wireless access. The controller can support up to 12, 24, 36, and 100 lightweight access points. The 4100 Series controller only has two 1000 Base-SX interfaces, while only one is active and one is standby. Cisco announced end-of-life of the 4100 Series wireless controller in September 2005 and will stop support of this platform in March 2011. The replacement product for the 4100 Series controller is the Cisco 4400 Series Wireless LAN Controller. The last supported controller image for the 2000 Series wireless controller is controller release 3.2.215. The Cisco 4100 Series wireless controller information can be viewed at www.cisco.com/en/US/products/ps6307/index.html.

The detailed Cisco 4100 Series controller end-of-life announcements can be viewed at www.cisco.com/en/US/prod/collateral/wireless/ps6302/ps8322/ps6307/prod_end-of-life_notice0900aecd803387a9.html.

B

Cisco Aironet Wireless Access Points

This appendix will provide a brief overview of the indoor wireless access points offered by Cisco for the Cisco Unified Wireless Network solution. The purpose of this appendix is to help you understand the differences between different models of Aironet access points and what environments and wireless applications they are best suited for.

Cisco Aironet 1250 Series Access Point

The Cisco 1250 Series access point is an 802.11a/g/n platform that is the most versatile and rugged of the Cisco access points with support for MIMO technology. The Cisco 1250 supports 802.11a and 802.11g radios and can also support up to 600 Mbps of bandwidth through the use of channel bonding technology with 802.11n on both the 2.4 and 5 GHz bands. The Aironet 1250 Series was introduced as Cisco's first 802.11n capable access point. Because of the 1250's rugged construction and antenna flexibility, you will often see 1250s in warehouse environments and areas with harsher conditions. The 1250 Series is also used in indoor environments with specific antenna requirements and coverage needs.

The Aironet 1250 Series access point is a modular platform that allows flexibility in both radio and antenna configuration. Through the use of RP-TNC antenna connectors, you can attach a variety of external antennas for proper RF coverage of a building or outdoor area. Table B-1 shows the different configurations and modules available for the Aironet 1250 Series access points. It is important to note that a single 1250 platform will not support two 802.11a/n or two 802.11g/n modules at the same time.

Figure B-1 is an image of an Aironet 1252 access point with an 802.11a/n and 802.11g/n radio installed. Notice the 6 RP-TNC connectors (3 for each radio) which allow the access point to support MIMO for 802.11n support in a 3x3 configuration.

TABLE B-1 Cisco Aironet 1250 Series Part Numbers and Description

Cisco Aironet 1250 Series Access Point	Description
AIR-AP1252AG-x-K9	2.4/5-GHz standalone/ IOS-based AP 802.11a/g/n (dual-band)
AIR-AP1252G-x-K9	2.4-GHz standalone/IOS-based AP 802.11g/n only (single-band)
AIR-LAP1252AG-x-K9	2.4/5-GHz Unified/controller-based AP 802.11a/g/n (dual-band)
AIR-LAP1252AG-x-K9	2.4-GHz Unified/controller-based AP 802.11g/n only (single-band)
AIR-AP1250=	Standalone/IOS-based AP Platform (no radio modules)
AIR-LAP1250=	Unified/controller-based AP Platform (no radio modules)
AIR-RM1252A-x-K9=	802.11a/n 5-GHz radio module
AIR-RM1252G-x-K9=	802.11g/n 2.4-GHz radio module
AIR-P1250MNTGKIT=	1250 Series Ceiling, wall mount bracket kit

FIGURE B-1 Cisco Aironet 1252 access point (Courtesy of Cisco Systems, Inc. Unauthorized use is not permitted.)

FIGURE B-2

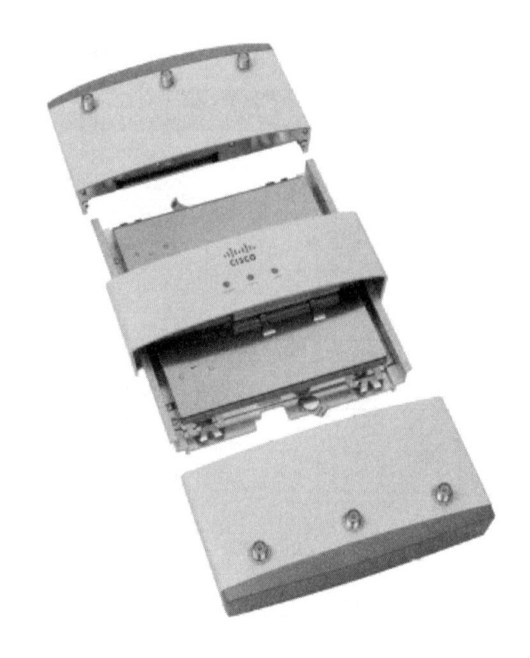

Cisco Aironet 1252 access point modules (Courtesy of Cisco Systems, Inc. Unauthorized use is not permitted.)

Figure B-2 is another image of the Aironet 1252 access point, which demonstrates how the 802.11a/n and 802.11g/n modules can be removed and changed.

One of the primary benefits of deploying the Cisco 1250 Series access point is the antenna flexibility that is inherent in the design. Table B-2 shows the available omnidirectional antennas that are available for the Cisco 1250 Series APs.

TABLE B-2 Cisco Aironet 1250 Antenna Options

Cisco Aironet 1250 Series Antenna Option	Description
AIR-ANT1728	2.4 GHz 2.2 dBi dipole antenna (802.11g/n)
AIR-ANT5135DG-R	5 GHz 3.5 dBi dipole antenna (802.11a/n)
AIR-ANT2430V-R	2.4 GHz 3 dBi omnidirectional antenna (802.11g/n)
AIR-ANT5140V-R	5 GHz 4 dBi omnidirectional antenna (802.11a/n)

FIGURE B-3

AIR-ANT1728 2.2 dBi dipole antenna (Courtesy of Cisco Systems, Inc. Unauthorized use is not permitted.)

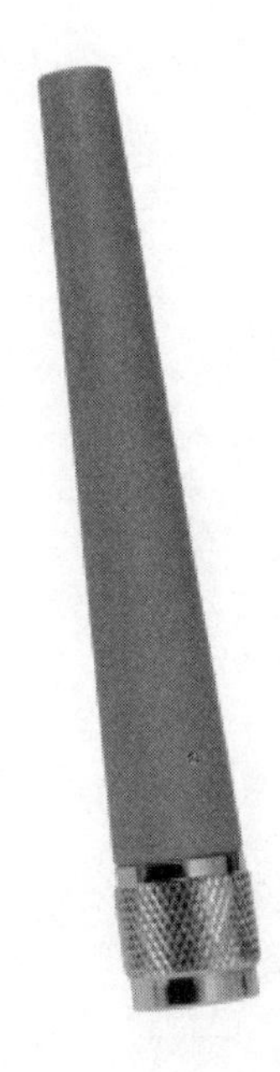

Figures B-3, B-4, B-5, and B-6 are images of the basic MIMO antenna options available for the Cisco Aironet 1250 Series access point. As you will see later in this appendix, there are other antennas that are also compatible with the 1250, but these are the ones most often seen. Note that the 5 GHz 802.11a/n antennas all have a blue dot or cable shield to help you identify which RP-TNC antenna ports to plug them into. It is also important to note that the dipole antennas in Figures B-3 and B-4 are installed on the 1250 Series in multiples of three, while the more cosmetic omnidirectional antennas in Figures B-5 and B-6 have three integrated antenna cables.

FIGURE B-4

AIR-ANT5135DG-R 3.5 dBi dipole antenna (Courtesy of Cisco Systems, Inc. Unauthorized use is not permitted.)

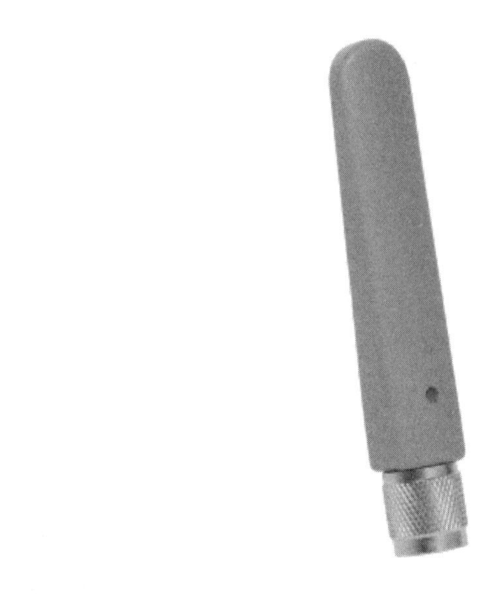

FIGURE B-5

AIR-ANT2430V-R 3 dBi omnidirectional antenna (Courtesy of Cisco Systems, Inc. Unauthorized use is not permitted.)

FIGURE B-6

AIR-ANT5140V-R 4 dBi omnidirectional antenna (Courtesy of Cisco Systems, Inc. Unauthorized use is not permitted.)

Cisco Aironet 1140 Series Access Point

The Cisco 1140 Series access point is an 802.11a/g/n platform that is targeted toward indoor enterprise and medium-sized office environments where there is a requirement for an easy-to-deploy wireless access point solution. Like the Cisco 1250 Series, the Cisco 1140 can support 802.11a and 802.11g and can support up to 600 Mbps of bandwidth through the use of channel bonding technology using 802.11n on both the 2.4 and 5 GHz bands. Along with the Aironet 1250 Series, the 1140 is part of Cisco's selection of 802.11n capable access points. Unlike the 1250, the 1140 Series is constructed of lighter weight plastics and has an integrated antenna for both the 802.11a/n and 802.11g/n radios. The 1140 Series is most often found in basic office environments and tends not to be as well-positioned for harsher industrial or warehouse environments. Table B-3 shows the different configurations available for the Aironet 1140 Series access points.

TABLE B-3 Cisco Aironet 1140 Series Part Numbers and Description

Cisco Aironet 1140 Series Access Point	Description
AIR-AP1142N-x-K9	2.4/5-GHz standalone/ IOS-based AP 802.11a/g/n (dual-band)
AIR-AP1141N-x-K9	2.4-GHz standalone/IOS-based AP 802.11g/n only (single-band)
AIR-LAP1142N-x-K9	2.4/5-GHz Unified/controller-based AP 802.11a/g/n (dual-band)
AIR-LAP1141N-x-K9	2.4-GHz Unified/controller-based AP 802.11g/n only (single-band)

Figure B-7 is an image of an Aironet 1142 access point with an 802.11a/n and 802.11g/n radio. The integrated antenna on the 1140 Series is an omnidirectional 3 dBi for the 2.4 GHz (802.11g/n) band and an omnidirectional 4 dBi for the 5 GHz (802.11a/n) band.

FIGURE B-7

Cisco Aironet 1142 access point (Courtesy of Cisco Systems, Inc. Unauthorized use is not permitted.)

Cisco Aironet 1240 Series Access Point

The Cisco 1240 Series access point is an 802.11a/b/g platform that is the more rugged of the current Cisco 802.11a/b/g capable access points. The Cisco 1240 can support up to 54 Mbps of bandwidth through the use of either its 802.11a or 802.11b radio. A direct replacement for the legacy 1200 Series, the 1240 supports the same antennas as the 1200 Series and provides a similar level of rugged construction and antenna flexibility seen in the 1200 and 1250 Series. Like the 1200 and 1250, the 1240 is often utilized in harsher environments but is also frequently found in traditional indoor office environments.

The Aironet 1240 Series access point has both an 802.11a and 802.11g radio integrated. Through the use of dual RP-TNC antenna connectors for each radio, you can also attach a variety of external antennas for proper RF coverage of a building or outdoor area. At the time of this publication, the 1240 has the advantage over the 1250 of also allowing the use of semidirectional patch and yagi antennas, which has proven useful in more challenging wireless deployments. Table B-4 shows the two configurations of the Aironet 1240 Series access points.

Figure B-8 is an image of an Aironet 1242 access point. Notice the four RP-TNC connectors (two for each radio) that allow the access point to support diversity for both the 2.4 and 5.8 GHz frequency bands.

As with the Cisco Aironet 1200 and 1250 Series, the Cisco 1240 Series access point provides for a great deal of antenna flexibility. Table B-5 shows the available omnidirectional antennas that are available for the Cisco 1240 Series APs. Note that the articulating (with the capability to bend up to 90 degrees in

TABLE B-4 Cisco Aironet 1240 Series Part Numbers and Description

Cisco Aironet 1240 Series Access Point	Description
AIR-AP1242AG-x-K9	2.4/5-GHz standalone/ IOS-based AP 802.11a/g (dual-band)
AIR-LAP1242AG-x-K9	2.4/5-GHz Unified/controller-based AP 802.11a/g (dual-band)

FIGURE B-8

Cisco Aironet 1242 access point (Courtesy of Cisco Systems, Inc. Unauthorized use is not permitted.)

either direction) dipole antennas can also be used on the Aironet 1250 Series. It is important to note that, in addition to these commonly used omnidirectional antennas, the 1240 supports a wide array of diversity and nondiversity antennas in a variety of different gains.

TABLE B-5 Cisco Aironet 1240/1200 Antenna Options

Cisco Aironet 1240 Series Antenna Option	Description
AIR-ANT4941*	2.4 GHz 2.2 dBi dipole antenna (802.11g/n)
AIR-ANT5135D-R*	5 GHz 3.5 dBi dipole antenna (802.11a/n)
AIR-ANT5959	2.4 GHz 2.0 dBi diversity omnidirectional antenna (802.11g)
AIR-ANT5145V-R	5 GHz 4.5 dBi diversity omnidirectional antenna (802.11a)
*Also compatible with the Cisco Aironet 1250 Series	

FIGURE B-9

AIR-ANT4941 2.2 dBi dipole antenna (Courtesy of Cisco Systems, Inc. Unauthorized use is not permitted.)

Figures B-9, B-10, B-11, and B-12 are images of the basic diversity omnidirectional antenna options available for the Cisco Aironet 1240 and 1200 Series access points. As mentioned before, the 5 GHz 802.11a/n antennas all have a blue dot or cable shield to help you identify which RP-TNC antenna ports to plug

FIGURE B-10

AIR-ANT5135D-R 3.5 dBi dipole antenna (Courtesy of Cisco Systems, Inc. Unauthorized use is not permitted.)

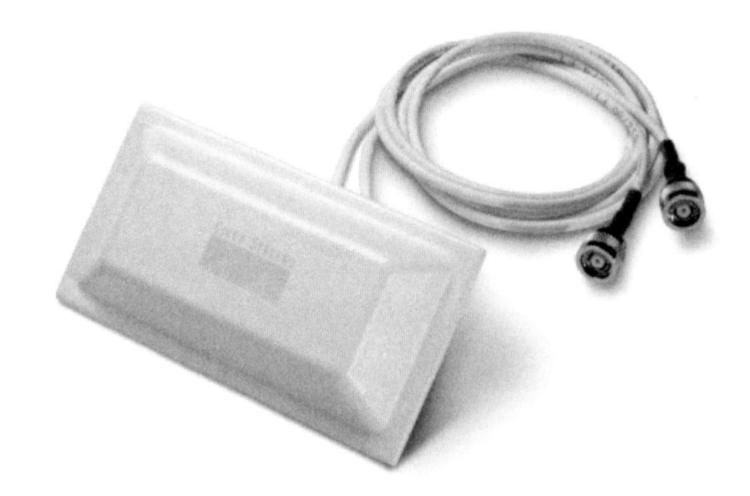

FIGURE B-11

AIR-ANT5959 2.0 dBi Diversity omnidirectional antenna (Courtesy of Cisco Systems, Inc. Unauthorized use is not permitted.)

them into. In the 1240 and 1200 Series, the dipole antennas (in Figures B-9 and B-10) are often installed on the 1240/200 Series in multiples of two for optional diversity, while the more cosmetic diversity omnidirectional antennas (in Figures B-11 and B-12) have two integrated antenna cables for built-in diversity.

FIGURE B-12

AIR-ANT5145V-R 4.5 dBi Diversity omnidirectional antenna (Courtesy of Cisco Systems, Inc. Unauthorized use is not permitted.)

Cisco Aironet 1130 Series Access Point

The Cisco 1130 Series access point is an 802.11a/b/g platform that, like the 1140 Series, is an easy-to-deploy platform targeted toward indoor enterprise and medium-sized office environments. Like the Cisco 1240 Series, the Cisco 1130 can support up to 54 Mbps of bandwidth in its integrated 802.11a or 802.11b radio. A direct replacement for the legacy 1100 Series, the 1130 Series is also very similar to the 1140 Series in that it is constructed of lighter weight plastics and has an integrated antenna for both its radios. The primary difference is that the 1130 Series does not support 802.11n in either band. Like the 1140 Series, the 1130 is found in basic office environments and is not typically used for harsher industrial or warehouse applications. Table B-6 shows the two configurations available for the Aironet 1130 Series access points.

Figure B-13 is an image of an Aironet 1131 access point with integrated 802.11a and 802.11b/g radio. The integrated antenna on the 1130 Series is an omnidirectional 3 dBi for the 2.4 GHz (802.11b/g) band and an omnidirectional 4.5 dBi for the 5 GHz (802.11a) band.

TABLE B-6 Cisco Aironet 1130 Series Part Numbers and Description

Cisco Aironet 1130 Series Access Point	Description
AIR-AP1131AG-x-K9	2.4/5-GHz standalone/ IOS-based AP 802.11a/g (dual-band)
AIR-LAP1131AG-x-K9	2.4/5-GHz Unified/controller-based AP 802.11a/g (dual-band)

FIGURE B-13

Cisco Aironet 1131 access point (Courtesy of Cisco Systems, Inc. Unauthorized use is not permitted.)

Discontinued Aironet Access Points

The following access points, which were common in deployments before and immediately after the Airespace acquisition by Cisco, are still seen in a substantial number of enterprise environments. The most common of these are the following models:

- Cisco Aironet 1200/1230 Series access point
- Cisco Aironet 1100/1120 Series access point
- Cisco Aironet 1000 Series access point

We will briefly discuss these models and their upgrade path.

Cisco Aironet 1200/1230 Series Access Point

The Cisco Aironet 1200/1230 Series access point is a rugged indoor access point that was replaced by the Aironet 1240 Series. The Aironet 1200 Series was introduced as an 802.11b access point before the wide adoption of the IEEE 802.11g standard and was among the first Cisco Aironet Access points to support both IOS standalone mode and unified/controller-based LWAPP mode. With an optional 802.11a module that could be fitted in a PCMCIA slot, the 1230 Series was Cisco's first dual-band capable access point and is still deployed in various environments today. Table B-7 shows the different configurations and modules that were available for the more recent Aironet 1230 Series access points.

Figure B-14 is an image of an Aironet 1232 access point with 802.11a radio and integrated antenna. The Aironet 1230 Series uses the ANT-4941 antenna (shown in the Aironet 1240 section of this chapter) and, in this image, has the unique AIR-MP21G integrated antenna which, when opened/raised is a 5 dBi omnidirectional antenna and, when flattened/lowered, behaves as a 9 dBi patch antenna.

TABLE B-7 Cisco Aironet 1230 Series Part Numbers and Description

Cisco Aironet 1230 Series Access Point	Description
AIR-AP1231G-x-K9	2.4 GHz standalone/ IOS-based AP 802.11g (single-band)
AIR-AP1232AG-x-K9	2.4/5-GHz standalone/ IOS-based AP 802.11a/g (dual-band)
AIR-LAP1231G-x-K9	2.4 GHz Unified/controller-based AP 802.11g (single-band)
AIR-LAP1232AG-x-K9	2.4/5-GHz Unified/controller-based AP 802.11a/g (dual-band)
AIR-MP21G	802.11a 5-GHz radio module (integrated antenna)
AIR-RM22A	802.11a 5-GHz radio module (RP-TNC connectors)

FIGURE B-14

Cisco Aironet 1232 access point with AIR-MP21G 802.11a module (Courtesy of Cisco Systems, Inc. Unauthorized use is not permitted.)

Cisco Aironet 1100 Series Access Point

The Cisco Aironet 1100 Series access point is a light plastic indoor access point that was replaced by the Aironet 1130 Series. The Aironet 1100 Series, like the 1200, was introduced as an 802.11b access point before the wide adoption of the IEEE 802.11g standard. The Aironet 1100 Series is strictly a 2.4 GHz access point and, along with the Aironet 1200/1230, was among the first Cisco Aironet Access points to support both IOS standalone mode and unified/controller-based LWAPP mode. That said, it is very rare to find 1100s deployed in controller-based LWAPP mode in most modern networks. Table B-8 shows the two configurations available for the Aironet 1100 Series access points.

TABLE B-8 Cisco Aironet 1100 Series Part Numbers and Description

Cisco Aironet 1100 Series Access Point	Description
AIR-AP1121G-x-K9	2.4-GHz standalone/ IOS-based AP 802.11g (single-band)
AIR-LAP1121G-x-K9	2.4-GHz Unified/controller-based AP 802.11g (single-band)

FIGURE B-15 Cisco Aironet 1121 access point with desktop stand (Courtesy of Cisco Systems, Inc. Unauthorized use is not permitted.)

Figure B-15 is an image of an Aironet 1121 access point. The 1100 is unique among Cisco APs in that it can only practically be wall-mounted and also has an optional cubicle mount or stand for desktop use (shown). The Aironet 1100 Series uses an integrated 2 dBi 2.4 GHz antenna.

Cisco Aironet 1000 Series Access Point

The Cisco Aironet 1000 Series access point is a directly inherited product from the Airespace acquisition. The 802.11a/b/g capable 1000 Series is unique in that it is the only Cisco access point that was capable of operating only in a unified/controller-based architecture. After the introduction of unified/controller-based capabilities in the Cisco 1100 and 1200 Series APs, the 1000 was quickly phased out as Cisco continued development of more unified/controller-capable access points. The Aironet 1000 Series features an unusual setup of integrated bidirectional antennas and, on the 1020 and 1030 models, also has RP-TNC connectors for the attachment of external antennas. The 1030 is distinguished from the 1020 by the fact that it is capable of operating as a remote edge access point (REAP). Table B-9 shows the various configurations of Cisco Aironet 1000 Series access points.

TABLE B-9 Cisco Aironet 1000 Series Part Numbers and Description

Cisco Aironet 1000 Series Access Point	Description
AIR-AP1010-x-K9	2.4/5-GHz Unified/controller-based AP 802.11a/g (dual-band) with Integrated antennas
AIR-AP1020-x-K9	2.4/5-GHz Unified/controller-based AP 802.11a/g (dual-band) with Integrated antennas and RP-TNC Connectors
AIR-AP1030-x-K9	2.4/5-GHz Unified/controller-based AP 802.11a/g (dual-band) with Integrated antennas and RP-TNC Connectors + REAP Capability

Figure B-16 is an image of an Aironet 1030 access point. The 1000 is unique among Cisco APs in that it does not use omnidirectional antennas by default. The Aironet 1000 Series uses an integrated bidirectional antenna, which produces 6 dBi bidirectional gain for the 2.4 GHz (802.11g) band and 5 dBi for the 5 GHz (802.11a) band.

FIGURE B-16

Cisco Aironet 1030 access point (Courtesy of Cisco Systems, Inc. Unauthorized use is not permitted.)

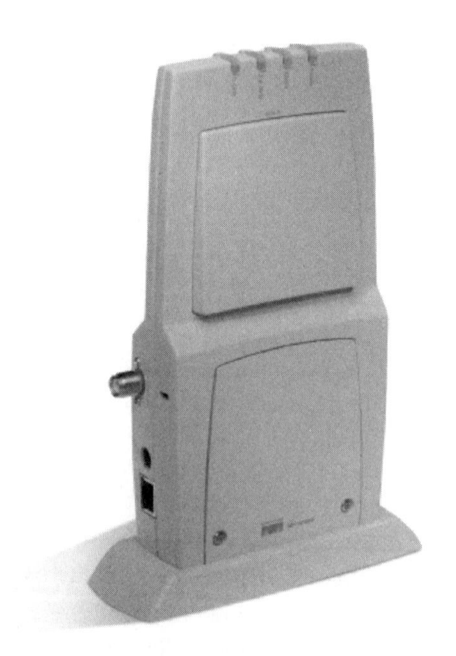

C

About the CD

The CD-ROM included with this book comes complete with MasterExam and the electronic version of the book. The software is easy to install on any Windows 2000/XP/Vista computer and must be installed to access the MasterExam feature. You may, however, browse the electronic book directly from the CD without installation. To register for the bonus MasterExam, simply click the Bonus MasterExam link on the main launch page and follow the directions to the free online registration.

System Requirements

Software requires Windows 2000 or higher, Internet Explorer 6 or above, and 20MB of hard disk space for full installation. The electronic book requires Adobe Acrobat Reader.

Installing and Running MasterExam

If your computer CD-ROM drive is configured to auto run, the CD-ROM will automatically start up upon inserting the disk. From the opening screen you may install MasterExam by clicking the MasterExam link. This will begin the installation process and create a program group named LearnKey. To run MasterExam, use Start | All Programs | LearnKey | MasterExam. If the auto run feature did not launch your CD, browse to the CD and click the LaunchTraining.exe icon.

MasterExam

MasterExam provides you with a simulation of the actual exam. The number of questions, the type of questions, and the time allowed are intended to be an accurate representation of the exam environment. You have the option to take an open book exam, including hints, references, and answers, a closed book exam, or the timed MasterExam simulation.

When you launch MasterExam, a digital clock display will appear in the bottom right-hand corner of your screen. The clock will continue to count down to zero unless you choose to end the exam before the time expires.

Electronic Book

The entire contents of the Study Guide are provided in PDF. Adobe's Acrobat Reader has been included on the CD.

Help

A help file is provided through the Help button on the main page in the lower left hand corner. An individual help feature is also available through MasterExam.

Removing Installation(s)

MasterExam is installed to your hard drive. For best results removing programs, use the Start | All Programs | LearnKey | Uninstall option to remove MasterExam.

Technical Support

For questions regarding the content of the electronic book or MasterExam, please visit www.mhprofessional.com or e-mail customer.service@mcgraw-hill.com. For customers outside the 50 United States, e-mail international_cs@mcgraw-hill.com.

LearnKey Technical Support

For technical problems with the software (installation, operation, removing installations), please visit www.learnkey.com, e-mail techsupport@learnkey.com, or call toll free at 1-800-482-8244.

Glossary

802.11 The IEEE standard that specifies carrier sense media access control with collision avoidance and physical layer specifications for 1 and 2 Mbps wireless LANs operating in the 2.4 GHz frequency band.

802.11a The IEEE amendment that specifies carrier sense media access control and physical layer specifications for wireless LANs operating in the 5 GHz frequency band using 802.11 a PHY and OFDM.

802.11b The IEEE amendment that specifies carrier sense media access control and physical layer specifications for 5.5 and 11 Mbps wireless LANs operating in the 2.4 GHz frequency band.

802.11e The IEEE 802.11 standard that defines a set of Quality of Service (QoS) enhancements for wireless LAN applications through modifications to 802.11 MAC.

802.11g The IEEE amendment that specifies carrier sense media access control and physical layer specifications for 6, 9, 12, 18, 24, 36, 48, and 54 Mbps wireless LANs operating in the 2.4 GHz frequency band using ERP-OFDM.

802.11i The IEEE amendment that specifies advanced wireless device security specifications for wireless LANs operating in both the 2.4 GHz and 5 GHz frequency bands.

802.11n The IEEE amendment that specifies increased bandwidth through spatial multiplexing using the new HT PHY and use of *multiple input/multiple output (MIMO)* antennas.

802.16 The IEEE standard for wireless broadband access. The standard is also known as WiMAX.

802.1X An IEEE standard for port-based access control. The wireless networking industry has utilized the 802.1X framework to enhance wireless security and allow user identity to be validated through an authority. Part of the 802.11i became recommendation for a robust security network (RSN).

access point (AP) A purpose built device that connects 802.11 wireless stations to a network infrastructure, which may be wired or wireless devices.

ad hoc mode An operation mode for Wi-Fi network that does not require use of an access point. This is also known as an independent basic service set (IBSS) mode.

Advanced Encryption Standard (AES) An encryption standard that uses block ciphers based on the Rijindael algorithm. AES has been adopted by the U.S. government for encryption purposes and is used by WPA2 or 802.11i standard devices.

antenna A transducer designed to transmit or receive electromagnetic waves through wireless and convert to electrical currents.

antenna diversity A technique that uses two or more antennas to improve the quality of signals of a wireless link against multipath. An access point will listen to multiple antennas of the same signal and receive from the one that provides the best signal. Antenna diversity requires additional antenna(s) to be deployed on an access point.

association The relationship that is created between a wireless client and the access point.

attenuation The diminishing in strength of an RF signal.

authentication The act of validating the identity of a device or user of a device.

authentication server The server in 802.1X authentication that validates client identities.

authenticator Enforces authentication in 802.1X authentication. The authenticator acts as a RADIUS proxy device that grants or denies access to a network.

basic service set (BSS) One or more wireless devices cooperating and communicating with one another identified by a BSSID. BSS is defined by the IEEE 802.11.

basic service set identifier (BSSID) An AP offering BSS services sends out a broadcast of a unique identifier to identify a unique network of Wi-Fi services. The BSSID is usually the radio MAC address of an access point.

Bluetooth A wireless protocol designed for short-range wireless communications in a wireless PAN. Bluetooth uses frequency hopping spread spectrum to send information over the 2.4 GHz spectrum and is widely used by personal electronic devices such as cellular phones, headsets, video game consoles, and video cameras. Bluetooth was originally published as the IEEE 802.15.1 standard but is now published and maintained by the Bluetooth Special Interest Group (SIG).

channel A defined frequency range used for wireless communications in an allowed spectrum.

co-channel interference Interference resulting from a radio system transmitting signals in the same channel and competing for the use of wireless medium.

Control and Provision of Wireless Access Point (CAPWAP) A mechanism defined by IETF RFC 5415 to provide a framework for communications used by wireless controllers and APs. CAPWAP allows wireless controllers to manage, configure, and monitor multiple access points from a centralized management interface. CAPWAP was based on LWAPP with additional features to protect control plane traffic, such as Dynamic TLS.

data rates The range of data transmission rates supported by a device. Data rates are measured in bits per second (bps). 1 Mbps represents one million bits per second.

dB (decibel) A unit used to measure sound, radio, or electrical signal level. In wireless, a decibel is a logarithmic representation used to describe the power of radio signals.

dBi (decibel isotropic) A ratio of decibels to a theoretical isotropic antenna that is commonly used to measure antenna gain. The greater the dBi value, the higher the gain, and the more acute the angle of coverage.

dBm (sometimes dBmW) Radio power measured in ratio of decibels in reference to one milliwatt (mW).

Direct Sequence Spread Spectrum (DSSS) A spread spectrum technique that is used in WLAN to transmit data over RF using bit sequence, or chipping code.

Distributed Coordination Function (DCF) The primary access method employed by 802.11 to provide wireless devices access to the medium to send traffic without colliding with other wireless device traffic. DCF employs the CSMA/CA algorithm, which is similar to the CSMA/CD used by 802.3 (Ethernet) standard, with optional virtual carrier sense using RTS/CTS control frames. PCF is an optional technique defined by 802.11 to provide similar functions and is rarely implemented.

distribution system (DS) A service defined by IEEE 802.11 that connects a set of BSSes to create an ESS. DS usually runs on an access point, and it allows stations within an ESS to communicate with one another indirectly.

diversity antennas A radio frequency instrument that uses two antennas to continually sense incoming radio signals and select the antenna with the best received signal. Diversity antennas may also be used to select the best antenna for transmitting signals.

European Telecommunications Standards Institute (ETSI) An organization legally and officially recognized by the European Commission as a European Standards Organization. ETSI, however, does not enforce the regulations within countries of the European Union.

extended service set (ESS) A collection of one or more BSSes connected to the same distribution system (DS) to form a larger WLAN.

Extensible Authentication Protocol (EAP) A protocol published by the IETF but widely used as an optional IEEE 802.1X security feature. EAP is ideal for organizations with a large user base and access to an EAP-enabled RADIUS server.

Federal Communications Commission (FCC) The regulatory authority in the United States responsible for regulating non–federal government use of radio and television broadcasting, interstate telecommunications (wire, satellite and cable), and international communications that originate or terminate in the United States.

frequency hopping spread spectrum (FHSS) A spread spectrum technique used in a WLAN that modulates the data signal with a narrowband carrier signal and switches in a patterned sequence from frequency to frequency over time in a wide band of frequencies.

Fresnel Zone An elliptical area immediately around an RF signal's beam from transmitter to receiver.

gain Increase in radio signal power. In RF, gain is typically referred to as ratio of the output power to the input power in decibels.

hertz (Hz) A unit of frequency defined as the number of complete cycles per second. In wireless LAN, gigahertz (GHz) is frequently used and is referred to one billion cycles per second.

hidden node In wireless LAN, client devices that are within range of an access point but not within range of one or more other client devices of the same access point.

Hypertext Transfer Protocol (HTTP) An application layer protocol in the OSI model widely used to distribute and share information using a client-server model on network. The client is typically a Web browser application running on a PC, and the server is a network device hosting and storing the information, also known as a Website. HTTP uses TCP port 80 by default and the Web browser uses "http://" to access information on an HTTP server.

Hypertext Transfer Protocol Secure (HTTPS) A protocol that implements the SSL/TLS protocol to provide a secure way of sharing information over HTTP. HTTPS uses the public key infrastructure (PKI) to create secure channels and to ensure confidentiality and integrity of communications between the client (Web browser) and the server (Website). HTTPS uses TCP port 443 by default and the Web browser uses "https://" to access information on an HTTP server.

Independent Basic Service Set (IBSS) As defined by the IEEE, the operation mode for an ad hoc wireless network.

Industrial Scientific Medical bands (ISM bands) Radio frequencies designated for industrial, scientific, and medical purposes. The FCC allows use of wireless LAN devices in the ISM frequencies defined in 47 CFR Part 15.5. The most commonly encountered ISM device is the microwave oven operating at 2.45 GHz; nonetheless, wireless LANs and cordless phones in the 915 MHz, 2450 MHz, and 5800 MHz bands also commonly use the ISM frequencies.

Institute of Electrical and Electronics Engineers (IEEE) One of the most influential standards bodies in the world for information technology. IEEE published the 802.11 standards and others related to wireless networking.

interference In wireless, the addition of two or more signals, which results in a new signal pattern that conflicts with the original signal.

Lightweight AP Protocol (LWAPP) A communications protocol used by Cisco Wireless LAN Controllers and Cisco lightweight APs. LWAPP allows a wireless controller to manage, configure, and monitor multiple access points from a management interface. LWAPP was first proposed by Airespace as a standard protocol, and Airespace was acquired by Cisco Systems in 2005. The IETF defined a similar mechanism in RFC 5415 using CAPWAP.

line-of-sight (LOS) The ability to see along a straight line between two points. In wireless LAN, RF LOS must consider the proper Fresnel Zone clearance to ensure proper communications between two end points.

loss The lowering of a signal's strength. Often measured in decibels in wireless networks.

Management Frame Protection (MFP)–infrastructure A security feature implemented on Cisco wireless LAN controllers to protect 802.11 management frames, such as authentication, association, deauthentication, disassociation beacons, and probes. When MFP–infrastructure is enabled, the wireless controller will send a unique key to the registered AP, and the AP will send out management frames with an additional message integrity check information element (MIC IE). The key is used to validate or generate the MIC IE.

Management Frame Protection (MFP)–client A Cisco security feature implemented to protect 802.11 class 3 management frames between clients and APs, such as disassociation, deauthentication, and QoS (WMM). Clients must support CCXv5 MFP and negotiate WPA2 with either TKIP or AES-CCMP. MFP–client clients encrypt management frames, and access points that support CCXv5 do not send any broadcast class 3 management frames.

mesh network A network having two or more paths to any node. Examples of possible mesh networks include Wi-Fi-based mesh, ZigBee, TCP/IP, and other routed protocols.

milliwatt (mW) One thousandth of a watt.

modulation Techniques for combining user information with a transmitter's carrier signal.

multipath The reflections created as a radio signal bounces off physical objects and arrives at a receiver in different paths and possibly different phases.

Multiple-Input/Multiple-Output (MIMO) An antenna technology that utilizes multiple transmitters and receivers to increase data bandwidth and to reduce errors.

open system authentication The default authentication for the IEEE 802.11 specification. Open system authentication is also used by 802.1X and EAP-based authentication in wireless LAN systems.

Open System Interconnection model (OSI model) Also known as the Open System Interconnection reference model or OSI reference model, an abstract conceptual description for layered communications and computer network design. The OSI model divides the computer communications into seven layers, including (from top to bottom) Application, Presentation, Session, Transport, Network, Data-Link, and Physical Layers. Physical, Data-Link, and Network layers are often referred to as Layer 1, Layer 2, and Layer 3, respectively, in a network. 802.11 is considered a Layer 1 and Layer 2 protocol, as it defines services and functions that span across Physical and Data-Link layers in the OSI model.

orthogonal frequency division multiplex (OFDM) A modulation technique used by IEEE 802.11a-compliant wireless LANs for transmission at 6, 9, 12, 18, 24, 36, 48, and 54 Mbps. OFDM is also used by 802.11g and 802.11n.

Point Coordination Function (PCF) An optional access method defined by the 802.11 standard to provide wireless devices access to the medium to send traffic without colliding with other wireless device traffic. Using PCF, a point coordinator (usually an AP) will poll each station and see if there's any traffic to be sent by the client station during a time period called contention free period. The AP will then control which client can transmit during any period of time. PCF is rarely implemented by wireless vendors.

polarization The physical orientation of the electromagnetic field of an antenna plane.

Power over Ethernet (PoE) A technology that allows standard Ethernet to carry electric power to enable a network device. IEEE has defined the 802.3af standard to describe sending electric (DC) power up to 15.4 watts over Category 5, 5e, or 6 cable in an Ethernet network.

power save mode One of two power management modes in which a wireless client may operate. In power save mode, the client wakes at predetermined intervals to check for packets intended for the client.

radio frequency (RF) A generic term for radio-based technology.

range A linear measure of the distance over which a transmitter can send a signal.

reflection A behavior of a wave signal, such as RF and light, where the signal bounces off a surface and changes direction.

refraction A behavior of a wave signal, such as RF and light, where the signal passes through a different medium and changes direction due to a difference in density of medium the signal travels in.

Remote Authentication Dial-In User Service (RADIUS) An authentication, authorization, and accounting (AAA) protocol frequently used by wireless LAN and other network services.

roaming The act of an access point that allows users to move through a facility while maintaining an unbroken connection to the LAN through connecting with other access points.

rogue access point An unauthorized access point.

Secure Shell (SSH) A network protocol that facilitates the exchange of information securely through a secure channel between two network devices. SSH uses strong authentication and encryption to ensure the confidentiality and integrity of information passing through the secure tunnel. It was designed to replace insecure remote shells, such as Telnet. SSH uses designed TCP port 22 for the SSH server.

service set identifier (SSID) A unique identifier that identifies a radio network; the SSID is used by stations to communicate with other stations or an access point.

shared-key authentication An alternate form of authentication for the IEEE 802.11 specification.

signal-to-noise ratio (SNR) The ratio of the wireless RF signal to the RF noise floor.

site survey A process used to capture a snapshot of the RF environment and to determine how to best implement or optimize a wireless network.

spread spectrum A radio transmission technique that spreads information over a wider bandwidth than otherwise required to improve interference tolerance and unlicensed operation

Telnet A network protocol designed to facilitate the exchange of information between two network devices. Usually Telnet is used to remotely access the command line interface (CLI) of a network device through virtual terminal session. Telnet sessions are usually sent in clear text; thus it is not considered a secure way to send information over the network. Telnet is a TCP-based application, and a Telnet server listens to TCP port 23 for connection requests.

Temporal Key Integrity Protocol (TKIP) A security protocol created to enhance WEP (Wired Equivalent Privacy). TKIP is required in the WPA certification.

transmit power The power level of a radio transmission.

Unlicensed National Information Infrastructure (U-NII or UNII) Regulations for radio frequency instruments operating in the 5.15 to 5.35 GHz and 5.725 to 5.825 GHz frequencies.

UNII-1 UNII devices operating in the 5.15 to 5.25 GHz frequency.

UNII-2 UNII devices operating in the 5.25 to 5.35 GHz frequency.

UNII-2 Extended UNII devices operating in the 5.470 to 5.725 GHz frequency.

UNII-3 UNII devices operating in the 5.725 to 5.825 GHz frequency.

Virtual Local Area Network (VLAN) A group of hosts logically sharing a broadcast domain. A VLAN provides Layer 2 functionality in the OSI reference model and offers logical segmentation of traffic. Traffic is usually switched by a switch, and there is usually a one-to-one relationship between a VLAN at Layer 2 and an IP subnet at Layer 3.

war driving Driving around using a vehicle and searching for Wi-Fi wireless networks. This is considered a passive wireless sniffing technique.

Wi-Fi Alliance An industry organization that creates certification programs to ensure interoperability among hardware from different vendors based on IEEE 802.11 technologies. This organization is responsible for issuing Wi-Fi, WPA, and WPA2 certifications.

Wi-Fi Protected Access (WPA) A certified security enhancement to WEP encryption defined by the Wi-Fi Alliance derived from and compatible with the

IEEE 802.11i standard. WPA leverages TKIP (Temporal Key Integrity Protocol) for data protection and 802.1X for authenticated key management. WPA is published and administered by the Wi-Fi Alliance.

Wi-Fi Protected Access version 2 (WPA2) A standards-based security enhancement to the WPA standard published by the Wi-Fi Alliance and compatible with the IEEE 802.11i standard; it supports both TKIP and AES.

WiMAX Primarily a wireless MAN-based technology not compatible with 802.11 or 802.15 based technologies; also refers to the WiMAX Forum's specifications for 802.16 wireless technologies.

Wired Equivalent Privacy (WEP) An optional security mechanism defined by the IEEE 802.11 standard and designed to make the link integrity of wireless devices equal to that of a wired device using RC4 encryption algorithm. WEP can be used with both 802.11 open system and shared key authentication.

wireless bridge A device that creates a point-to-point or point-to-multipoint wireless connections between separate networks. The device will perform a bridging function only by forwarding traffic between network segments.

Wireless Control System (WCS) A Cisco network management solution that provides centralized management to monitor and maintain the Cisco Unified Wireless Network solution.

Wireless LAN Controller (WLC) A Cisco device that provides centralized wireless LAN management by collapsing large numbers of managed end-points into a single, unified system.

WPA and WPA2 Enterprise mode A mode defined in the WPA and WPA2 program by the Wi-Fi Alliance, where an 802.1X authentication and use of an EAP method are required. The 802.1X and EAP method ensures that the client and the infrastructure properly authenticate each other. Upon successful authentication, a dynamic key is generated, and TKIP and AES are used for encryption purposes to ensure confidentiality. WPA and WPA2 Enterprise mode is considered to be more secure than the WPA/WPA2 PSK mode.

WPA and WPA2 PSK (Pre-Shared Key) mode A mode defined in the WPA and WPA2 program by the Wi-Fi Alliance, where a predefined static key, similar to a static WEP key in the 802.11 standard, is used by the wireless client and AP to securely communicate with each other. The pre-shared key may be entered in the form of either a string of 64 hexadecimal digits or a passphrase of 8 to 63 ASCII characters. When used in WPA and WPA2 PSK mode, an 802.1X authentication server is not required. TKIP and AES are used for encryption purposes to ensure confidentiality. PSK mode is vulnerable to password cracking if a weak passphrase is used.

ZigBee A wireless mesh network communication protocol used by short range radio systems based on the IEEE 802.15.4 standard for wireless PAN. ZigBee devices are usually low-cost and low-power, including temperature meters, home automation, and electrical meters. The ZigBee standard is maintained and published by the ZigBee Alliance.

INDEX

NUMBERS

B

M

N

O

P

Q

R

S

LICENSE AGREEMENT